Software Radio

Springer
London
Berlin
Heidelberg
New York
Barcelona
Hong Kong
Milan
Paris
Singapore
Tokyo

Enrico Del Re (Ed)

Software Radio

Technologies and Services

With 209 Figures

Enrico Del Re, Professor
Dipartimento di Elettronica e Telecomunicazioni, V.S. Marta 3, 50139 Firenze, Italy

ISBN 1-85233-346-4 Springer-Verlag London Berlin Heidelberg

British Library Cataloguing in Publication Data
Software radio : technologies and services
1.Radio
I.Del Re, Enrico, 1947-
621.3'84
ISBN 1852333464

Library of Congress Cataloging-in-Publication Data
A catalog record for this book is available from the Library of Congress

Printed in Great Britain

Typesetting: Camera ready by contributors
Printed and bound by Athenæum Press Ltd., Gateshead, Tyne & Wear
69/3830-543210 Printed on acid-free paper SPIN 10772861

Preface

Mobile communications technologies and services are world wide considered one of the key factors for the economic development of industrialised as well as less-developed countries in the near and mid-term future.

Next generation mobile communications presently are in the standardisation phase and will likely employ different techniques and standards. The implementation of as many as possible transceiver functionalities in software appears to be the most effective solution to cope with the multiplicity of communications alternatives. Software Radio technology will provide flexible mechanism for the configuration and management of mobile networks.

The concept of Software Radio, dating back to 1991, originally attracted commercial interest due to the possibility that transmission layer functions could be fully software-defined. During the years the concept and the technical interest extended to protocols of the higher layers too, thus conceiving a programmable hardware to implement the functionalities of several layers protocols by resident software or software downloaded from the network.

Software Radio is world-wide recognised as an approach that will be able to greatly improve technical and service performance of next generation mobile communications systems and we can expect that its introduction will deeply characterise and be the distinguishing feature of the mobile communications of the future.

This volume, containing the contributions presented at the 12th Tyrrhenian International Workshop on Digital Communications focused this year on *Software Radio Technologies and Services*, includes the updated state-of-the-art of the enabling technologies and the prospective services of software radio implementations for future mobile communications. The contributions come from leading experts and researchers in the field from Europe, Japan and USA.

Part 1 *Software Radio for Universal Wireless Internet Access* deals with wireless access scenario and the role of software radio is expanded beyond the air interface to enable new high-speed, packet-switched, IP-based wireless network.

Part 2 *Software Radio for Multimedia Communications* deals with Implementation stage of Software Radio for third generation mobile communication including capacity consideration and adaptive scheme selection for Software Radio multimedia communications and application to surveillance networks.

Part 3 *Software Radio Architecture* deals with the architecture evolution for military Software Radio and commercial development for future wireless systems.

Part 4 *Software Radio Technology towards Pervasive Appliance* deals with applications for mobile appliance, for Intelligent Transport Systems, for Air Traffic Control application and an overview of Software Radio technology in Japan.

Part 5 *Network Architecture, Protocols and Services* deals with new developments at the higher levels, protocol stacks and open platforms and service reconfigurability.

Part 6 *Enabling technologies* shows different front end architectures for cellular mobile systems, and future alternative for a CDMA satellite modem.

Finally I would like to express my sincere and grateful appreciation to the session organisers whose dedicated and enthusiastic effort has made the Workshop not only possible but more importantly of scientific value and to thank the authors for their updated contributions.

Enrico Del Re
September 2000

Acknowledgements

The editor wishes to express his sincere thanks to the members of the Technical Committee of the 2000 edition of Tyrrhenian International Workshop on Digital Communications: Software Radio Technologies and Services, namely *Anthony Acampora* from the University of California San Diego, USA, *Enrico Buracchini* from CSELT, Italy, *Testu Ikegami* from NTT and University of Aizu, Japan, *Joseph Mitola III* from MITRE Corporation, USA, *Ramjee Prasad*, from the Aalborg University, Denmark, *Silvano Pupolin* from the University of Padova, Italy, for their valuable contribution to the organisation of the sessions of the Workshop and to the publication of this volume.

The Workshop has benefited of the support of the following sponsors, whose contribution is gratefully acknowledged :

cnit

Italian National Consortium for Telecommunications

Acknowledgements

Table of Contents

Part 1. Software Radio for Universal Wireless Internet Access

Part 2. Software Radio for Multimedia Communications

Part 3. Software Radio Architecture

Part 4. Software Radio Technology Towards Pervasive Appliance

Part 1

Software Radio for Universal Wireless Internet Access

Bluetooth: "Last Meter" Technology for Nomadic Wireless Internetting

M. Gerla[1], P. Johansson[2], R. Kapoor[1], and F. Vatalaro[3]

[1] University of California Los Angeles, USA
[2] Ericsson Radio Systems AB, Sweden
[3] Università di Roma Tor Vergata, Italy

Abstract: New technologies are being introduced for broadband Internet access, both wired and wireless. The variety of "nomadic" access scenarios clearly shows the need to provide sophisticated interworking functions to ensure efficient, seamless, ubiquitous and flexible support to the user. Bluetooth plays an important role in the functional integration of systems and networks. After a brief Bluetooth overview, this paper discusses the issue of indoor access/distribution in which Bluetooth can provide the "last meter" interconnection between the user and existing and forthcoming "last mile" technologies. Then, the paper provides simulation results for TCP and Voice-over-IP traffic. Simulation results for TCP traffic confirm that Bluetooth provides predictable and robust performance, and that it can favorably compare with other wireless local area networks. It also allows better performance to Voice-over-IP in heavy load situations.

1. Introduction

With the increasing dependence on the Internet in many aspects of our daily life, users demand ubiquitous, high performance Internet access whether they are at work, at home, or on the move. Until now, the predominant way to access the Internet has been the wired access, from fixed terminals. However, due to the increasing bandwidth of the multimedia contents offered by the Internet, traditional local loop phone lines may be inadequate to support the needs of residential users. New technologies are emerging for broadband residential access, both wired and wireless. Table 1 accounts for some of the most promising ones.

However, the larger market for communications services and applications is expected not to be for the residential user but for the user on the move. Personal mobile users are estimated to largely exceed 100% of the population, and some estimates are in the order of 400%, or more, when also appliances and vehicles are taken into account as "users" of telecommunication networks. Very likely in the future a variety of mobile devices will operate as thin clients of much larger information services throughout the networks. The Internet

Table 1. List of acronyms/systems

ADSL: asymmetric digital subscriber line
CATV: cable television
DECT: digital enhanced cordless telecommunications
GEOS: geostationary earth orbit satellites
GPRS: general packet radio service
HDSL: high speed digital subscriber line
HFC: hybrid fiber coaxial
IMT-2000: international mobile telecommunications 2000
LAN: local area network
LEOS: low earth orbit satellites
LMDS: local multipoint distribution service
MMDS: multichannel multipoint distribution service
PLT: power line telecommunications
PWT-E: personal wireless telecommunication - enhanced
UMTS: universal mobile telecommunications system
VDSL: very high speed digital subscriber line
WATM: wireless asynchronous transfer mode
WLAN: wireless local area network

itself may be the "ultimate" personal digital assistant (PDA), and Internet access will be nomadic and supported by wireless technology.

The emerging IMT-2000 third generation cellular technologies (UMTS, etc.) are designed to integrate traditional voice communications with data and Internet access. However, cellular infrastructures are limited in their ability to efficiently handle situations such as local interconnection of personal devices, as well as of intelligent, networked appliances at home and in the office. Individual users may need to establish "ad-hoc" networks to communicate and collaborate with each other as a team. Furthermore, over large distances, such as sparsely populated areas, or by air or sea, only satellites can provide Internet access.

The variety of expected nomadic Internet access scenarios clearly shows the need for flexible support to the user on the move. The concept of "smart spaces" has been introduced [1] to handle the complex situations arising in a mobile scenario with a large variety of multimedia services and technologies and their seamless integration. We can view a smart space as a collection of objects equipped with embedded computing and communications devices. These objects interact with the environment via sensors and actuators, and communicate with each other. The higher the degree and promptness of interaction, the "smarter" the space. For example, an individual may carry many personal devices, such as a cellular phone, a PDA, earphones, etc. As a whole, these are parts of a "subspace", i.e. the mobile individual. Many individuals can step into an environment (a room, a bus, etc.), which in turn becomes an object of a larger space (a building, a highway, etc.), and so on.

The Bluetooth system concept [2] turns out to provide a near-term answer to this need of different systems and networks integration. It can be considered as one first powerful building block of the smart space concept and is emerging as a de-facto open standard for short-range wireless networks. For instance, when in an indoor environment a terminal equipment through Bluetooth can act as an interphone or as a wireless local loop (WLL) terminal, while outdoor it can become a cellular phone, or a multimedia terminal. Also, printers, computers, keyboards, mice, and whatever digital equipment can be wirelessly connected in a Bluetooth network.

The paper is organized as follows. After a brief Bluetooth overview (Section 2), the issue of indoor access/distribution is discussed (Section 3) in which the Bluetooth system can play the role of "last meter" technology to interconnect the user to existing and forthcoming "last mile" technologies. Section 4 focuses on a few representative indoor service scenarios to provide simulation results on throughput for TCP connections established through Bluetooth, and for Voice-over-IP applications. Finally, Section 5 collects some paper conclusions.

2. Bluetooth Technology Overview

The Bluetooth system operates in the worldwide unlicensed 2.4 GHz Industrial-Scientific-Medical (ISM) frequency band. To make the link robust to interference, it employs a Frequency Hopping - Spread Spectrum (FH-SS) technique, with carrier frequency changed at any packet transmissions. To minimize complexity and to reduce the cost of the transceiver, it adopts a simple binary Gaussian frequency shift keying modulation. To allow efficient wideband data transmission the bit rate is 1 Mbit/s.

Two or more Bluetooth units sharing the same channel form a piconet, see Fig.2.(a). Within a piconet a Bluetooth unit can be either master or slave. Within each piconet there may be only one master (and there must always be one) and up to seven active slaves. Any Bluetooth unit can become a master in a piconet. Furthermore, two or more piconets can be interconnected, forming what is called a scatternet, see Fig.2.(b). The connection point between two piconets consists of a Bluetooth unit that is a member of both piconets. A Bluetooth unit can simultaneously be a slave member of multiple piconets, but master in only one, and can only transmit and receive data in one piconet at a time, so participation in multiple piconets has to be on a time division multiplex basis.

The Bluetooth system provides full-duplex transmission built on slotted time division duplex (TDD), where each slot is 0.625 ms long. Master-to-slave transmissions always start in an even-numbered time slot, while slave-to-master transmissions always start in an odd-numbered time slot. An even-numbered time slot and its subsequent odd-numbered time slot together are called a frame. There is no direct transmission between slaves in a Bluetooth piconet, only between master and slave, and vice versa.

The communication within a piconet is organized such that the master

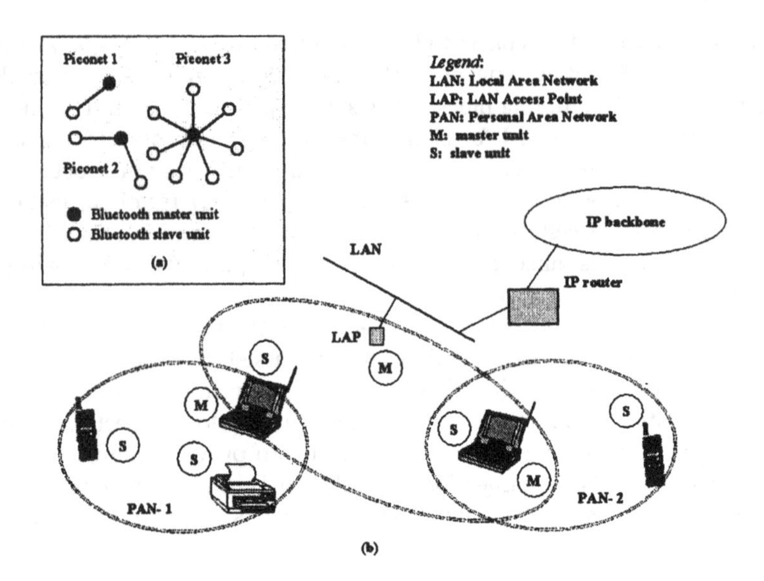

Fig. 1. Examples of: (a) Bluetooth piconets; (b) a Bluetooth scatternet

polls each slave. A slave is only allowed to transmit after the master having polled it. The slave will then start its transmission in the slave-to-master time slot immediately following the packet received from the master. The master may or may not include data in the packet used to poll a slave. It is however possible to send packets that cover multiple slots. These multi-slot packets may be either three or five slots long.

Each Bluetooth unit has a globally unique 48-bit IEEE 802 address. This address is permanently assigned when the unit is manufactured. In addition to this, the master of a piconet assigns a local active member address (AM_ADDR) to each active member of the piconet. The AM_ADDR is three bits long, is dynamically assigned and reassigned, and is unique only within a single piconet. The master uses the AM_ADDR when polling a slave in a piconet.

Bluetooth packets can carry either synchronous data on synchronous connection oriented (SCO) links mainly intended for voice traffic, or asynchronous data on asynchronous connection-less (ACL) links. Depending on the type of packet that is used, to ensure reliable transfer of data an acknowledgment and retransmission scheme is used, except for SCO links. In addition, a forward error correction (FEC) scheme may be used to further improve reliable packet transmission.

Fig. 2 shows the standard format of a Bluetooth packet, although there are exceptions for certain control packets. The AM_ADDR is located in the packet header followed by some control parameters (e.g., a bit indicating acknowledgment or retransmission request of the previous packet, when applicable) and a header error check (HEC).

ACCESS CODE	HEADER	PAYLOAD

Fig. 2. Standard Bluetooth packet format

Scatternets may be used for a number of reasons [3]. First, this is a way to extend the rather limited coverage that Bluetooth provides. The line-of-sight behavior may be improved if multihops between piconets are provided, allowing traffic paths around obstacles. Second, a scatternet may be the best solution to handle certain traffic and configuration patterns. For instance, when multiple Bluetooth piconets connect with a LAN through a LAN access point (LAP), the LAP is preferably chosen as the master. Since a node can be master of one piconet only, a simple architecture consists in connecting the remaining piconets to the "root" piconet in a tree-shaped scatternet. Third, the partitioning of a network into several scatternets may give an overall gain in capacity for the same area. For instance, if two slave nodes in different piconets have a sustained traffic between them, it may be a better choice to create a new piconet containing only the two nodes, and the resulting scatternet has improved performance. In fact, the FH-SS access system to the channel makes Bluetooth very robust against interference. Therefore, new overlaid piconets gain substantially more capacity than it is lost due to the increased interference between piconets.

3. Interfacing "Last Mile" and "Last Meter" Technologies

Bluetooth will be integrated with a large set of wireless and wired technologies, such as: wired/wireless LANs, WLL systems, HFC systems, xDSL systems (x = A, H, V), and third generation cellular systems. Wireless access to the Internet and, in general, to any kind of telecommunication networks, both fixed and mobile, is among Bluetooth most appealing applications. In this picture, Bluetooth can cover the role of "last meter" technology for which two kinds of environment can be envisaged:

- Indoor wireless universal coverage, where the "last mile" access can be provided through wired or wireless technologies (see Table 1)
- Mobile (e.g., vehicular) link to the personal or automata user, where the access function is provided through the IMT-2000 family of wide-band standards.

In the indoor scenario we consider in this paper, ad-hoc Bluetooth networks provide simple and cost-effective intercommunication of various mobile/personal devices. An integration scenario of Bluetooth and last mile technologies is the direct access to the building through a WLL technology, (DECT, PWT-E, MMDS/LMDS), as well as through GEOS, LEOS [5], [4]. The interfacing of

Bluetooth and WLL technologies involves aspects spanning from the physical layer to the network and transport layers, and in general, interface means are needed to ensure proper interworking.

In-building interconnection through the TV coax cable system is an example of a cable infrastructure of interest for interactive multimedia services delivery. This is especially useful in those countries where the TV distribution networks are mainly based on VHF/UHF radio links, instead of CATV. In these countries (e.g., Italy) often the TV reception is centralized with a tree coax distribution system. The existing building cable network can thus be reused to provide a building backbone for the interconnection of Bluetooth access ports among each other, as well as to provide a two-way low-cost access network for WLL and satellite systems. Therefore, adding and integrating Bluetooth to the coax system can provide an indoor low-cost hybrid wired/wireless LAN architecture. To extend the building cable system to this WLL access function we face some of the problems already considered for the HFC access to the home [4].

The above networking scenario involves access to the building with a WLL technology (case of small cells) or a GEOS/LEOS technology (case of large cells); transport within the building with the TV cable; access/transport in indoor picocells with the Bluetooth. The advantage of this architecture is reuse of an existing infrastructure (i.e., the cable), no need for civil works in the building, low cost and flexibility of the network through Bluetooth ad-hoc networking (Fig. 3).

The proper use of the building coax cable asks for a regenerative and, in the future, reconfigurable (e.g. through a software radio approach, [6]) common unit (CU) that must take care of all the interworking functions. The CU can be located at roof level, and should transparently deliver the TV signals to the home. Adaptation of the Bluetooth signals to the cable bandwidth implies the need for a simpler interworking unit at each cable end, the individual unit (IU). In addition, the ability of Bluetooth to establish piconets and scatternets can be made easier if local access points are made available to connect nearby rooms with very limited civil works, see Fig. 3. A possible solution is to incorporate simple Level 2 switching functions in the IU's (e.g. LAN bridges) and Level 3 routing functions in the CU.

4. Simulation Experiments

In this Section we present simulation results based on Internet access from Bluetooth terminals with TCP file transfer and Voice-over-IP applications. We assume that the environment is a single room and evaluate achievable traffic throughput taking into account the interference between coexisting piconets, as well as external noise sources such as microwave ovens. We use NS-2 as the simulation environment [7]. We augmented NS-2 with a Bluetooth ACL link protocol package both for piconets and scatternets. Work is in progress to implement the SCO link function.

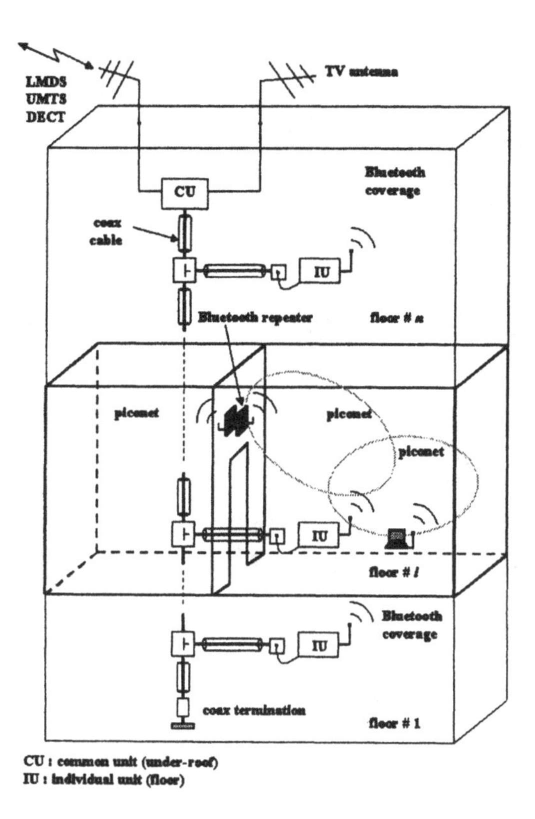

Fig. 3. Indoor interconnection of Bluetooth and last mile technologies

In our TCP experiments we vary the number of sources. The destination hosts are placed on a wired 10 Mbit/s Ethernet, beyond the Internet access point. The sources are engaged in file transfers over TCP connections. Different configurations can be used to connect the Bluetooth terminals to the wired Internet via an access port (see Fig. 4):

a) In the first configuration, the Internet access point has one Bluetooth port. In this case, all the TCP sources belong to the same piconet and the access point is the master.

b) In the second configuration, the Internet access point has as many access points as the TCP transactions. Each TCP source belongs to a different piconet and each piconet includes one master and one slave only.

c) In the third configuration, a single Internet access point is equipped with multiple Bluetooth cards and a single terminal can connect to the Internet via "parallel" links if it is equipped with multiple Bluetooth cards.

Using multiple cards, there is no limit to the number of piconets that can coexist in a room, except for the interference among them. For the first two cases above (the third is equivalent to the second) we measure the effective throughput achieved by Bluetooth. First, we examine the single piconet case

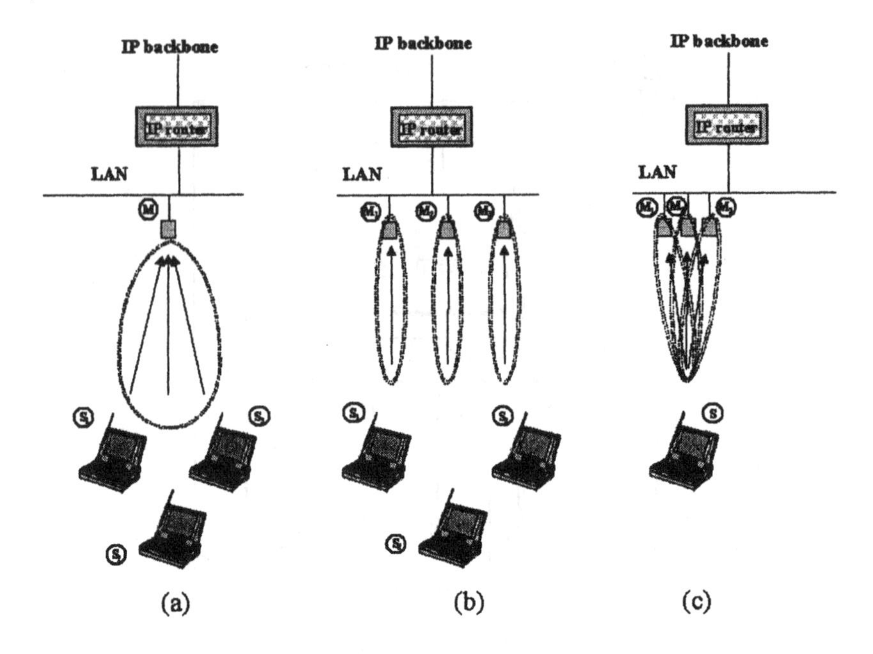

Fig. 4. Bluetooth TCP simulation scenarios

(see Fig. 4(a)). The results are reported in Fig. 5. Note that the total throughput in a single piconet (sum over all connections) is equal to 0.7 Mbit/s for any number of connections. In the same figure we show the total throughput for the scenario in Fig. 4(b). The total Bluetooth TCP throughput now increases linearly with the number of connections. This is due to the modest interference level between the different piconets.

Another interesting aspect is the throughput fairness across different TCP connections. Fig. 6 reports the individual TCP throughput values for an experiment with ten TCP connections and multiple piconets (Fig. 4(b)). For comparison, we show the uneven throughput achieved by the IEEE 802.11 WLAN 10 Mbit/s maximum data rate TCP connections. In fact, for the case of WLAN only four out of the ten TCP connections manage to transmit traffic. This capture effect was discovered before in similar situations with TCP running on IEEE 802.11. The main cause is the interplay between TCP timeouts and IEEE 802.11 timeouts. In contrast, Bluetooth provides very uniform performance by virtue of the fair sharing enforced by the polling MAC protocol.

All the experimental results reported above do not include external noise sources. It is well-known that microwave ovens can be a main source of interference in the ISM band (the Klystron source radiates around 2.45 GHz). Interference effects and measurements have been reported in the recent literature (e.g. [8]). Therefore, we investigated the effect of microwave interference on Bluetooth under two different assumptions:

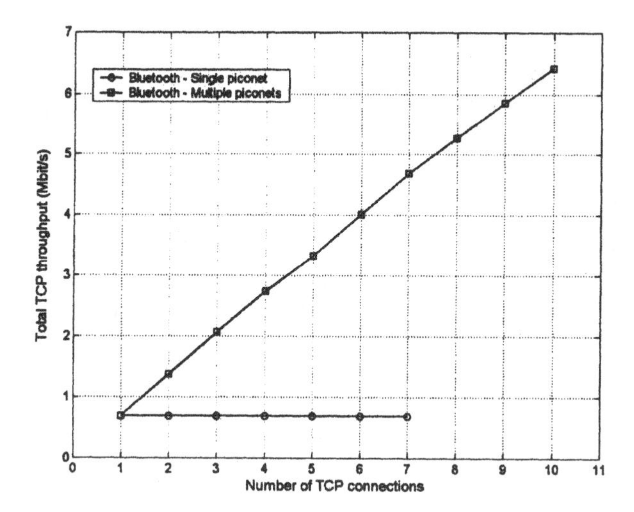

Fig. 5. Total TCP throughput as a function of TCP connections

- The health hazard federal radiation standard for microwave ovens produced since 1971 requiring that radiation leakage be limited to less than five milliwatts per centimeter squared at two inches from the oven over the lifetime of the unit
- Typical radiation characteristics as measured on residential microwave ovens (Table 2) [8]

The interference scenario assumes a microwave oven located 3m from the Bluetooth receiver and a transmitter to receiver distance variable between 2m and 18m. The link budget parameters and results are reported in Table 3.

Based on this link analysis a set of simulations with NS-2 has been also performed to evaluate the effect on packet loss. For the case of maximum radiation, packets are systematically lost when the interference is active and it hits the instantaneous frequency band of the hopping packet. However, due to both the interfering oven duty cycle and the frequency hopping gain, the actual reduction in capacity can be estimated around 5%, and this result was confirmed through simulations.

Another application in the indoor wireless environment could be IP telephony in which voice is transmitted as IP traffic and is carried by the ACL link. We evaluated the performance of a mix of voice and TCP sources. The experiments consisted of four interfering piconets. Each piconet supports one TCP connection and six voice connections, each between the master and a different slave. Therefore, we have four TCP and 24 voice connections. The TCP data connections are always active, with 500-byte packets. The voice connections are modeled according to the Brady model [9]. In particular, the voice connections are "on-off" sources, with "on" time equal to 1 s and "off" time

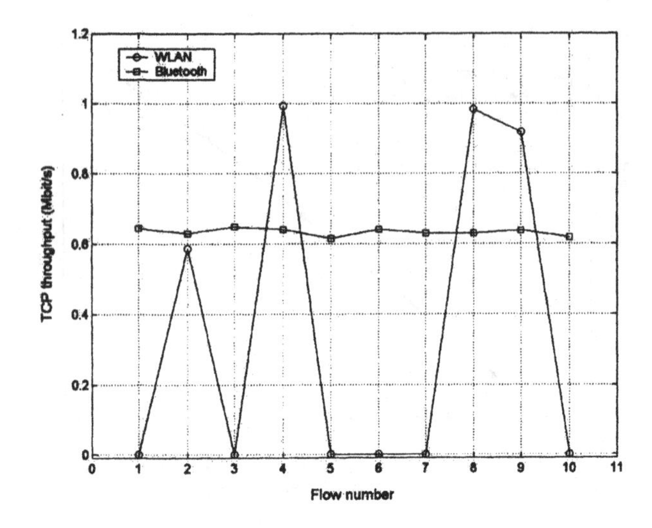

Fig. 6. Throughput of individual TCP flows

Table 2. Microwave oven characteristics

Frequency	
Center frequency:	2450 to 2460 MHz
Bandwidth:	2 to 6 MHz
Time	
Active time:	8 ms
Repetition time:	20 ms
Power	
Maximum EIRP:	+16 to +33 dBm

Table 3. Link budgets

Parameter	**Value**	
	Typical	**Worst-case**
Terminal transmitted power (dBm)	1	
Transmit antenna gain (dB)	3	
Receive antenna gain (dB)	3	
Coding gain (dB)	3	
Transmitter to receiver distance, D (m)	2...18	
Free space loss (dB)	$20 \log_{10} D + 40.4$	
Thermal noise power density (W/Hz)	$4 \cdot 10^{-21}$	
Signal bandwidth (Hz)	10^6	
Interference power density (W/Hz)	$1.3 \cdot 10^{-16}$	$3.2 \cdot 10^{-11}$
Signal to noise ratio (dB)	31.4...12.3	$-22.4... -41.5$

1.35 s, with "on-off" times exponentially distributed. The voice coding rate used is 8 kbit/s and the packetisation period is 20 ms, which gives a payload size of 20 bytes. Header compression is assumed for voice packets in Bluetooth and the total packet size is 30 bytes. Voice packets are sent using RTP over UDP. Each experiment lasts 5 seconds of simulation time. The results show that each TCP connection manages to fully load the piconet capacity left over by voice, achieving a throughput of approximately 0.31 Mbit/s. This is uniform over all piconets. Moreover, each piconet delivers full throughput since the probability of collision among four piconets is still fairly low. Voice end-to-end delays are acceptable, see Fig. 7(a). They range between a mean delay of

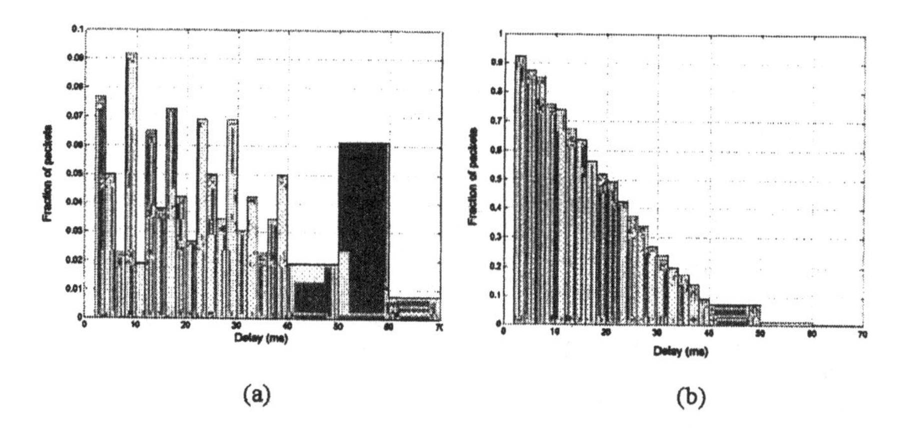

Fig. 7. IP telephony delay distribution: (a) probability density function; (b) complementary cumulative distribution function

about 25 ms and a maximum delay of about 60 ms. From the tail distribution in the complementary cumulative distribution graph, see 7 (b), we note that a packet loss ratio of less than 5% can be obtained for a play-out buffer of about 50 ms.

5. Conclusions

This paper considered the role of Bluetooth in the forthcoming telecommunications scenario in which a variety of emerging "nomadic" Internet access media and technologies will be available and there will be a need to provide rerouting and/or interworking between such technologies to ensure efficient, seamless, ubiquitous, and flexible support to the user on the move.

Several solutions for the deployment and use of Bluetooth as a "last meter" interconnect technology were outlined. Moreover, a simple indoor service scenario was studied and simulation results were provided under different traffic and topology situations. We evaluated the throughput performance in the presence of competing TCP sessions for different Bluetooth network layouts. In all, the simulation results confirm that Bluetooth performance is very predictable and robust. We have also compared Bluetooth with the IEEE 802.11 WLAN in terms of fairness. Bluetooth cannot match the single TCP connection throughput of the WLAN, but can actually surpass WLAN in aggregate throughput, measured over many connections in overlapped piconets. The effect of microwave oven interference has also been considered. In principle, a capacity reduction due to interference exists as the maximum radiated power allowed by health hazard considerations is too penalizing for Bluetooth. In spite of this, even under these extreme conditions, the loss of capacity is in the order of a few percent. The paper also showed that Bluetooth also guarantees a better performance to Voice-over-IP in heavy load situations.

References

[1] M. Gerla, "Mobile, wireless access to the Internet: from personal 'bubble' to satellites", CSELT internal report, Turin, Italy, Sept. 1999.

[2] J. Haartsen, "BLUETOOTH - the universal radio interface for ad hoc wireless connectivity", Ericsson Review, N. 3, 1998, pp. 110-117.

[3] P. Johansson, N. Johansson, U. Krner, J. Elg, G. Svennarp, "Short range radio based ad-hoc networking: performance and properties", ICC'99, Vancouver, 1999, pp. 1414-1420.

[4] M. Gagnaire, "An overview of bradband access technologies", Proc. IEEE, Vol. 85, N. 12, Dec. 1997, pp. 1958-1972.

[5] C. C. Yu, D. Morton, C. Stumpf, J. E. Wilkes, M. Ulema, "Low tier wireless local loop radio systems - Part 1: Introduction", IEEE Comm. Mag., March 1997, pp. 84-92.

[6] J. Mitola, "The software radio architecture", IEEE Comm. Mag., May 1995, pp.26-38.

[7] http://www-mash.cs.berkeley.edu/ns/ns.html

[8] A. Kamerman, "Microwave oven interference on wireless LANs operating in the 2.4 GHz ISM band", Proc. 8th IEEE PIMRC Symp., 1997, pp. 1221-1227.

[9] P.T. Brady "A model for generating on-off speech patterns in two-way conversation", Bell System Technical Journal, Sept. 1969, pp. 2445-2471.

The Pitfalls of Proxy Mechanisms Outside of the Research Domain

Bruce Zenel

Juno Online Services, Inc., New York NY 10036, USA

Abstract: The use of proxies in today's network environment has become commonplace; they are relied upon by Internet ready cell phones, wireless personal digital assistants, and ISP users. Generally, proxies are used to process data flowing between two endpoints using an intermediary. More specifically, they can be used to filter or process traffic flowing to and from a network-limited host. Benefits include more efficient use of network resources, reduced cost, and increased security. Along with these benefits come a host of problems and associated solutions. In this paper we will take a retrospective look at various proxy related research, identify the pitfalls associated with proxy mechanisms outside of the research domain, discuss why the general proxy model does not work well in industry, and present a case study of a contemporary proxy.

1 Introduction

Proxy based systems are nearly ubiquitous in today's heterogenous network environment. In general, a proxy takes the form of an intermediary placed between two communicating endpoints such as a client and server. The purpose of the intermediary is to improve the quality of the network as perceived by the client along some dimension which is constrained (e.g., connectivity, cost, security). Examples include firewalls [11], application specific proxies [1], programmable routers [24], and gateways that reduce traffic over low bandwidth or heterogeneous network links [4,17].

As part of our research on mobile computing, we spent two years building a general purpose proxy system [28,27]. Our work was general purpose in that it provided a mechanism for downloading and executing proxy programs (called "filters"), interposing filters into the middle of client-server connections, and dynamically controlling filter behavior. The filters may drop, delay, or transform data moving to and from the mobile host.

This work was done from 1995 to 1997; at that time proxies were relatively new and it was unclear where the long term evolution of proxy environments would lead. We envisioned a world where a general purpose proxy mechanism using application and protocol specific filters would be considered valuable and accepted over time. To our surprise, the general proxy mechanism fell by the wayside while a number of narrow focus proxy solutions (e.g., Palm.Net [1]) were developed and widely accepted. In this paper we will take a retrospective look at this research and identify the pitfalls associated with proxy mechanisms outside of the research domain. We will discuss why the general proxy model does not work well in industry, review a contemporary proxy system, and discuss the reasons for its success.

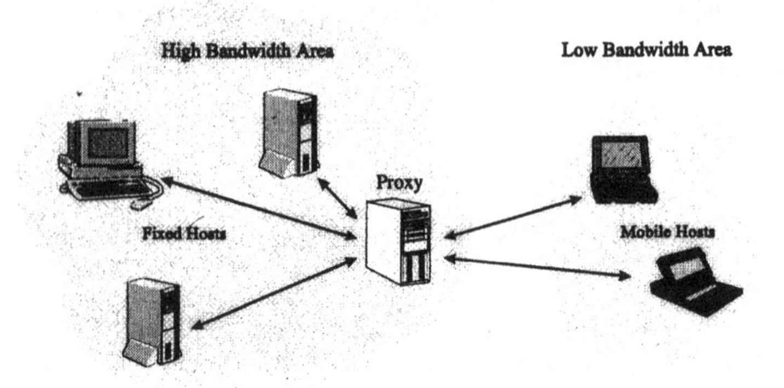

Figure 1. Proxy Computing Environment

This paper is organized as follows. Section 2 briefly describes our system, so as to establish a vocabulary and a concrete context for discussing issues. Sections 3 and 4 describe the architecture of the system and introduce the various filters used for the evaluation; Section 5 provides a summary of our quantitative analysis. Section 6 discusses the underlying problems associated with this architecture and other proxies in general. Section 7 provides a case study Palm.net, a wireless service provider that relies heavily on proxy technology. Summary and conclusions are provided in Section 8.

2 Our Proxy

The environment for which our work was intended is shown in figure 1. In this figure a pair of mobile hosts are communicating with a number of fixed hosts. To simplify the discussion, we assumed that an application (or "client") runs on the mobile host and communicates with a server on a fixed host, but the proxy mechanism can work with any two communicating endpoints located anywhere, including two mobile hosts.

In our model of the environment, we assumed that the "last link" to the mobile host is the likely problem area. This link will typically be slower, more costly, less reliable, and perhaps less secure than typical WANs or LANs. Accordingly, a proxy is placed somewhere on the opposite side of this link from the mobile host, and the two work together to overcome or attenuate the link's limitations. The proxy need not be physically attached to this last link.

We applied our proxy system to problems arising in the mobile environment. Specifically, our work focused on adaptation to heterogeneous network environments. Mobile hosts exacerbate the heterogeneity problem because they may move unpredictably through areas whose networks have very different speed, cost, security, and loss rate. Few existing distributed systems react well to sudden drastic changes in network characteristics. So in our setting the purpose of filters were to allow dynamic run-time, situation-specific, adjustment of design-time decisions about the nature and amount of traffic across the problem link.

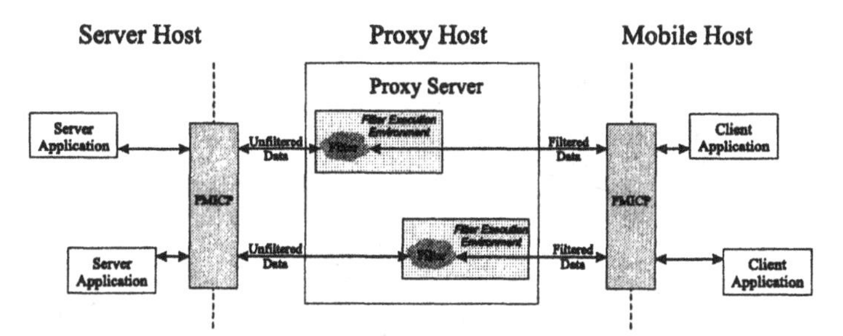

Figure 2. Proxy Server/PMICP Interaction

Actions a proxy filter might take include:

- Selectively drop structured data. Examples are frames in an MPEG [18] data stream, and superfluous header information in Email.
- Compress data instead of dropping it. Compression occurs at the proxy, rather than "end to end" at the server because the need for compression arises from link characteristics.
- Use a different transport protocol, or different parameter settings of the same protocol.
  Studies have shown that, in mobile and wireless environments, small changes in the use of UDP and TCP can result in large performance variations [5,6,10]. Therefore, it is desirable to use a protocol suited to the current network environment.

The filters executing on the proxy may be loaded from a number of locations; they may be built into the proxy, or downloaded from another site, such as the client or a filter repository. A filter may receive control messages (e.g., information about packet loss rate) from the application or from anywhere in the surrounding environment, allowing it to dynamically adapt to its environment. Filter control may be coarse (such as turning a filter on or off) or fine (such as directing a filter to ignore some portion of a message).

3 Architecture

The architecture of the system can be broken down into two components:

1. Proxy Server: Filtering unit.
2. PMICP: Protocol which moves data to and from the Proxy Server.

The PMICP protocol and the Proxy Server interact as shown in figure 2. The Proxy Server is sandwiched by the PMICP protocol, which guarantees that all traffic moving to and from the Mobile Host will pass through the Proxy Server. These two components are described in the following sections.

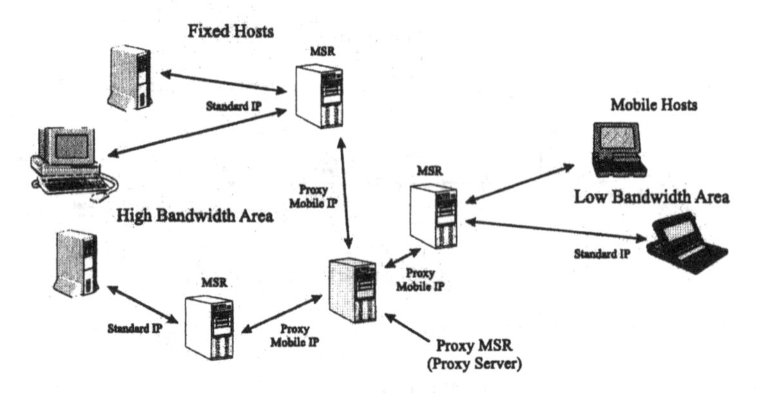

Figure 3. PMICP Environment

3.1 PMICP

In order to facilitate proxy filtering, all traffic traveling to and from a particular mobile host must pass through a single gateway (e.g., the Proxy Server). Without this assertion, data may find some other path around the gateway, and thus could not be filtered. In the world of stationary hosts, a simple solution can be provided by using static routes that force data between the two hosts to follow a strict path.

When mobile hosts are considered however, the problem becomes complicated. The Columbia Mobile IP Protocol (MICP) [16] supports host mobility by changing the path that the data takes depending on the location of the mobile host (MH). This is accomplished by keeping track of the location of the MH and using Mobile Support Routers (MSR) to direct data to its current location. Given this, the problem of assuring that data traveling to and from the MH pass through a single gateway becomes non-trivial. This is due mainly to the fact that as the MH moves through the network, MSRs will dynamically change the path taken by the data.

We have augmented the Columbia Mobile IP protocol in order to solve this problem. The new protocol, Proxy Mobile Internetworking Control Protocol (PMICP), allows a mobile host to choose an MSR to be its Proxy MSR (PMSR). Once chosen, the PMICP protocol guarantees that all traffic traveling to and from the MH will pass through the PMSR. This allows data to be filtered or otherwise processed at this point.

A more complicated view of the environment we are interested in is depicted in figure 3. In this figure we see a pair of mobile hosts communicating with a number of fixed hosts. The mobile hosts are using PMICP not only for host mobility, but to direct all traffic through a particular Proxy MSR (where the Proxy Server resides).

3.2 Proxy Server

The primary component of the filtering architecture is the *Proxy Server* (PS), which has been realized as a process executing on the PMSR. It provides a dynamic execution environment for filters, which may be permanently resident within the Proxy

Server, or downloaded on demand from a Mobile Host or filter repository elsewhere on the wired infrastructure. These filters have access to data streams traveling through the PMSR and are controlled primarily by applications executing on the Mobile Host.

4 Filters

In order to evaluate the system we created filters to be applied to application, transport, and network layer protocols. In particular, we looked at applications and protocols that perform poorly in the mobile environment. These include:

- HTTP: Hypertext Transfer Protocol.
  The HTTP [8] protocol is used by most HTML [7] browsers to transfer data from an HTTP server to a browser. In an HTTP transaction, the client (browser) makes a request for a document, which is returned by the server using the same connection. In general, HTTP transactions (messages) consist of two sections, a text header, followed by an optional body that may be text or binary in nature. The HTTP protocol is a good candidate for filtering since text in the header and body is highly compressible. The filter for this protocol compresses the header and body of HTTP messages destined for a browser on the Mobile Host.
- SMTP: Simple Mail Transfer Protocol.
  The SMTP [19] protocol is used to transfer email from one host to another. It has a well defined structure that lends itself to filtering. An email header can contain various types of information, some of which may be considered superfluous. There are many image and voice standards, such as MIME [9], that are used to encode voice/image data within an email message. To the mobile user, these images may be considered a waste of bandwidth if the image data is nothing more than a picture of the sender. Since these voice/image encoding methods are somewhat standard, they can be effectively filtered within the Proxy Server.
- NFS: Network File System protocol.
  The NFS [22,3] protocol is a UDP based RPC [21] protocol used for stateless file transfer. Within the environment that our system exists, an NFS client on the Mobile Host makes NFS requests to servers within the wired network. Most requests are NFS READ requests, which return file data to the MH. These replies are good candidates for filtering since they will often contain compressible text. Accordingly, we have designed and implemented an NFS filter specifically to handle NFS READ requests and replies.
- TCP: Transport Control Protocol.
  A number of problems (and solutions) with the TCP protocol in the weakly connected environment have been raised [4,26,10]. Most of the issues are related to how the congestion control algorithm used by TCP is not valid in the weakly connected environment. Most solutions improve TCP performance at the cost of breaking TCP's end-to-end semantics. An alternative solution investigated by Balakrishnan et. al. [2] at Berkeley propose the use of an intermediary that "snoops" TCP conversations, but does not break the end-to-end semantics. The

main purpose of the intermediary is to cache packets headed towards the Mobile Host and to perform local retransmissions when packet loss between the intermediary and Mobile Host is detected. This has the effect of speeding up the retransmission process and hiding packet loss from the sender side of the connection. The latter effect reduces the risk that TCP's congestion control algorithm will be activated accidentally. We took the basic mechanism they developed and implemented it using our system.

5 Evaluation

We performed a quantitative analysis directed towards the overall performance of the system including the filters. Specifically, we wanted to measure how well the filters performed under varying conditions. We chose three filters to be evaluated, HTTP, NFS, and TCP. These were chosen primarily because their performance could be readily measured. The HTTP and NFS filters use "lossless" compression, so overall data throughput measurements can be made for transfers of varying sizes. The TCP filter improves performance over a lossy link, which can also be measured for varying packet loss rates. We will only give a summary of the results from the quantitative evaluation; the complete results have been published previously and can be found in [28] and [27].

Overall the quantitative results were very positive. In nearly all the cases the filtering system performed somewhat poorly in environments with little heterogeneity, but performed very well when the level of heterogeneity was high. Since our target environment was one with a high degree of heterogeneity, we felt that we had achieved one of our primary goals.

6 Problems

While the quantitative evaluation of our proxy provided some promising results, a more qualitative review exposed a number of problems with our architecture. In this section we will address some of these problems and how they impact the application of our research. Where possible, we will also discuss how these issues affect other proxy architectures.

6.1 Invalid Assumptions

General versus Specific

One of our underlying assumptions was that a general purpose mechanism would be inherently more valuable than a specific one (whether specific to a device, application, or protocol). What we did not consider, or at least failed to address, was that general solutions are often harder to implement, deploy, and gain acceptance than solutions for a specific problem. This can be readily seen today in that nearly all widespread proxy based solutions (e.g., Palm.Net, WAP [25] technology, etc.) are specific to a particular technology or vendor. In the long run it is clearly more expensive to create a series of per-domain proxy solutions than one general purpose proxy (which is perhaps more complicated than any per-domain proxy), but apparently the need for a general purpose solution has yet to arise.

New Protocols to Address Specific Problems
We created a mechanism that would improve the quality of service for the end user by operating on existing protocols in seamless way. This is a general solution allowing many end users to benefit from a relatively small number of filters within a fully deployed filtering environment. While this is valuable we failed to consider the acceptance of new protocols that address specific problems with current protocols or particular application domains (e.g., a disconnected environment such as Internet telephony). Given the historical lack of interest/adaptation/integration of new IP based protocols (e.g., IPv6 [12]), we did not expect new protocols that deal with specific issues to become widely implemented and accepted. An example that contradicts our view is WAP, which was developed to deal with the specific issues of wireless Internet use and has been implemented and installed by a number of cell phone producers and Internet service providers (ISPs). The success of this and similar domain specific solutions is likely to continue in the future due to the acceptance of newer technologies which may all but replace older ones.

6.2 Deployment

In our original research we glossed over the issue of deployment, which has a clear impact on long term acceptance. Our mechanism required access to both the client (i.e., the end users' host) and the proxy being used. We required modifications to the operating system on the both the client and proxy host and in some cases modifications to particular end user applications. This is not practical in a real world situation. In addition, in order to ensure that the proxy was involved in every interaction, we developed and required the use of the PMICP protocol. In actuality this would involve a large scale deployment onto many hosts on the Internet, which once again is not practical.

Going beyond a discussion of our particular proxy it is interesting to examine the proxy deployment problem in general. A given entity (e.g., end user, content portal, Internet service provider) usually controls one side of an interaction. For example, an end user has control over their local PC and to some extent the applications used; a content provider has control over the server machines used to deliver content. In order for a proxy to be effective, the entity interested in increasing the end user quality of service (be it the end user or ISP) must either have the ability to deploy a proxy, or an infrastructure must exist such that a proxy is inherent to the network. While this is not practical in the general case (which includes our proxy mechanism), there are a number of areas where proxy deployment is very practical. Examples include:

- Large scale Internet service providers. Internet service providers such as America Online that own or have control over large portions of their network are able to deploy proxies very effectively.

- Palm.Net. The Palm.Net network used by the Palm VII personal digital assistant is an ideal environment for proxy deployment since all traffic traveling to and from *every* Palm VII device must traverse this network.
- Cell-phone based wireless Internet service providers. Similar to Palm.Net, wireless Internet service providers have an advantage over wired providers in that traffic must traverse their wireless access points. These provide ideal locations for proxy deployment.

6.3 Semantic Filtering Issues

We created a series of filters meant to span the standard network protocol stack. For example, we developed a low level filter to improve TCP throughput in high loss environments, a medium level filter to reduce NFS bandwidth requirements, and a high level filter to reduce SMTP message length. One issue that was not discussed in our original work is the fact that as one moves upward in the network protocol stack (e.g., from TCP towards SMTP), filtering solutions must be more semantic in nature. Specifically, our TCP filter was very general in that it acted on all TCP data, regardless of content. This is in contrast to the SMTP filter which required an understanding of the email content in order to remove superfluous header information. Not only does this problem result in high level filters becoming less general, but it also raises a semantic problem that is best explained by example. Let us assume that a particular high level filter reduces HTTP transaction time by processing image content in some way (e.g., distillation [13] or some other transformation). While this is likely to improve the quality of service by reducing the time required to receive an HTTP reply, it may *reduce the perceived quality of service* by disturbing the resulting layout of the content (possibly resulting in navigation problems). Unfortunately this is a very general problem and can only be dealt with effectively by building an infrastructure that supports filtering as an inherent feature (e.g., the Palm VII).

6.4 Mobility

In our domain of interest, mobility and wireless go hand in hand. While our research can be applied to any heterogenous network, our primary intent was for use in a wireless environment. We envisioned a world where clients could roam and their proxies would have the ability to move if appropriate. Since the proxy contains some amount of client side state, it would be advantageous to allow it to move within the network. For example, there are advantages to having the proxy physically attached to the "last link"; e.g., it is easier to gather information about link characteristics, which the proxy could use to alter its behavior. So as the mobile host moves through the network, the proxy would have the ability to follow it. This turned out to be a problem of such complexity that we decided to work around it by enhancing Columbia's mobile IP protocol (MICP) to force data through our proxy regardless of the client's location. Any proxy solution (whether general or restricted) that makes use of client side state information will face this problem and choose a solution that best suites the application environment.

6.5 Economics

While discussing the application of our research in industry it would be remiss to not discuss the economic aspects of the work. Capitalism suggests that the only way to gain widespread acceptance of a particular product is to slot it into an existing niche or create and exploit a new niche. As far as we can tell, the appropriate sized niche for a general purpose proxy never opened. At the time that we concluded our research there appeared to be a lack of strong interest in general proxy technology, although at least one product did appear on the market. In 1997, IBM released its WebExpress [15] product, which had a number of features very similar to our work. It was not truly *general*, but it did address a number of problems associated with standard client/server operation in a disconnected environment. Their proxy acted upon HTTP streams, reducing the amount of bandwidth consumed and the number of interactions required by a fixed server and client residing on a mobile host. The product was well received by critics [23] but it never gained a strong foothold in the market. It is unclear whether this was a result of the complexity of the application, poor execution of the business model, or the lack of interest from the general public.

7 Case Study of Palm.Net

While general purpose proxy mechanisms have not gained a strong foothold within industry, there are numerous narrow focus proxies which been widely deployed and accepted. One such proxy architecture is Palm.Net which will be discussed in the following sections.

7.1 Introduction

The Palm VII Connected Organizer [1] is a personal digital assistant (PDA) developed and marketed by Palm Computing Inc. (part of 3Com Inc.). It follows an evolutionary line of Palm Pilot devices, the closest relative being the Palm III. The most striking point of differentiation from its predecessors is the Palm VII's integrated wireless functionality. The device comes equipped with a radio transmitter and receiver that are designed to be used within Palm.Net, a wide area wireless network. Currently, Palm.Net is comprised solely of the BellSouth Wireless Data Network, which covers more than 260 population centers within the United States. The Palm VII owner can subscribe to this network service for either a flat monthly fee or a pay-by-the-kilobyte metered rate. The Palm.Net model allows new wireless networks (possibly using a different technology) to be added as the product matures.

7.2 Hurdles and Solutions

The designers of the device were presented with a number of problems. Once the underlying data network was chosen, an architecture had to be developed that allowed interoperability between the Palm VII device and the rest of the Internet. The approach taken by the Palm was to use proxy technology to translate standard Internet

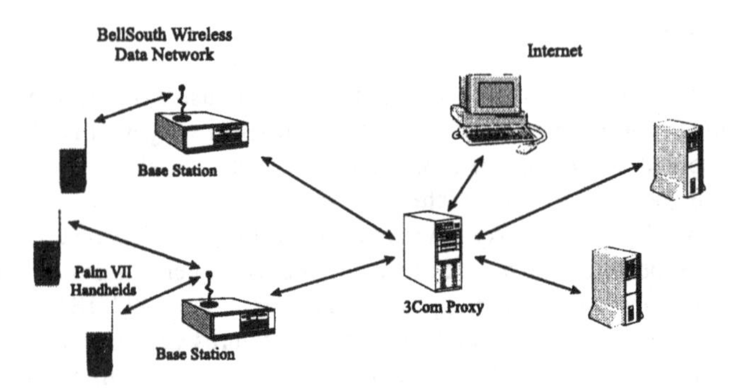

Figure 4. Palm.Net Proxy Architecture

protocols into ones that were more suitable to the Palm VII's operating environment (e.g., disconnected, low bandwidth, etc.). The Palm.Net proxy architecture is shown in figure 4. This figure shows Palm VII hand held devices communicating with base stations within the BellSouth Wireless Data Network. In order to reach the Internet, data travels from this network, through the 3Com Proxy, and out into the Internet at large; data returning to Palm VII devices follows the reverse path. The purpose of the 3Com proxy is to translate protocols that are native to Palm.Net into standard Internet protocols and visa versa.

The proxy specifically addresses the following protocols:

- HTTP. The 3Com proxy filters HTML data within an HTTP stream in order to make it more compatible with the Palm VII device; all remaining data is compressed.
- SSL. While the SSL [14] protocol suite is widely accepted as being extremely secure, it is an expensive protocol in that the use of large keys (crucial to strong security) results in long message lengths, slow response times, and decreased battery life. In contrast, the Palm VII uses ECC [20] technology, which is arguably as strong as SSL, but uses shorter keys. The 3Com proxy takes an SSL encrypted stream, decrypts it, re-encrypts it using ECC technology, and forwards this data to the Palm VII device.
- TCP. As discussed in Section 4, TCP is not very well suited for the weakly or disconnected environment. The Palm.Net designers deal with this by doing away with TCP altogether. Instead only UDP is supported between the Palm VII device and the 3Com proxy. It is the 3Com proxy's duty to turn TCP streams into UDP "streams" that have similar features as TCP (e.g., reliable, in-order delivery, etc.) without the overhead associated with TCP.

The Palm VII takes another novel approach to reduce interactions between the Palm VII device and the rest of the Internet. Palm advocates the use of "web clipping" applications on the Palm VII device. These are web applications that have been partitioned such that part of the functionality resides on the Palm VII device and the rest remains on a web server as before. This partition restricts the interaction between client (i.e., Palm VII) and server to a model resembling a transaction. The client sends a well-defined query to the server, which replies with a well-defined response. An example of this would be a stock quote application. The application would be partitioned such that the stock symbol entry form would reside on the Palm VII; the only interaction between client and server would be the transmission of the symbol to the server and the associated result (formatted in an expected way). Another valuable asset of this model is that standard HTML can be used on the server, allowing applications to be more universal in nature.

7.3 Reasons for Success

The success of the 3Com proxy and associated Palm.Net architecture can be attributed to a number of factors.

- Constrained domain. The goals associated with the 3Com proxy and Palm.Net are limited and well understood. Within the disconnected operating environment of the Palm VII device, the primary goal was to allow HTTP like transactions using the limited capabilities available. This allowed the designers to come up with solutions that were more optimal than ones which would be required for a general purpose system. This resulted in an improved quality of service, measured both quantitatively and qualitatively.
- Deployment. As discussed in Section 6.2, one of the main problems preventing widespread acceptance of a proxy system is deployment. The Palm.Net architecture has the advantage of having control over the proxy, the network used by the client, and to some extent the client itself. The design of Palm.Net places the 3Com proxy at the interface point between the BellSouth Wireless Data Network and the Internet; this is an ideal location. In addition, since 3Com controls both the proxy and the operating system running on the client (i.e., Palm VII), technology can be put in place that best handles operation in a disconnected environment (e.g., UDP based protocols, ECC encryption, application partitioning).
- Success of the Palm VII device itself. Last but not least, the success of the 3Com proxy has no doubt been bolstered by the success of not only the Palm VII, but of the series of Palm Computing devices that preceded it. There is some "early adoption" risk associated with the Palm VII due to its new wireless functionality, but beyond that the device is basically a Palm III (a mature and widely accepted PDA). Where 3Com excelled was in its integration of the wireless technology into the existing interface. The Internet enabled Palm VII applications have functionality associated with both standard Palm applications and HTML browsers, providing an integrated end user experience.

8 Conclusions and Summary

Our original research showed that a general purpose proxy mechanism could be successfully applied to problems arising from mobility. Unfortunately, as can be concluded from Section 6, the quantitative reasons alone are not compelling enough to build and deploy such a complex system. We believe that future proxy architectures will be of a specific nature, built to solve a particular problem or application domain. The relative success of current domain specific proxies versus the lack of general purpose ones will continue and the gap between the two is likely to widen considerably.

In this paper we have taken a retrospective look at some proxy related research and discussed the pitfalls associated with proxy mechanisms outside of the research domain. The design and capabilities of a general purpose proxy were presented, along with sample filters that address "real world" protocols. A quantitative evaluation of the system was provided along with a discussion of the problems associated with both general and domain specific proxies. We concluded the paper with a case study of Palm.Net, a contemporary, widely deployed proxy architecture.

References

1. 3COM, INC. Palm VII Connected Organizer. URL: http://www.palm.com/pr/palmvii/-7whitepaper.pdf, 1998.
2. AMIR, E., BALAKRISHNAN, H., SESHAN, S., AND KATZ, R. H. Improving TCP/IP Performance over Wireless Networks. In *Proc. First Annual International Conference on Mobile Computing and Networking* (November 1995), ACM, pp. 2–11.
3. B. CALLAGHAN AND B. PAWLOWSKI AND P. STAUBACH. NFS Version 3 Protocol Specification. RFC 1813, IETF Network Working Group, June 1995.
4. BADRINATH, B., BAKRE, A., MARANTZ, R., AND IMIELINSKI, T. Handling Mobile Hosts: A Case for Indirect Interaction. In *Proc. Fourth Workshop on Workstation Operating Systems* (October 1993), IEEE, pp. 91–97.
5. BADRINATH, B., AND SUDAME, P. To Send or Not to Send: Implementing Deferred Transmissions in a Mobile Host. In *Proc. Sixteenth Intl. Conf. on Distributed Computing Systems* (May 1996), IEEE, pp. 327–333.
6. BAKRE, A., AND BADRINATH, B. Handoff and System Support for Indirect TCP/IP. In *Proc. Second Symposium on Mobile and Location-Independent Computing* (April 1995), USENIX, pp. 11–24.
7. BERNERS-LEE, T., AND CONNOLLY, D. Hypertext Markup Language - 2.0. RFC 1866, IETF Network Working Group, November 1995.
8. BERNERS-LEE, T., FIELDING, R. T., AND FRYSTYK, H. Hypertext Transfer Protocol - HTTP/1.0. IETF HTTP Working Group Draft 02, Best Current Practice, August 1995.
9. BORENSTEIN, N., AND FREED, N. MIME (Multipurpose Internet Mail Extensions) Part One. RFC 1521, IETF Network Working Group, September 1993.
10. CACERES, R., AND IFTODE, L. Improving the Performance of Reliable Transport Protocols in Mobile Computing Environments. *IEEE Journ. Selected Areas of Communication 13*, 5 (June 1994).
11. CHAPMAN, D., AND ZWICKY, E. *Building Internet Firewalls*. O'Reilly and Associates, 1995.
12. DEERING, S., AND HINDEN, R. Internet Protocol, Version 6 (IPv6) Specification. RFC 2460, IETF Network Working Group, December 1998.
13. FOX, A., GRIBBLE, S. D., BREWER, E. A., AND AMIR, E. Adapting to Network and Client Variability via On-Demand Dynamic Distillation. In *Proc. Seventh International Conference on Architectural Support for Programming Languages and Operating Systems* (October 1996), ACM, pp. 160–170.

14. FREIER, A., KARLTON, P., AND KOCHER, P. The SSL Protocol, Version 3.0. Internet Draft, IETF Transport Layer Security Working Group, November 1996.
15. HOUSEL, B. C., AND LINDQUIST, D. B. Web*Express*: A System for Optimizing Web Browsing in a Wireless Environment. In *Proc. Second Annual International Conference on Mobile Computing and Networking* (November 1996), ACM, pp. 108–116.
16. IOANNIDIS, J., DUCHAMP, D., MAGUIRE, JR., G. Q., AND DEERING, S. Protocols for Supporting Mobile IP Hosts. Internet-Draft, ftp://software.watson.ibm.com/pub/mobile-ip/columbia-draft-june-92, June 1992.
17. KOJO, M., RAATIKAINEN, K., AND ALANKO, T. Connecting Mobile Workstations to the Internet over a Digital Cellular Telephone Network. In *Proc. 1994 Mobidata Workshop* (November 1994), Rutgers University.
18. LE GALL, D. MPEG: A Video Compression Standard for Multimedia Applications. *Communications of the ACM 34*, 4 (April 1991).
19. POSTEL, J. B. Simple Mail Transfer Protocol. RFC 821, IETF Network Working Group, August 1982.
20. SCHNEIER, B. *Applied Cryptography, Protocols, Algorithms, and Source Code in C.* John Wiley and Sons, Inc., 1996.
21. SUN MICROSYSTEMS, INC. RPC: Remote Procedure Call Protocol Specification, Version 2. RFC 1057, IETF Network Working Group, June 1988.
22. SUN MICROSYSTEMS, INC. NFS: Network File System Protocol Specification. RFC 1094, IETF Network Working Group, March 1989.
23. TALLEY, B. Artour Gateway and WebExpress use smart caching. *InforWorld 19*, 20 (May 1997).
24. TENNENHOUSE, D., AND WETHERALL, D. Towards an Active Network Architecture. *ACM Computer Communications Review 26*, 2 (April 1996), 5–18.
25. WIRELESS APPLICATION PROTOCOL FORUM LTD. Wireless Application Protocol White Paper. WAP Forum White Paper, May 1995.
26. YAVATKAR, R., AND BHAGAWAT, N. Improving End-to-End Performance of TCP over Mobile Internetworks. In *Proc. Workshop on Mobile Computing Systems and Applications* (December 1994), ACM/IEEE, pp. 146–152.
27. ZENEL, B. *A Proxy Based Filter Mechanism for the Mobile Environment.* PhD thesis, Columbia University, May 1998.
28. ZENEL, B., AND DUCHAMP, D. A General Purpose Proxy Filtering Mechanism Applied to the Mobile Environment. In *Proc. Third Annual ACM/IEEE International Conference on Mobile Computing and Networking (Mobicom '97)* (September 1997), ACM, pp. 248–259.

Analysis of File Transfer Protocol Over Bluetooth Radio Link

Diego Melpignano[1] and Andrea Zanella[2]

[1] Philips Research Monza, G.Casati,23, 20052 Monza (Italy),
Phone +39 039 2037809, Fax. +39 039 2037800,
mailto: Diego.Melpignano@philips.com

[2] Department of Electronic Engineering, University of Padova,
Via Gradenigo 16/A, Padova (Italy),
Phone +39 049 8277662, Fax: +39 049 8277699
mailto: andrea.zanella@cnit.it

Abstract: TCP is the current dominant transport protocol, mainly used in fixed networks. It is well known that TCP performance may degrade over paths that include wireless links, where packet losses are often not related to congestion, but to the unreliability of the transmission medium. In this paper, we examine this problem considering a wireless link based on Bluetooth radio equipment. Bluetooth (BT) is a low-cost system in the unlicensed 2.4GHz band. It provides a reliable data transmission using fast frequency hopping technique and Stop-and-Wait ARQ scheme. In our experiments, we have studied the performance of a heavy file transfer over a BT link, with different environmental conditions and BT radio packet formats. Results show that the best FTP performance in a wide range of radio channel conditions is obtained by using long non-FEC-protected radio packets. Nevertheless, in particularly hostile situations, the intermediate-length packet format appears more suitable. Furthermore, analysis has focused the possibility of inefficiency due to bad interaction between TCP and BT retransmission mechanisms.

1. Introduction

The increasing popularity of wireless technologies and the continuing advancements in portable computing are flowing together in a raising demand for wireless networks. This consideration is forcing researchers to study the performance of the most common data applications over different radio equipment. In particular, much work has been made about the behaviour of Transmission Control Protocol (TCP) [1,2] over wireless networks, thanks to its primary role in the Internet protocols stack. It is well know that TCP is tuned to work well in wired networks, where packet losses and unusual delay are primarily due to congestion. Unfortunately, communication over wireless links suffers from sporadic high bit error rates that produce packet losses not related to congestion. These events may trigger the congestion reaction mechanisms on TCP sender [3], resulting in an unnecessary reduction in the end-to-end throughput and hence, sub-optimal performance.

A method proposed to alleviate these problems consists in hiding the unreliability of the wireless link from the TCP sender by using local retransmissions and forward error correction (FEC) schemes [4,5]. In this way, the lossy link appears as a reliable link with a reduced effective bandwidth. Unfortunately, the TCP sender may not be fully shielded from wireless losses. Indeed, the link layer retransmissions could generate sporadic long delay on segment delivery, causing the TCP sender timer to expire. These spurious timeouts trigger unnecessary segment retransmission and start the congestion control mechanisms, leading to a waste of available capacity in the wireless link and a significant degradation of end-to-end throughput. Besides, another problem is the low radio link capacity that could produce a buffer overflow at the interface between fixed and radio parts, resulting in a loss of performance.

In spite of these problems, many proposals to support wireless Internet access are based on link layer approach. Some examples can be found in [6] and [7], where a radio connection based respectively on GSM and DECT technology is analysed. In this paper we consider a new link layer solution, based on the emerging Bluetooth technology [8,9] to supply radio link connectivity.

The Bluetooth (BT) radio technology allows users to make wireless connections between various communication devices, such as phones, desktop and notebook computers. It supports both symmetric and asymmetric connections, with up to six radio packet formats characterised by different values of packet length, time duration and error protection. With the current specification, up to seven *slave* devices can be set to communicate with a *master* radio in one device, establishing a so-called *piconet*. Several of these piconets can be linked together in *ad-hoc scatternets* to allow communication in flexible configurations. Furthermore, the use of unlicensed ISM band has a good impact on the cost of BT systems. All these aspects make BT suitable for the realisation of an *ad-hoc* wireless LAN, especially for short-range indoor environments.

In our experiments, we have tested the behaviour of a typical TCP/IP application, such as an FTP session, through a BT wireless connection. The experiments aim to explore the impact of single-hop radio link configuration on the end-to-end FTP performance. The results obtained show that the best FTP performance in a large range of radio channel conditions is obtained by using long non-protected radio packets. Nevertheless, in particularly hostile situations, the intermediate-length packet format appears more suitable. Furthermore, analysis has focused the possibility of inefficiency due to bad interaction between TCP and BT retransmission mechanisms.

This paper is organised as follows. Section 2 provides a brief overview of Bluetooth system and outlines the mechanisms used by TCP to react against congestion. Section 3 introduces the methodology used in carrying out the tests and defines the target metrics considered. Section 4 describes the measurement platform. Section 5 presents measurement results and their analysis. Finally, Section 6 presents a summary of results and conclusions.

2. Background

This section gives a brief overview of Bluetooth technology and of the TCP features of primary interest for our purposes. A complete description of such topics can be found in e.g. [9] for BT and [1,2] for TCP.

2.1 The Bluetooth Radio System

Bluetooth (BT) is a low-cost system for medium bit-rates wireless *ad-hoc* networks in the unlicensed 2.4 GHz ISM band. Two up to eight BT units sharing the same channel form a piconet. In each piconet, one unit acts as master, controlling the channel access in order to avoid collisions, based on a centralised polling scheme. The channel is represented by a pseudo-random hopping sequence spanning 79 (or 23 for France, Spain & Japan) 1-MHz RF channels. The hopping sequence is unique for the piconet and is determined by the Bluetooth (BT) device address of the master, whereas the BT clock of the master determines the phase of the hopping sequence. The time is divided into time slots where each slot corresponds to a RF hop frequency. The nominal hop rate is 1600 hops/s, corresponding to a slot duration of 625μs. Full duplex is obtained with a slot-based Time-Division Duplex (TDD) scheme.

Both Synchronous Connection Oriented (SCO) and Asynchronous Connection Less (ACL) links can be established among BT terminals, respectively for coded voice and best-effort data traffic (symmetric and asymmetric), with an available bitrate of up to 721kb/s. The master maintains the SCO link by using reserved slots at regular intervals (circuit-switched connection). SCO packets are exclusively one slot long and they are never retransmitted. In the slots not reserved for SCO links, the master can exchange ACL packets with any slave on a per slot basis (packet-switched connection). A slave is permitted to return an ACL packet in the slave-to-master slot if and only if it has been addressed in the preceding master-to-slave slot, following the polling scheme mentioned above. A baseband packet can extend over **one, three** or **five** consecutive time slots. When a multiple slot packet is used, the transmitter frequency remains unchanged for the whole packet duration, thus reducing the efficiency loss due to the PLL settling time (~220μs), occurring each time a new frequency is used.

Each packet starts with a 72-bit **access code** (AC), used for synchronisation and piconet identification. At the beginning of each reception slot, the BT unit correlates the sequence received to the expected AC. The incoming packet is accepted only if the correlator output exceeds a certain threshold, elsewhere it is discarded. The AC is followed by an 18-bit packet **header** field, coded with a 1/3 FEC code (three-time repetition of every bit), resulting in a 54-bits header. A packet received with an unrecoverable header is immediately discarded. A **payload** field completes the baseband data packet. Payload is covered by a 16-bit CRC code, used by a simple Stop-and-Wait ARQ scheme to verify the integrity of received data. After reception and CRC control, a baseband packet is immediately acknowledged by using the ARQN bit in the header of the reply packet. An optional 2/3 FEC code can be applied to the packet payload to reduce the number of retransmissions. Formats that include FEC are denoted with DMn where *n* indicates the packet extension in number of slots, whereas DHn indicates packets without FEC. Table 1 synthesises the main characteristics of the six different ACL packet formats.

Table 1. ACL packet fields

Pck Type	DM1	DH1	DM3	DH3	DM5	DH5
Number of Slot	1	1	3	3	5	5
Payl Head (bit)	8	8	16	16	16	16
Payl Data (bit)	136	216	968	1464	1792	2712
FEC	2/3	no	2/3	no	2/3	no
Fw Max Rate (kbit/s)	108,8	172,8	387,2	585,6	477,8	723,2
Rv Max Rate(kbit/s)	108,8	172,8	54,4	86,4	36,3	57,6
Tot Pck Size(bits)	366	366	1626	1622	2871	2870

2.2 TCP Congestion Recovery Algorithms

Modern implementations of TCP include four intertwined algorithms for flow control: slow start, congestion avoidance, fast retransmit, and fast recovery. This section summarises only the main characteristics of these four algorithms, a complete description can be found in [3].

Slow Start and Congestion Avoidance
The flow control scheme used by TCP is based on a "sliding window" concept. The allowed window (*awnd*) is the minimum of the sender's congestion window (*cwnd*) and the receiver's advertisement window (*rwnd*). During the *slow-start phase* the value of *cwnd* is increased by one segment for each received ACK, whereas, in congestion phase, the *cwnd* is incremented by 1 full-sized segment per round-trip time (RTT). Slow start phase ends when *cwnd* exceeds a specific threshold, called *ssthresh,* after that congestion avoidance phase is executed. When congestion is detected, the slow start phase is restarted with a *cwnd* set to one 1 full-sized segment and *ssthresh* set to half of the amount of outstanding data in the network.

Exponential Backoff and Round Trip Estimation
TCP sender dynamically adapts the retransmission timeout (RTO) to the end-to-end delay, by using an estimation of mean (*SRTT*) and standard deviation (*D*) of the round-trip time (RTT). These values are adjusted each time an ACK is received, on the base of the following formulas

$$SRTT=SRTT+0.125\cdot(RTT-SRTT) \quad (1)$$

$$D=D+0.25\cdot(|RTT-SRTT|-D) \quad (2)$$

The retransmission timeout (RTO) is computed as

$$RTO=SRTT+4\cdot D \quad (3)$$

When a timer goes off before receiving the acknowledgement for the relative segment, the segment is retransmitted and the value of the timer is doubled (exponential backoff).

Fast Retransmit and Fast Recovery
The TCP may generate a duplicate acknowledgement (DACK) when an out-of-order segment is received ([10]). If three or more duplicate ACKs are received in a row, it is a strong indication that a segment has been lost. TCP then performs a retransmission of what appears to be the missing segment, without waiting for a retransmission timer to expire (**fast retransmission**). After this retransmission, congestion avoidance, but not slow start is performed (**fast recovery**).

3. Methodology of Analysis

The main focus of our analysis is to study the performance of different BT packet formats and the potential protocol interactions between TCP and radio link protocol (RLP). We are only interested in stable connections that last long enough to allow for all TCP sender state variables (e.g. retransmission timer, slow-start threshold, *etc.*) to converge from their initial values to a stable range operation. We have therefore performed a series of large bulk data transfers (roughly 1 Mbytes) between a fixed FTP server and a nomadic FTP client. Experiments have been made in different real-world situations, obtained by moving the nomadic station around our research laboratory. Unfortunately, we have no method to precisely determine the signal strength in each situation. So, the environmental hostility has been classified on the basis of practical considerations, as the distance between BT devices, the presence or the absence of Line-of-sight (LOS), *etc.* This empirical classification is consistent with the measured packet drop probability (PDP), *i.e.* the probability that a packet is received with unrecoverable AC or Header fields. Since these fields are identical for all the packet formats, the PDP depends only on the radio channel condition. Figure 1 shows the PDP (continuous line) for the five environments considered (the others lines depicted in the figure will be discussed in Section 5), whereas Table 2 gives a synthetic description of the environments. Note that each environment has been assigned a specific mark, as shown by the legend in the figure.

Table 2. Environments description

Env1	2mt distance between terminals on different desks
Env2	3mt distance on different desktops and microwave oven 4m away beneath a wall
Env3	4mt distance, wall in the middle with metal whiteboards on both sides
Env4	8mt distance on the same laboratory, with many obstacles between BT devices
Env5	3mt distance on different desktops with microwave oven between them

Many tests have been carried out in every environment, using different BT packet types. More precisely, we have only changed the packet formats in forward direction (from master to slave), whereas in reverse direction we have always used single slot unprotected packets (DH1), because they appear ideal to carry the low traffic in this direction (mainly due to TCP acknowledgement). From data collected we evaluated the radio link "goodput", defined as the average number of data bits successfully transmitted in the unit of time. The goodput is determined by baseband packets that are not retransmitted, that is packets that arrive with correct AC, HEC and CRC fields and that are successfully acknowledged. Instead, all sent data packets determine "throughput", defined as the total amount of data bits sent per time unit. In practice, the throughput represents the maximum link capacity obtainable with a given packet format, whereas goodput gives the effective link capacity perceived by the upper level protocol.

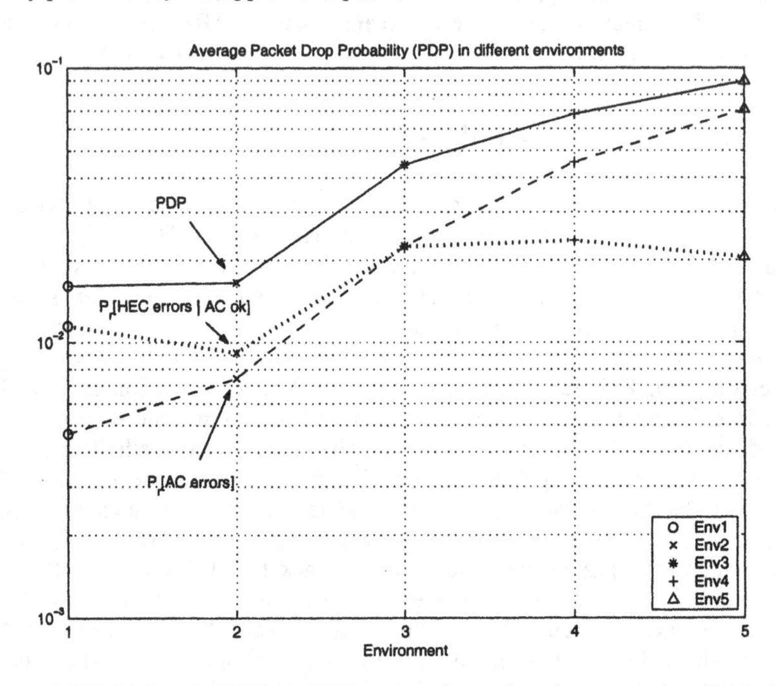

Figure 1. PDP, PAC and PHEC in different experimental environments (see Table 1 for environments description)

Besides goodput, we considered another metric, called the "Segment Service Time" (SST) defined as the time that BT entities employ to transmit a TCP segment through the radio link. Note that SST is only a part of the end-to-end delay, since it does not include the buffering time. SST cannot be precisely measured observing the radio channel only, since we are not able to distinguish radio packets belonging to the same TCP segment. However, we can estimate its mean and standard deviation. Indeed, assuming that every packet transmission experiments the same packet retransmission probability (PRP), then the number of transmission attempts until positive acknowledgment is a modified geometrically distributed random variable, with parameter 1-PRP. Consequently, if a TCP segment is split into N radio packets, the SST is given by the sum of N independent identically distributed (i.i.d.) random variables. Thus it results a random variable with modified Pascal distribution, having mean and standard deviation (expressed in number of slots) given by

$$\text{Mean}: ASST = \frac{N}{1-PRP}(n+1) \quad (4)$$

$$\text{Std dev.}: DSST = \frac{\sqrt{N \cdot PRP}}{1-PRP}(n+1) \quad (5)$$

In the previous expressions, $n+1$ is the total number of slots needed for packet transmission and acknowledgement (n slot per packet in the forward direction, one slot per ACK in reverse direction).

4. Measurements Platform and Tools

The architecture of the system that we have used for measurement collection is depicted in Figure 2. An FTP server is connected to a router through a 10Base-T Ethernet. The router interfaces fixed and radio parts establishing a BT piconet with an FTP client. The piconet has been configured with *master* on the router and *slave* on the client to maximizing the link capacity in the forward direction. Router and

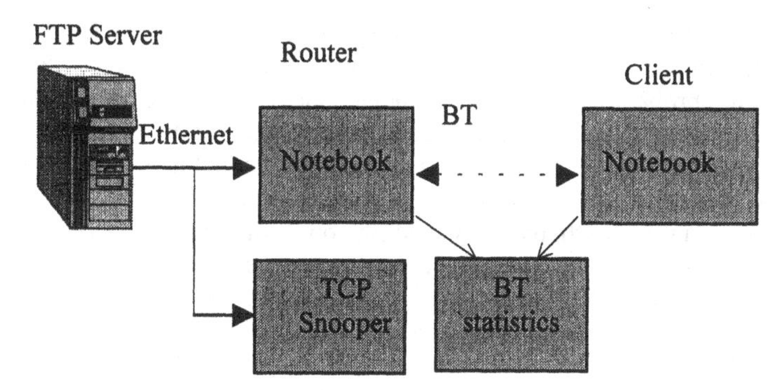

Figure 2. Measurement platform

FTP client are running on two Pentium II notebooks, clocked to 200MHz and using the Windows 98 operating system. The radio interface runs a Bluetooth DigiAnswer firmware, release 4.02, where BT baseband processing occurs. Winsock2 TCP/IPv4 Windows implementation has been used in tests. The "TCP Snooper" entity depicted in Figure 2 uses TCPDUMP and LIBPCAP 4.05 under Linux operating system to trace the TCP behaviour. The related plots are prepared with TCPTRACE and XPLOT. Traces are collected also at the link layer exploiting the DigiAnswer Bluetooth Application Programmer's Interface to collect statistics regarding an active Bluetooth link in terms of AC, FEC and CRC errors. Unfortunately, at this stage of our work, the tracing of radio link behaviour suffers of some drawbacks. For instance, the probing programs work independently on master and slave devices, then correlating master transmissions to slave receptions, and *vice versa,* is not easy. Furthermore, the probing time is not always constant and sporadically can assume very high values. The result is that we are able to determine only the trend of radio connection, and not its step-by-step history. Nevertheless, the information obtained permits to draw some interesting observation about the BT performance.

5. Measurement Results

In this section we present and analyse the results obtained by using different BT packet formats in our test environments. Note that the values obtained by measurements must be considered only as indicative, because even repeated stationary measurements in an identical location often yield different results. Anyhow, our aim is to determine the most suitable packet format in each situation and this information can be extracted with sufficient clearness from the performed measurements.

In the following, we present the comparison between protected (DMn) and unprotected (DHn) formats, and successively we proceed comparing long and short packet formats. Finally we present some preliminary results about TCP behaviour over a BT radio link.

5.1 Protected vs Unprotected Packet

Protecting payload with FEC produces two opposite effects: on one hand, the FEC theoretically improves the payload error probability and lowers the packet retransmission probability (PRP), on the other hand, the code overhead reduces the maximum bit rate achievable. Thus a trade-off between goodput realised by DMn and DHn packets can be expected. In reality, we have found that unprotected overrun protected formats in almost all the situations considered.

This fact can be partially explained by observing the two dashed curves depicted in Figure 1. These curves represent respectively the AC error probability (PAC) and the Header Checksum error conditioned probability (PHEC), *i.e.* the probability of having an HEC failure given that the AC field is good. We can note that, while the AC error probability grows rapidly passing from Env1 to Env5, the PHEC remains roughly of the same order of magnitude. This is due to the fact that the HEC is considered only when the packet has a good AC field and hence the signal strength is sufficiently high. Since the radio channel is slowly time-variant, with high probability it will remain good also during header reception. This consideration can be extended to the payload field. Indeed, if the AC and HEC fields are good, then with high probability the radio signal is strong enough to guarantee a low error probability in the payload field, and consequently the FEC results useless. Measurements confirm this observation, at least for packets with short and medium length. Instead, long packets show an effective improvement of PRP due to FEC, especially in dynamic environment. The waste of capacity caused by FEC overhead is not compensated by the code benefit on PRP; thus, at least in the considered situations, unprotected packet format results preferable protected ones.

5.2 Short vs Long Packet

In this subsection we compare the performance obtained by using packet of different length. On the basis of the results presented in the previous subsection, we consider only unprotected packet formats. As already mentioned in Section 2, during the transmission of a multi-slot packet, the carrier frequency remains unchanged for the whole packet duration, thus reducing the efficiency loss due to the PLL settling time. On the other hand, if an error occurs anywhere in the payload field, the entire packet must be retransmitted; thus in the same channel condition, long packets have higher retransmission probability than short ones. The combination of these two effects determines the effective link capacity as seen by the upper layer.

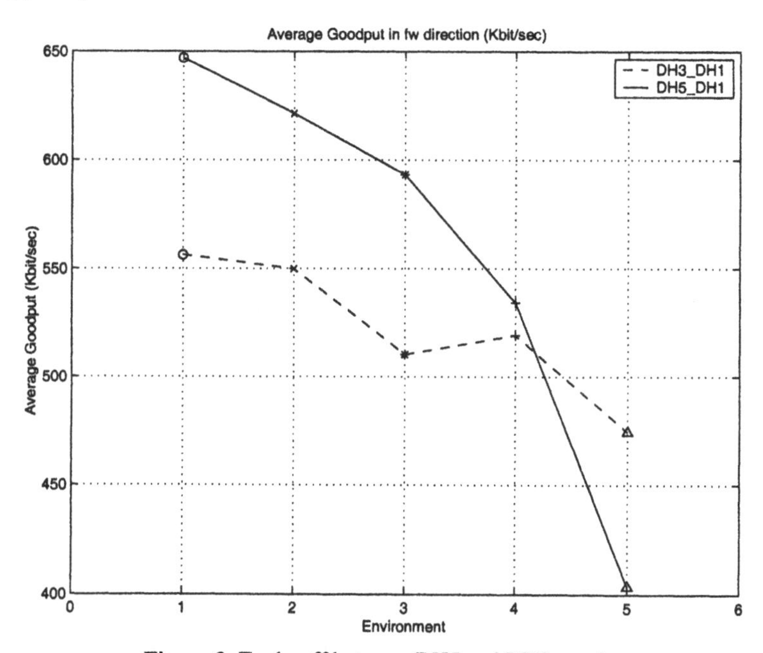

Figure 3. Trade-off between DH5 and DH3 goodput

Figure 3 shows the goodput with DH5 and DH3 packets in our five test environments. For the sake of clarity, we have not plotted the curve relative to DH1 goodput, because it is much lower than the others. This is due to the very low payload capacity of DH1 packet that is not compensated by its better retransmission probability.

Curves depicted in Figure 3 show an interesting trade-off between goodput obtained with DH5 and DH3 packets. As expected, performances of both packet types progressively worsen with the increase of environment hostility; however, DH5 degrades faster than DH3 so that, in particularly unfavourable conditions, the second outruns the former.

Besides goodput, the end-to-end performance is determined also by the link latency. For this reason, in Figure 4 we report the average and the standard deviation (in number of slots) of the Segment Service Time (SST) that has been introduced in Section 3. The curves are obtained considering the default value of TCP Maximum Segment Size (MSS), corresponding to an IP datagram size of 576 bytes. Each of these IP datagrams is split in 4 DH3 packets or in 2 DH5 packets.

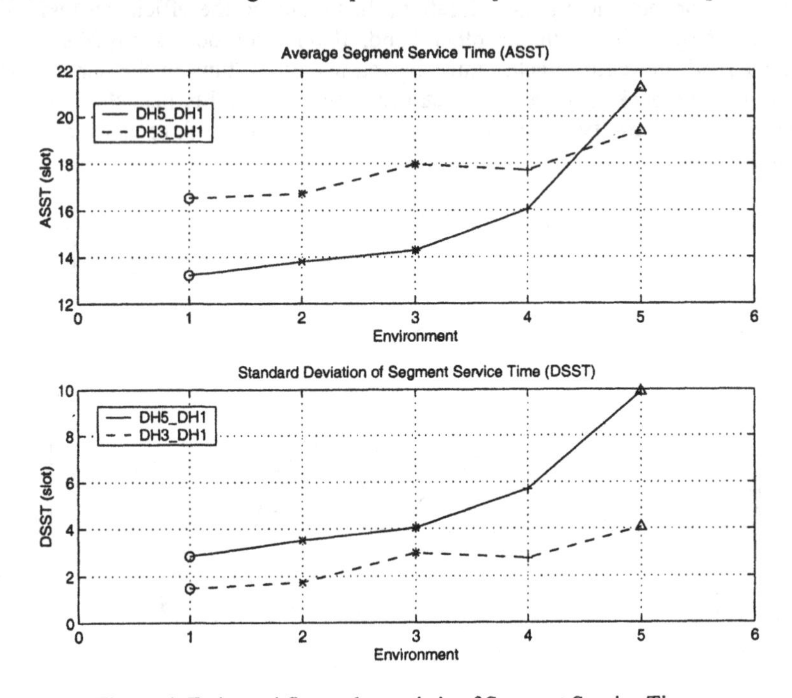

Figure 4. Estimated first order statistic of Segment Service Time

As we can observe from the upper part of Figure 4, the average Segment Service Time (ASST) follows the same trend observed for goodput. This fact does not surprise since both metrics are closely related to the average link bandwidth. More interestingly, the trend of the estimated SST standard deviation (DSST), shown on the bottom part of Figure 4, must be noticed. Indeed, we can observe that DSST is greater for DH5 than for DH3 in all the situations considered. This can be explained by the fact that each time a DH5 packet must be retransmitted a delay of 6 slots is incurred, while for DH3 packet retransmission only 4 slots are needed. In other words, the transmission time of long packet has a thick granularity compared with that of short packet, resulting in a greater deviation of segment delay around its mean. Note that the end-to-end delay could be much longer and variable than the estimated SST. Indeed, due to the low radio link capacity, segments could be buffered on the router and their delay consequently grows with the queue length. This, in turn, could lead to long latency and large round trip time deviations at the TCP sender.

5.3 TCP over BT Link: Preliminary Results

In Figure 5, an example of a trace of an FTP connection running on Bluetooth is presented, for a mobile host that moves in areas characterized by different received signal quality. It can be noticed that, under normal operation, TCP Reno adapts to the latency variations due to the Bluetooth baseband retransmissions, as expected. However, under critical channel conditions, the TCP sender timeout is triggered, resulting in unnecessary retransmissions of TCP packets and in congestion control mechanisms being invoked, which reduce the connection throughput. In the graphic, the black diamonds represent packets sent by the FTP server, while the lower and the upper lines show respectively the acknowledgement value and the transmission window size advertised by the client. After a timeout occurs at ~62.3s, packets are retransmitted by the sender (R in the figure) without the whole advertised window being exploited (slow-start phase) and the abnormal situation persists for about 1 second. In addition several duplicate acknowledgements are also generated (the small lower vertical lines).

In general, retransmissions often result in a cascade-effect whose negative impact extends for seconds.

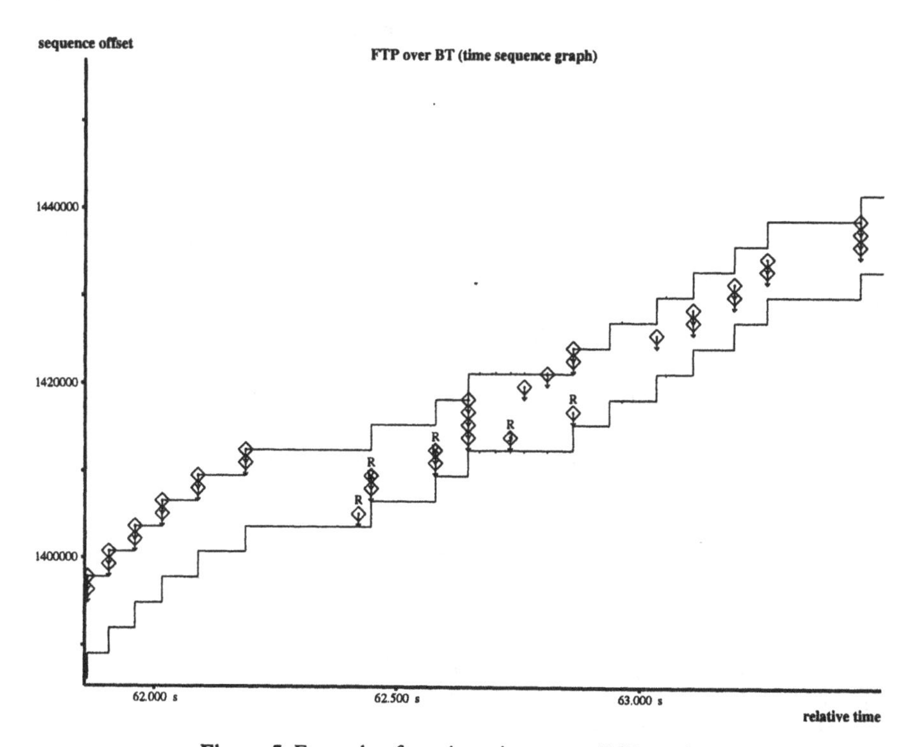

Figure 5. Example of spurious timeout on TCP sender

6. Conclusions

In this paper we have presented the results of a set of measurements relative to an FTP connection over a single-hop BT radio link. Our measurement-based approach has given us the opportunity to observe the system behaviour in real-world situations and thus to analyse the performance obtained by using different Bluetooth packet formats. From the collected bulk data, we can conclude that BT protected packet formats suffer from the inefficiency of FEC overhead that is not compensated by the improvement of packet retransmission probability. Furthermore, in almost all the considered situations, DH5 packets appear more suitable than DH3 in terms of goodput and average SST; whereas in particularly hostile environments, the intermediate-length packet format seems better. Finally, in all the considered situations, the DH5 packet format is characterised by an estimated value of SST standard deviation greater than that obtained for DH3, due to the thick granularity of long packet transmission time. The effect of this delay variation on the end-to-end performance must be investigated in more detail.

Future work will focus on analysing the possible impact of BT latency on end-to-end performance, and on deriving models for the wireless link that capture the aggregate of real-world effects like noise, interference and fading.

7. References

[1] W. R. Stevens, «TCP/IP Illustrated, Volume 1: The Protocols,» Addison-Wesley, 1994.

[2] G. R. Wright, W. R. Stevens, «TCP/IP Illustrated, Volume 2: The Implementation», Addison-Wesley, 1995.

[3] W. Stevens, «TCP Slow Start, Congestion Avoidance, Fast Retransmit, and Fast Recovery Algorithms», RFC 2001 Jan.1997.

[4] A. DeSimone, M. C. Chuah, and O. C. Yue. «Throughput Performance of Transport-Layer Protocols over Wireless LANs,» *Proc. Globecom '93*, December 1993.

[5] A. Chockalingam, M. Zorzi and V. Tralli, «Wireless TCP Performance with Link Layer FEC/ARQ,» *Proc. of International Conference on Communications*, pp.1212-1216, Vancouver, Canada, June 1999.

[6] S. Hoff, M. Meyer and A. Schieder, «A Performance Evaluation od Internet Access via the General Packet Radio Service of GSM,» *Vehicular Technology Conference (VTC)*, pp.1760-1764, Ottawa, Canada, May 1998.

[7] P. Wong and F. Halsall. «Data Applications for DECT». *Proc. of ICC'93*, pp. 1274-1278, Geneva, Switzerland,. May 1993.

[8] Specification of Bluetooth System, ver. 1.0, July 1999

[9] The Bluetooth Special Interest Group, Documentation available at http://www.bluetooth.com/.

[10] V. Jacobson, «Modified TCP Congestion Avoidance Algorithm», end2end-interest mailing list, April 30, 1990. ftp://ftp.isi.edu/end2end/end2end-interest-1990.mail.

Role of Software Defined Radio in Wireless Access to the Internet

Anthony Acampora, Joseph Soma Reddy, Haipeng Jin
and Ralph Gholmieh

Center for Wireless Communications, University of California, San Diego, USA

Abstract: The convergence of the Internet and Cellular networks will result in a new high speed, packet switched, IP based wireless network. We examine some of the technical issues associated with such a system. Smart antenna based media access protocols, better scheduling algorithms in interference limited environments and a micromobility protocol are presented. The role of software radio in expanded beyond the air interface to enable these new technologies.

1. Introduction

Over the past decade, two unprecedented events have radically transformed the nature of modern telecommunications. The first of these is the meteoric rise and spread of the Internet, resulting in a new world wide telecommunication infrastructure based upon high speed, packet based, bandwidth upon demand as needed to support a vast array of new multimedia service offerings and applications. The second is the equally impressive emergence of cellular radio based systems which enable, in essence, wireless voice communications to small battery operated cell phones.

At the vanguard of modern telecommunications is the convergence of these two megatrends, the result of which will be a high speed, packet based wireless network capable of delivering to small handheld terminals the same grade and range of Internet baesd services now available only to desktop computers. Characterizing this trend are packet switching(as opposed to circuit switching), broadband(as opposed to voiceband) and on-demand bandwidth as needed to deliver multimedia content filled data files between wireless terminals and web sites.

In this paper, we consider several of the many technical issues associated with broadband packet based wireless access to the Internet. In particular, we focus on three particular technologies, and examine the relationship between these technologies and Software Defined Radio(SDR). Historically, SDR has often been viewed in a radio limited sense: since different cellular systems conform to different air interface standards, SDR has been considered as a means for re-programming a cell phone to operate with any of several possible air interfaces, thereby enabling a multiterminal capability.

As we converge on the wireless Internet era, the range of options which characterize high rate, packet-based wireless access extends well beyond modulation and coding as used on the air interface, and SDR must, more generally, be viewed as a technology which allows a wireless terminal to operate in a variety of different environments. In the following, we examine the roles of smart antenna technology and efficient media access protocols, smart scheduling algorithms for multiple traffic classes in an interference limited environment and mobility management strategies as needed to ensure access of the Internet Protocol(IP) based, content-rich web sites by small, mobile terminals. As will be shown, all extend the scope and definition of Software Defined Radio.

2. Smart Antennas

2.1 Introduction

There is an increasing interest in architectural integration between Software Defined Radio (SDR) base stations and the adaptive processing needed for smart antennas. With wireless data service becoming a major component in the telecommunications world, the users are demanding anytime, anywhere access to a wide spectrum of information by handheld mobile devices. This requires a ubiquitous high capacity wireless network and provides strong motivation for considering smart antenna applications in the base station (and possibly in the mobile terminals). SDR provides the flexibility to support multiple air interfaces and signal processing functions at the same time, and thus could easily incorporate smart antenna ability when needed.

By dynamically adjusting the antenna array weights, a smart antenna could change its array pattern as a function of the location of the desired user and the interfering users. It has the ability to both combat co-channel interference and counteract multipath delay spread. This could be exploited to extend the coverage range, increase system capacity, improve quality of service and achieve a higher date rate.

One very important issue which needs to be addressed before smart antennas could be used in a wireless system is the design of new Media Access Control (MAC) protocol. Originally intended for bursty data traffic, the MAC protocol seeks to insure an orderly sequencing of packets from the various users onto the shared channel, with a minimum of time lost to collisions. It delivers bandwidth-on-demand, meaning that a user having a greater volume of packets to send contends for the channel more frequently than one having fewer to send. For wireless access to the Internet, the efficiency of the MAC protocol becomes an important design consideration, especially insofar as guaranteed quality of service is concerned, and the software defined radio must be capable of supporting a multiplicity of traffic classes and operating environments.

When applied to cellular radio systems, a MAC protocol must cope with the various impairments suffered on the radio link such as shadowing, multipath delay spread and co-channel interference from other users. This is especially troublesome in a cellular system, since the users are always geographically separated and uncoordinated, and in consequence not all receivers will hear all transmissions with the same intensity, making access cooperation among the users more difficult to achieve. In addition, the upstream/downstream traffic may be both highly dynamic and highly asymmetric, implying that the full bandwidth channel must be shared by the base station and all remote units within the cell.

The use of smart antenna technology to abate radio link impairments further exacerbates the MAC protocol problem. In a packet access environment, the base station does not know which mobile station might need to send a packet. Consequently, even if the base station has stored a correct set of antenna weights to receive from each user, it would have no way of knowing which user's weights to use, or even if any user was trying to access. Furthermore, since a given user may not have sent anything for some prolonged period of time, the currently stored weights corresponding to that user might not be valid. This would also be true if the base needs to send anything to the user.

Packet switching essentially requires that the antenna array be tuned slot-by-slot, as the set of users transmitting in a given slot is no longer determined in advance, but rather depends on the random traffic dynamics. This means that a fast and accurate mechanism must be provided to enable proper channel estimation. Therefore, training fields are needed in the radio burst. During the training period, the base station runs certain algorithms (e.g. Recursive Least Squares algorithm) on the training sequences to get a set of correct weights.

Fundamentally, the advantage of smart antennas is needed not only on the up-link transmission but, even more importantly, on the down-link as well. The principle of reciprocity implies that the channel is identical on the up-link and down-link as long as the channel is measured at the same frequency and at the same time instant. For a Time Division Duplex (TDD) system where the designated bandwidth is shared among both links, when the turn-around time is sufficiently small, the channel can be assumed stationary and the reciprocity holds. The weights obtained from up-link training can be used on the down-link transmission to preprocess the signal, so that the strength is enhanced for the intended user and the interference to other users is kept small at the same time.

Also, some means to identify the active users at a given time is necessary. Each user is assigned a pseudo-noise (PN) code to distinguish it from others. Furthermore, the base station needs to know which PN code it should use to get the correct weights. One possibility is to let the base station poll the users, and thus the base station knows when the user will transmit. Another possibility is to let the user make reservations beforehand and thus the base station knows when to expect which user.

2.2 Smart Antenna based MAC Scheme

We studied a TDMA/TDD reservation based scheme[1] for using smart antennas in a packet access cellular system. In a Time division Duplex (TDD) system with slow fading, where the time between receiving and transmission is small compared to the channel coherence time, the reciprocity principle holds. Also, TDD flexibly enables instantaneous and asymmetrical division of the radio spectrum between the up-link and down-link directions as may be needed to support multimedia applications.

The frame structure of the MAC protocol is shown in Fig. 1. It consists of three parts: reservation and grant period, up-link transmission period and down-link transmission period.

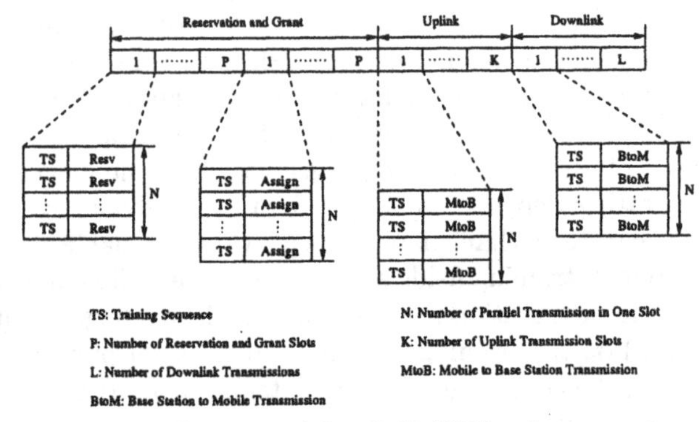

Frame Structure for Reservation Based MAC Protocol

Fig. 1.

The up-link period is composed of multiple slots. In each slot, a number of users will transmit data to the base station simultaneously. The data transmissions are preceded by training sequences, which enable the base station to acquire the proper combining weights for the smart antennas for each user. The training field here provides the base station with information on the interference environment that the user will experience during this slot. We assume that synchronization is achieved between neighboring cells within some small guard time, so that the training field will contain all interference and the interference situation will not change in the duration of one slot.

The down-link period is also composed of multiple slots. In each slot, those users that are going to receive information from the base station will first send their training sequences to the base station, enabling the base station to acquire correct antenna combining weights. Then the base station will use those weights to send information to these users.

In the reservation and grant period, the users are all assigned a fixed slot both in the reservation part and in the grant parts. A user will transmit its request to the base station in the slot assigned to it in the reservation part, and the base station will inform the user about the information transmission slots assignment in the corresponding grant slot. Note that the users are only assigned to fixed slots in the reservation and broadcast period, the information transmission slots in the up-link and down-link period are still assigned dynamically based on the users' demand.

Since each user is assigned to fixed slots in the reservation and grant period, it is possible to take advantage of the strength of smart antennas during this period. Several users can transmit their requests or receive their assignment information simultaneously. Every slot starts with a training field, during which the users send their training sequences to the base station. The base station computes the correct antenna weights for each user at the end of the training field. Then it can either receive requests or transmit assignment information to the users using the corresponding weights.

2.3 Results

The performance of the MAC protocol has been studied[1] in a cellular system with log-normal shadowing and flat Rayleigh fading. SINR statistics study shows that smart antennas can be used effectively to combat the co-channel interference from users in the same cell and in the adjacent cells. Also under a slow fading assumption, the combining weights obtained on the up-link training can be used to improve the down-link transmission, SINR statistics shows good performance.

The maximum achievable utilization efficiency and mean delay were also derived. Compared with a polling scheme, the overhead for training in the reservation scheme is much less and thus it can provide higher efficiency and accommodate more users.

Extra components are needed to incorporate a smart antenna in the SDR base station. For example, multiple antenna elements are required in the antenna segment of the software radio architecture[2]. Also, appropriate base station architectures are needed to provide the adaptive beam-forming capability and to accommodate cost effective production. A wide-band phase array architecture was proposed in[3] and an N-channel beam-former architecture was proposed in[4].

The tight integration of smart antenna functions with the other functionality of the base station enables the processing unit to implement not only air interface demodulation and beam forming, but also advanced signal processing ability to mitigate co-channel and inter-symbol interference. This offers great improvement in system performance and simplification in system design.

3. Scheduling Algorithms

3.1 Introduction

Software defined radio (SDR) is the future technology for wireless terminals and base stations[5, 2, 6]. In particular, complex transceiver structures will be configured, modified and enhanced "on demand" at SDR terminals and mobile stations. The early focus on the radio interface has given way to a more general approach that applies software radio principles to every function in the radio from the physical level up to the application level. In that light, proprietary network solutions may be deployed at base stations where connecting user terminals must be programmed with some enhanced features; a standardized download solution like the one described in[7] allows this reconfigurability across different networks.

We have studied the potential increase in the allowable number of data terminals per cell on the forward channel attainable by using an interference measurement-based scheduling algorithm at the base stations and a pilot tone measurement function at the mobiles. Such a pilot tone measurement function may be programmed into SDRs that roam into the system. SDR based base stations may also be easily upgraded to run our scheduling algorithm for data transmission. Our results have shown that a large increase in capacity is obtained by power efficient scheduling when compared to a system that does not provide such a function.

Further, a SDR configured, for example, for a CDMA-2000 HDR (high data rate) system[8] may be reconfigured to allow our type of scheduling. HDR requires a mobile based transmission rate selection. In the following we propose an alternative base station based transmission rate selection that requires the remotes to report all pilot tone measurements to the base station and also allows base station coordination.

3.2 Scheduling Algorithm

Fig. 2 shows the system we are studying: each mobile terminal roams between cells and connects to that base station, BS_i, whose pilot tone (T_i in the figure) is received by that mobile with the strongest average power. Further, the mobile terminal tracks and estimates the received power of other measurable base station pilot tones. This tracked information is relayed back to the terminal's own serving base station. The serving base station uses the pilot tone power measurements and the actual transmission power of surrounding base stations to schedule transmissions to the users experiencing lower interference than their individual long term average interference. In our scheme, every base station estimates and commits *apriori* to a maximum transmission power for the next forward channel frame; the chosen interference is calculating by consulting the expected state of the queue in the next frame at the base station. The details of the power selection algorithm will be described briefly later. Note that we assume that the intelligent parts of the base stations in general, and base station schedulers in particular, are co-located in a base station controller (BSC) or that some backbone network interconnects the base stations so that information between base stations may be exchanged in a timely manner.

The air interface is a DS-CDMA synchronous slotted CDMA system. One frame is formed of one time slot in our work and a base station may use several codes to transmit at a higher rate to any given terminal.

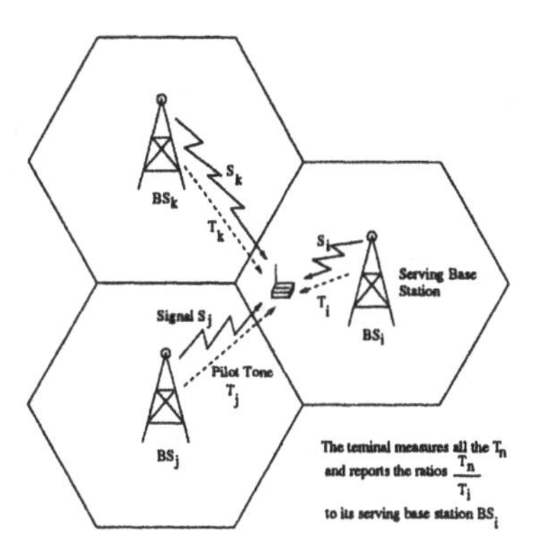

Fig. 2. Cellular System

Individual mobiles receive batch arrivals that are segmented into packets where one packet may be transmitted during one time-slot and using a single code. The job of the scheduler is to select how many packets are to be simultaneously transmitted for each batch per slot.

In this setting, every batch has a target delivery time that provides base stations with some scheduling flexibility. Every queued batch of packets has a unique deadline timestamp calculated as follows:

$$T^d = T^a + K * S \tag{1}$$

where T^d is the delivery deadline of the batch, T^a is the time of arrival of the batch at the queue of the destination terminal, K is a proportionality factor in seconds per kilobit and S is the size of the batch in bits.

The number of packets that the scheduler attempts to schedule for a batch is equal to the ceil of the priority of that batch. The priority of a batch is given by:

$$Priority = \frac{N_b}{T^d - T} InterferenceFactor \tag{2}$$

where N_b is the number of remaining packets in the batch, T^d is the delivery deadline, T is the current time in the system, and the InterferenceFactor is greater than 1 if the destination mobile is experiencing better than average interference. Whenever a batch is considered for scheduling, the scheduler attempts to schedule $\lceil Priority \rceil$ packets per slot for that batch if enough power is available.

When the estimated outer-cell interference at the destination receiver of a batch decreases, more packets may be scheduled for that batch given a fixed transmission power. To conserve power and benefit from the improved channel conditions for a particular terminal, the $InterferencFactor$ may be set to the ratio of the number of possible transmissions in the current predicted interference conditions at the receiver, to the number of possible transmissions at the average interference level at the destination receiver at the same allocated transmission power. When forward link transmissions are orthogonal, the number of transmissions that may be scheduled for a given user is inversely proportional to the out of cell interference at the terminal antenna and thus we choose:

$$InterferenceFactor = \overline{OOC}/OOC \tag{3}$$

where OOC is an estimate of the interference at the user given the pilot tone measurements at that particular user and given base station power transmission information at the scheduler. $\overline{OOC}$ is the long-term average measured interference at the destination terminal.

Batches are sorted in the decreasing order of the priority. Packet transmissions are scheduled for the batches in the order of the highest priority until the available power is exhausted. After this scheduling, the base station selects its next slot transmission power by assuming that all currently scheduled transmissions are successful and determining an estimate of the power required to schedule all remaining queued packets that will need transmission in the next slot.

We identified the following possible cases:

1. Base stations have no knowledge of the value of the fade for either scheduling or power control. In this case the target decoding SNR should be increased to provide a "fade margin" that reduces the number of retransmissions.
2. Base stations are able to track the changes in the Rayleigh fade at the mobile through closed loop feedback and/or channel estimation. We assume that the power requirements of the packets with the highest priority are fulfilled. Two sub-cases are identified: (a) the fade information can only be used for power control purposes, and no special interference-based scheduling is done, (b) the fade information can only be used for power control purposes but long-term shadow fading and path loss measurements are used to prioritize the packets, and (c) the small scale fading value is known *apriori* and may be used for scheduling the packets.

3.3 Results

We ran simulations to test the improvement brought by our scheduling algorithm. In our simulations, the available bandwidth is 4.96 Mbps, the spreading gain is equal to 128, and the duration of a slot is $20ms$; thus, a packet is formed of 750 bits. Transmissions use a rate 1/2 convolutional code and QPSK modulation. One batch arrives for each user every 5 seconds on average and the interarrival time is exponentially distributed. The size of a batch in packets is geometrically distributed with an average of 213 packets. Thus the data arrival rate per user is 32 Kbps. Fig. 3 shows our results in the presence of fast, flat multipath fading and assuming that transmissions from any given base station are orthogonal.

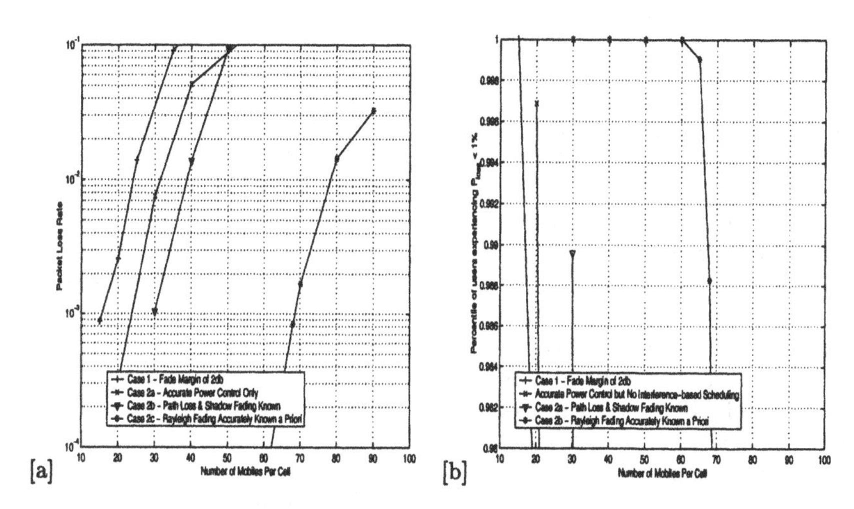

Fig. 3. Capacity improvement due to better scheduling

The performance for case 1 is always worse than that of case 2 as can be seen in Figure 3. Figure 3a shows the average packet drop rate vs. the average number of accepted terminals per cell for cases 1, 2a, 2b and 2c above. We assumed that the thermal noise can be ignored, *i.e.*, the system is interference limited. Mobiles/base stations monitor interfering base stations whose pilot tone is received with a power of at least 20% at the mobile when compared to the pilot tone of the serving base station. For case 2a, the $InterferenceFactor$ in equation 2 is always set to 1, emphasizing the effect of improved scheduling on the obtained results. For an acceptable average packet loss value of 10^{-2}, if the $InterferenceFactor$ is always set to 1 then the acceptable number of terminals per cell is 31 (case 2a). This number increases to 38 when shadow fading and path loss are taken into account (Case 2b) and further increases to 78 when multipath fading is known accurately at the time of scheduling (Case 2c). Thus, as the accuracy of the small scale fade estimate increases, the acceptable average number of terminals per cell increases from 38 to 78. The improvements are larger if the requirement is that 99% of the terminals experience a packet loss of less than 10^{-2} as seen in Fig. 3b. The acceptable number of terminals per cell increases from 20 when basic scheduling is done ($InterferenceFactor$ set to 1), to 30 when path loss and shadow fading information are used for scheduling, and finally that number increases to 67 when shadow fading is known accurately at the time of scheduling.

This scheme clearly benefits users which are experiencing fast but tractable multipath fading. Once such a user is identified by a base station, the base station should reconfigure or reprogram the SDR-able terminal to track and report pilot tone measurements from the strongest interfering base stations.

4. Mobility Management

4.1 Introduction

Due to the growing demand for providing packet data services to the mobile world, future cellular networks are expected to be based on IP protocols. Packets would be routed upto the base stations using IP routers and then transmitted over the air to the mobile host. Thus the cellular infrastructure would be just an extension of the Internet. However, the Internet was designed as a network of stationary hosts. The IP address identifies the point of attachment of the host rather than the host itself. Hence, a mobile host would need to change its IP address whenever it moved to a new base station, causing all existing connections to break down.

The Mobile IP[9] protocol was designed to enable routing of packets to mobile hosts. It allows a mobile host to maintain a constant IP address even when it changes its point of attachment. However, while Mobile IP works well for nomadic hosts, its high latency and overhead make it unsuitable in situations requiring fast and frequent handoffs, such as might occur within a cellular network. Thus, a hierarchical approach is needed with a micromobility protocol that can manage host mobility within a large domain (intradomain mobility), and a macromobility protocol to handle mobility between domains (interdomain mobility). We divide the global wireless infrastructure into domains each of which spans an area large enough so that mobile handoffs between them are infrequent. Also, each domain is under the administrative control of a single operator. Mobility within a domain is managed by a micromobility protocol while handoffs from one domain to another are handled by the macromobility protocol. (See Fig. 4)

While Mobile IP is a natural choice for the macromobility protocol, several proposals[10, 11, 12] exist for the micromobility protocol. It is likely that each operator will implement a protocol of its own choice for micromobility within its domain. If a single handset is to operate on a global level, it must be capable of detecting which micromobility protocol is used in the current domain and execute accordingly. A software radio based handset, with the capability to execute one of several different protocols, would be ideal for such a role. Also, if the particular micromobility protocol is not available with the handset, it should be capable of downloading the required software components over the air. Thus a single handset would be able to roam on a global scale.

In the following section, we present one possible micromobility scheme.

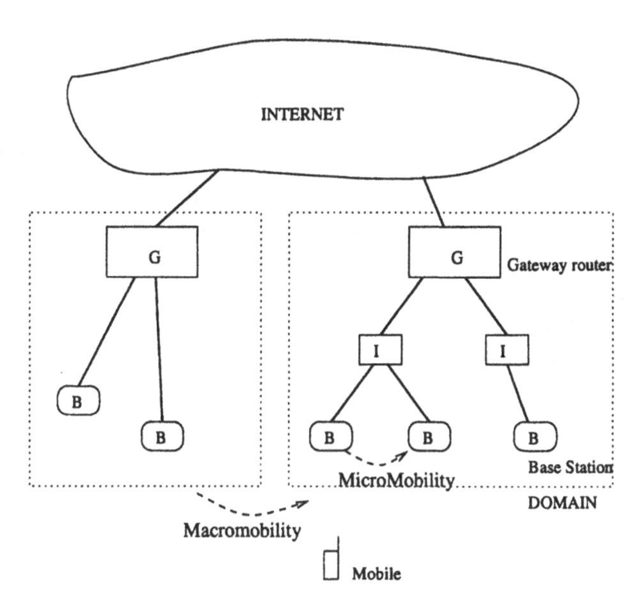

Fig. 4. Global Mobility

4.2 Micromobility Protocol

As illustrated in Fig. 4, a domain consists of a Gateway router, several base stations and some intermediate routers. It is connected to the Internet through the Gateway router. A mobile host that is admitted to the domain is loaned an IP address that can be used as long as the mobile is connected to a base station within the domain. Packets addressed to the mobile's IP address reach the gateway router (through the network id portion of the IP address) and are routed to the base station and transmitted to the mobile. As the mobile host moves from one base station to another within the domain, the micromobililty protocol ensures that packets addressed to it are routed to the current serving base station and delivered to the mobile.

Each domain owns a pool of global IP addresses and all packets sent to these addresses on the Internet are routed to the Gateway router based on the network id portion of the address. In addition, each base station is assigned a set of private IP addresses and the intermediate routers are configured to route all packets addressed to these private IP addresses to the respective base stations. The configuration of the intermediate routers may be done manually or through standard routing protocols.

When a mobile is admitted to a domain, it is loaned a global IP address. The mobile uses this address in it communications with other hosts on the Internet as long as it connected to a base station within the domain. It is also assigned a set of private IP addresses, each of which belongs to a particular base station.

Packets addressed to the mobile's IP address reach the Gateway router (based on the network id portion of the address). The Gateway encapsulates the packet within another IP packet and sets the destination address to the private IP address of the mobile corresponding to its current serving base station. The encapsulated packet is routed to the current serving base station and transmitted over the air to the mobile where it is deencapsulated.

When the mobile performs a handoff between base stations, a control message is sent by the new base station to the Gateway router informing it of the change in the identity of the current serving base station. The Gateway router correspondingly changes the private IP address it uses for encapsulating packets destined to the mobile. Thus packets destined to the mobile are always routed to the correct base station.

The tunnelling of packets between the Gateway router and the mobile is transparent to the transport and application layers in the mobile, which are only aware of the global IP address loaned to the mobile and which communicate with other hosts on the Internet using that address.

The Gateway router maintains tables relating each mobile with the global IP address it has been loaned and its current serving base station. The table entry for each mobile is created when the mobile is admitted to the domain and the current serving base station field is updated every time the mobile performs a handoff. Each base station maintains a table relating each mobile connected to it with its link layer address. The entry for each mobile is created when the mobile is connected to the base station.

4.3 Example

We will use the private address space $10. * . * .*$ for our private addresses. Let each base station have an identifier b_i of, say, 8 bits length. Similarly, let each mobile have an identifier m_i of, say, 16 bits length. These identifiers are meaningful and unique only within a domain (*i.e.*, they are reused by base stations and mobiles in other domains).

Let the pool of private addresses $10.b_i. * .*$ be assigned to base station b_i. Thus all packets with destination address $10.b_i.*.*$ would be forwarded to base station b_i. Mobile m_i would need to be assigned a set of private addresses, one from each base station's pool. Let $10. * .m_i$ be the set of private addresses assigned to the mobile m_i. Then, mobile m_i, when connected to base station b_i, can be reached with the address $10.b_i.m_i$

The Gateway router maintains a mapping between each global IP address (belonging to the domain's pool) and the identifier m_i of the mobile it has been loaned to. It also maintains a mapping between a mobile's identifier m_i and the identifier b_i of its current serving base station. When it receives a packet destined to a global IP address (belonging to the domain's pool), it constructs the current private IP address of the mobile using the two mappings and tunnels the packet to that address.

Each base station maintains a mapping between the mobile identifier m_i and its link layer address. Upon receiving a packet, it derives the mobile identifier from the destination IP address, looks up the link layer address corresponding to it and delivers the packet.

4.4 Results

Our micromobility scheme achieves fast handoffs with no involvement of the mobile at each handoff. There are no signalling messages exchanged over the air at each handoff. This reduces overhead on the wireless link. The handoff latency is reduced since message exchange over the air typically takes greater amount of time due to the slower speed of the wireless link. Also, standard IP routers can be used for intermediate routers and, with little extra software, for Gateway router.

It is very likely that several micromobility protocols will be in use by different networks. A software radio based handset would be an ideal choice in such a situation so that a single handset is capable of roaming globally.

5. Conclusion

As we can see, enabling universal wireless access to the Internet involves a multiplicity of issues relating to software defined radio. In this paper, a sampling of such issues were identified and addressed. Clearly, in the wireless Internet era, software defined radio must play an expanded role beyond the air-interface and it is hoped that this paper will stimulate additional discussion and attention with regard to these issues.

6. References

[1] Technical Report, "MAC Protocol Design for Systems Using Space Time Processing", Center for Wireless Communications, Univ. of California, San Diego, June 2000

[2] Joe Mitola, "The Software Radio Architecture", IEEE Communications Magazine, May 1995, p26-38

[3] Dimitrios Efstathiou and Zoran Zvonar, "Enabling Components for Multi-Standard Software Radio Base Stations", Wireless Personal Communications 13, 2000, p145-166

[4] Joseph Kennedy and Mark C. Sullivan, "Direction Finding and Smart Antennas Using Software Radio Architecture", IEEE Comm., May 1995, p62-68

[5] J. Mitola III, "Software radios: Survey, critical evaluation and future directions," IEEE Aerospace and Electronics Systems Magazine, vol. 8, no. 4, pp. 25-36, Apr 1993.

[6] J. Mitola III, "Technical challenges in the globalization of software radio," IEEE Communications Magazine, vol. 37, no. 2, Feb 1999.

[7] M. Cummings and S. Heath, "Mode switching and software download for software defined radio: the SDR forum approach," IEEE Communications Magazine, vol. 37, no. 8, pp. 104-106, Aug 1999.

[8] P. Bender, P. Black, M. Grob, R. Padovani, N. Sindhushyana, and A. Viterbi, "CDMA/HDR: a bandwidth efficient high speed wireless data service for nomadic users," IEEE Communications Magazine, vol.38, no.7, pp.70-77, July 2000.

[9] Charles Perkins, editor, "IP Mobililty Support," Internet RFC 2002, October 1996.

[10] A. G. Valko, "Cellular IP - A New Approach to Internet Host Mobility," ACM Computer Communication Review, January 1999.

[11] Ramachandran Ramjee et. al.,"HAWAII: A Domain-based Approach for Supporting Mobility in Wide area Wireless Networks," International Conference on Network Protocols, ICNP'99.

[12] Claude Castelluccia and Ludovic Bellier, "A Hierarchical Mobility Management Framework for the Internet," IEEE Intl. Workshop on Mobile Multimedia Communications, Nov. 1999.

Energy/Latency/Image Quality Tradeoffs in Enabling Mobile Multimedia Communication

Clark N. Taylor, Sujit Dey, and Debashis Panigrahi

Department of Electrical and Computer Engineering
University of California, San Diego
La Jolla, California, USA
{cntaylor, dey, dpani}@ece.ucsd.edu

Abstract: Future wireless applications, such as cellular video telephony, wireless LANs and PANs, home networks, and sensor networks, point towards a growing demand for multimedia content in wireless communication. However, mobile multimedia communication has several bottlenecks including bandwidth requirements, low-power constraints, and channel noise. In this paper, we propose a method to overcome the energy and bandwidth bottlenecks by adapting to the varying conditions and requirements of mobile multimedia communication. We focus on source coding, which can have significant impact on both the computation and communication energy consumption of the multimedia radio, as well as the Quality of Multimedia Data transmitted and the Quality of Service (latency of transmission) achieved. In particular, we study the effect of varying some parameters of the JPEG image compression algorithm (a type of source coding) on energy, latency, and image quality. We present a methodology to enable selection of the appropriate image compression parameters to implement the energy/latency/image quality tradeoff in mobile multimedia radios.

1. Introduction

With the growing popularity of new mobile multimedia applications such as cellular video telephony, wireless internet access, wireless LANs and PANs, home networks, and sensor networks, there will be a growing demand for fast, low-energy, mobile multimedia communication. Figures 1 and 2 illustrate the dramatic growth of mobile multimedia communication in one of these applications, wireless internet access. Figure 1 records and predicts a dramatic growth in the number of wireless internet users from 1996 to 2001. Figure 2 shows how the quantity of multimedia traffic is predicted to increase rapidly compared with voice traffic. The growth in the need and ability for wireless access to the internet, together with other applications requiring mobile multimedia communication, is fueling the need for wireless multimedia communication.

However, before high-quality mobile multimedia communication can be achieved, there are several bottlenecks which need to be addressed. First, the bandwidth requirements of multimedia communication can be very high. This bottleneck will be addressed in the future by new "3G" cellular systems including WCDMA[1, 2], CDMA2000[3], and HDR standards[4], as well as non-cellular standards such as Bluetooth[5] and IEEE 802-11a(b)[6].

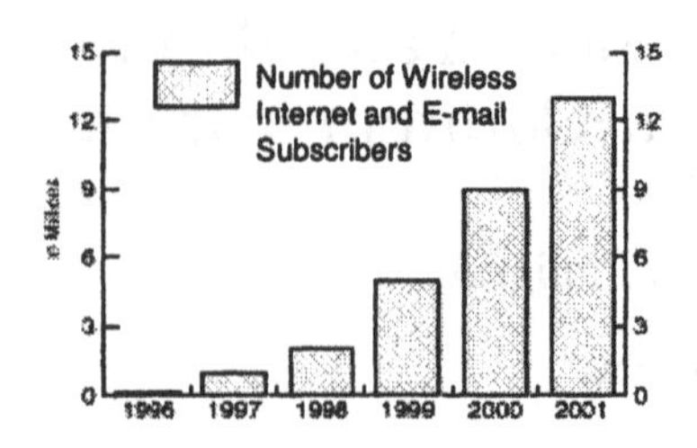

Fig. 1. Number of wireless users increasing exponentially (source phone.com)

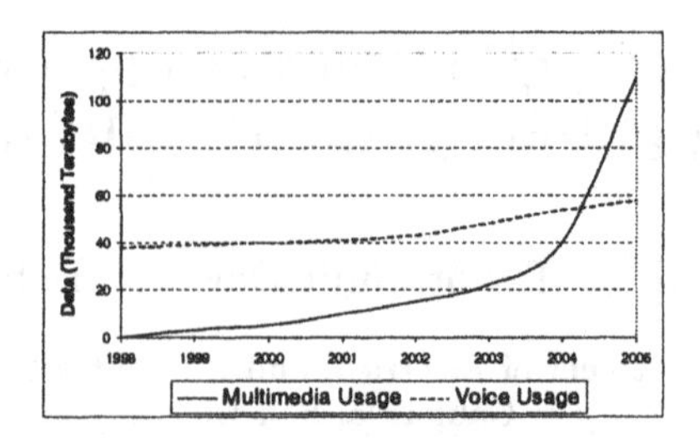

Fig. 2. Type of data transmitted over internet (source Analysys Ltd.)

In addition to bandwidth, energy consumption presents a significant bottleneck to mobile multimedia communication. As radios built for mobile multimedia communication will be powered primarily by battery, the energy consumed must be minimal. In addition, a large amount of information is needed to represent multimedia data. Therefore, both the computation energy (energy consumed in processing information to be transmitted) and communication energy (energy consumed in wirelessly transmitting information) can be very high.

Another bottleneck to mobile multimedia communication is channel noise. As the number of mobile users increase, the interference between users will also increase, causing more channel noise. While many methods, such as retransmission and channel coding, exist for overcoming the effects of channel noise, more bandwidth and energy is required to implement these methods.

A characteristic of wireless communication which can be used to overcome the bandwidth and energy bottlenecks is that the conditions and requirements of wireless multimedia communication vary. Variations in channel conditions may be due to user mobility, changing terrain, etc. For example, in [7], the Signal to Interference Ratio (SIR) for cellular phones was found to vary by as much as 100dB for different distances from the base-station.

Moreover, the Quality of Service (QoS) – such as transmission latency or bit error rate (BER) – and Quality of Multimedia Data (QoMD) – including image/video quality – required during multimedia communication changes depending on the current multimedia service. For example, the QoS (latency) and QoMD (image quality) requirements of transmitted data are different between video telephony and web browsing.

One way to design a multimedia radio is to assume the worst-case conditions and requirements. However, in this paper we show that by designing a mobile multimedia radio to adapt to the varying conditions and requirements of the communication system, we can overcome the bottlenecks of mobile multimedia communication.

For example, in Figure 3, we show three different cases of wireless communication of the same image. Because of the type of multimedia communication, the quality of the image is not very important, but the latency is constrained to 0.1 seconds. In case 1, the channel conditions are good, so the wireless radio handset receives a high quality image within 0.1 seconds. The bandwidth available in case 2, however, is lower, causing a high quality image transmission to take 1 second. However, this violates the latency requirements of the desired service. Case 3 shows that by adapting the image compression to the available bandwidth, the same latency as case 1 may be achieved. By sending a lower quality image than in cases 1 and 2, case 3 achieves a latency of 0.1 seconds, meeting the requirements of the multimedia communication.

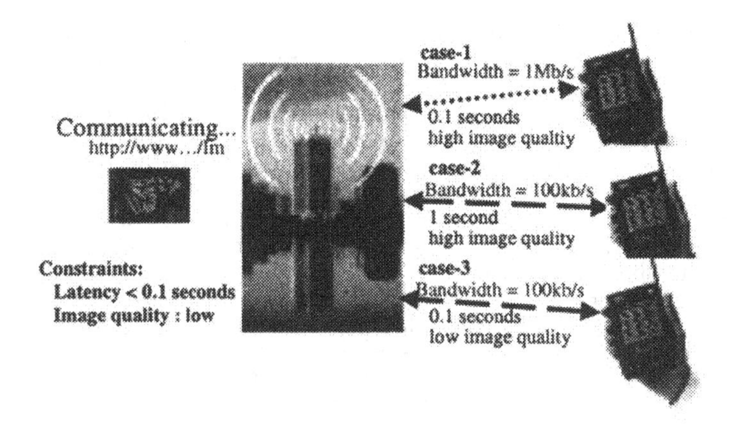

Fig. 3. Example adaptation of image quality to meet latency constraints

There has been active research on adapting to current channel conditions to save energy and bandwidth. For example, in [8], the authors adapt the channel coding parameters used to match current channel conditions, thereby increasing the average bandwidth available. In [9], an algorithm is proposed to modify the broadcast power of a power amplifier to meet QoMD requirements, thereby lowering energy consumption. In [10], the authors change channel coding and power amplifier settings to adjust for current conditions, thereby lowering energy consumption.

In this paper, we present an approach to help overcome the bandwidth and energy bottlenecks in enabling mobile multimedia communication. By adapting a mobile multimedia radio to the current communication conditions and constraints, we can execute the right algorithms, with the right parameters, at the right time and cost, rather than consistently communicating as if in the worst case. This will maintain the necessary performance of the system while lowering the energy and bandwidth requirements.

An important component in mobile multimedia communication is the source coder. By changing the source coder algorithms and parameters, we can implement energy, latency, and QoMD tradeoffs according to the current system conditions and requirements to minimize energy and bandwidth. In this paper, we focus on one example of source coding, the JPEG image compression algorithm. We study the effects of varying parameters of the JPEG image compression algorithm in mobile multimedia communication. We also introduce a methodology to select the optimal image compression parameters to effect an energy/latency/image quality tradeoff.

This paper is organized as follows. Section 2 discusses how source coding can be used to adapt to communication conditions and constraints. Section 3 presents how changing JPEG image compression parameters effects the energy, latency, and image quality in mobile multimedia communication. Section 4 presents a methodology for selecting image compression parameters which minimize the energy consumption of a multimedia radio given the latency and image quality constraints of the communication. Section 5 concludes the paper.

2. Flexible Multimedia Data Compression and Communication

One of the issues that makes mobile multimedia communication difficult is the large amount of data that needs to be transmitted wirelessly. However, multimedia data can be compressed in a lossy (as opposed to lossless) manner, leading to smaller compressed representations of the multimedia data than is available with traditional data compression. Therefore, source coding (compression) plays an important role in communicating multimedia information.

The flexibility enabled by lossy compression can be exploited to enable tradeoffs in mobile multimedia radios. Traditionally, when configuring the source coder, only the tradeoff between the quality of multimedia data (QoMD) and the number of bits to be transmitted was considered. However, with wireless communication, source coding affects more than just QoMD and bandwidth. Figure 4 illustrates the effects of configuring the source coder through choosing different compression algorithms and parameters. The choice of source coding algorithms and parameters affects not only the bandwidth required and the QoMD, but also the Quality of Service (QoS) and energy consumption during wireless communication.

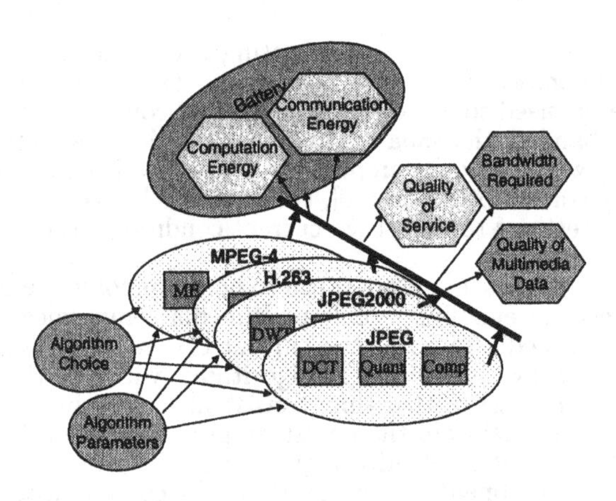

Fig. 4. Effects of source coding algorithms and parameters

For example, in Table 1, some of the tradeoffs between JPEG[11, 12] and JPEG2000[13, 14], two different image compression algorithms, are illustrated. For JPEG2000, the computation stage (source coding) consumes more energy and takes longer than when using the JPEG algorithm. However, because JPEG requires more bits to achieve the same image quality, the communication energy and latency for JPEG is greater than for JPEG2000.

To leverage the advantages of configuring a mobile radio to current communication conditions and requirements, we propose the mobile multimedia radio architecture shown in Figure 5. Our new architecture includes a traditional radio transceiver (unshaded), consisting of a speech/data coder, channel coder, RF modulator, and power amplifier. Additionally, it will include an adaptive image/video source coder and Network Aware Operating System (NAOS), indicated by the shaded regions. The NAOS is responsible for understanding the

Table 1. Effect of different image compression algorithms on energy and latency requirements

Which Algorithm	Computation Energy	Computation Latency	Communication Energy	Communication Latency
JPEG2000	More	More	Less	Less
JPEG	Less	Less	More	More

current conditions and requirements of the radio and network, and configuring the adaptive image/video source coder accordingly. The adaptive image/video coder must be designed to handle different multimedia data compression algorithms, with their parameters, so that the NAOS can optimally configure it. With this architecture, we can leverage the variations in communication conditions and requirements to help overcome the bottlenecks to mobile multimedia communications.

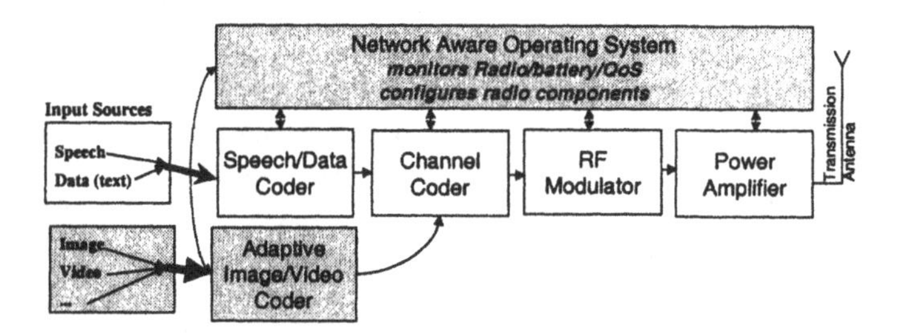

Fig. 5. Our Proposed Radio Architecture

In the next section, we investigate the effects of modifying the parameters of an existing source coding algorithm, JPEG image compression, on mobile multimedia communication.

3. Effects of Varying JPEG Image Compression Parameters on Energy, Latency, and Image Quality

As discussed in section 2, source coding affects not only the Quality of Multimedia Data (QoMD) transmitted and bandwidth required for communication, but also the energy and Quality of Service (latency) required in wireless multimedia communication. To better understand the effects of source coding on QoMD, bandwidth, energy, and latency, we present the results of varying the parameters to a commonly used source coding algorithm, JPEG image compression.

Figure 6 shows a basic flow diagram of the JPEG image compression algorithm. To implement JPEG, the input image is divided into blocks of pixels of size 8 pixels by 8 pixels. Each of these 8x8 pixel blocks is transformed by a Discrete Cosine Transform (DCT) into its frequency domain equivalent. After the transform stage, each frequency component is quantized (divided by a certain value) to reduce the amount of information which needs to be transmitted. These quantized values are then encoded using a Huffman encoding-based technique to reduce the size of the image representation.

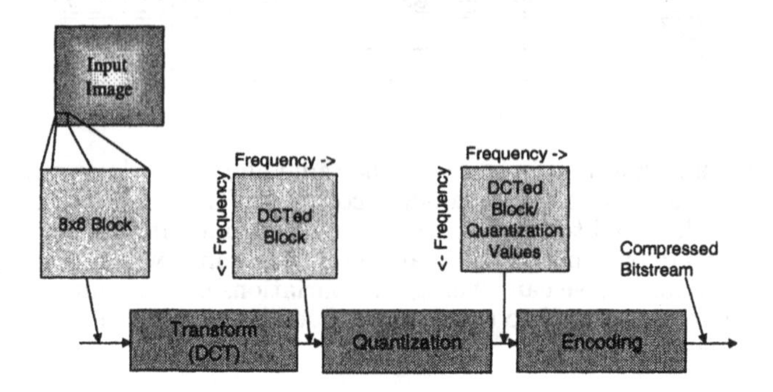

Fig. 6. Basic flow of JPEG image compression algorithm

To investigate the possible tradeoffs between energy, latency, and image quality, we selected two parameters of the JPEG image compression algorithm to vary. The first parameter we chose is the scaling of the quantization values used in the quantization step of JPEG. The JPEG standard defines some default quantization tables which can be scaled up or down depending on the desired quality of the final image. As the quantization level decreases, the image quality increases, but more information needs to be transmitted, causing more bits to be transmitted.

The second parameter whose effects we study is Virtual Block Size (VBS). This parameter affects the DCT portion of JPEG as first introduced in [15]. To implement VBS, the DCT still inputs the entire 8x8 block of pixels, but outputs a VBSxVBS amount of frequency information rather than an 8x8 block. Figure 7 shows an example of setting the VBS to 8 and 5. When the block size is 8, all information frequency information is computed. When the block size is 5, all frequency data outside the 5x5 block is set to 0. By setting the components outside the VBSxVBS block to zero, less computation energy is required for smaller VBSs because the elements set to zero do not have to be computed or quantized. In addition, zeros require less information bits to transmit, lowering the computation energy as the VBS values decrease.

In the subsections that follow, we discuss the effect of varying the selected parameters (quantization level and VBS) on energy, latency, and image quality. We conducted our experiments using the Independent JPEG Group's C code [12] modified to implement VBS. All numbers presented are an average across four different images (monarch, peppers, sail, and tulips). Image quality is represented by Peak Signal to Noise Ratio (PSNR), while computation energy is estimated by the number of operations needed to compress an image. Communication energy and latency is measured by the number of bits needed for image communication. We have assumed the computation energy needed for compressing a full-color, 704x512 size image, with VBS=8, to be the same as the communication energy required in transmitting 280kb (35kB).

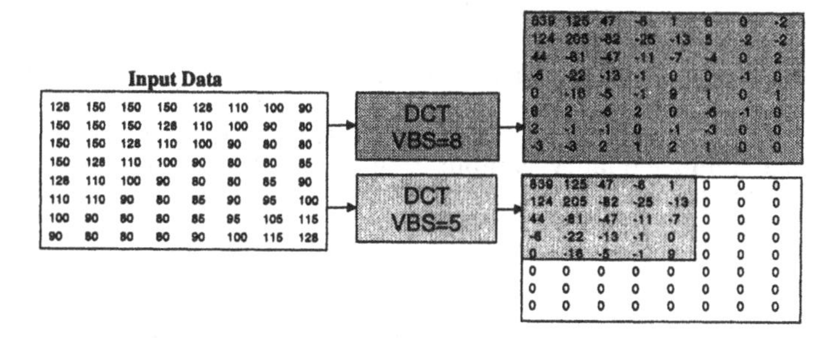

Fig. 7. An Example of virtual block sizes 8 and 5

3.1 Effects of Varying Quantization Level

Varying the quantization level of the JPEG algorithm has several effects on the mobile multimedia communication. First, increasing the quantization level reduces the image quality. Second, increasing the quantization level decreases the number of bits to be sent. This leads to a decrease in communication energy, latency and bandwidth required to send the image. Figure 8 illustrates how increasing the quantization level leads to a decrease in the image quality (PSNR) and communication energy (number of bits transmitted). The quantization values are linearly interpolated between quantization levels 0, 50, and 100, as defined by the IJG [12] (where 0 is no quantization and 100 is maximum quantization). The number of bits transmitted are normalized to compare against transmitting with no quantization.

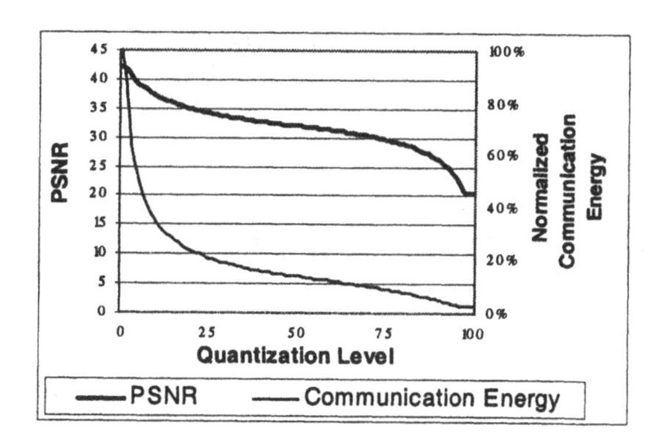

Fig. 8. Effect of varying the quantization level on image quality and communication energy (number of bits transmitted)

3.2 Effects of Varying Virtual Block Size

As mentioned before, the effect of Virtual Block Size (VBS) on computation energy and image quality has been studied before [15]. In this subsection, we present the effect of VBS on communication in addition to computation so that we can have a comprehensive knowledge of the tradeoffs present in using VBS.

As VBS decreases, the quality of image and the energy used in the DCT and quantization portion of the JPEG algorithm decreases, while the energy consumed in the encoding portion remains the same. In Figure 9 we show the decrease in image quality, computation energy, and communication energy as the VBS decreases. The amount of energy expended is normalized against when the VBS is 8.

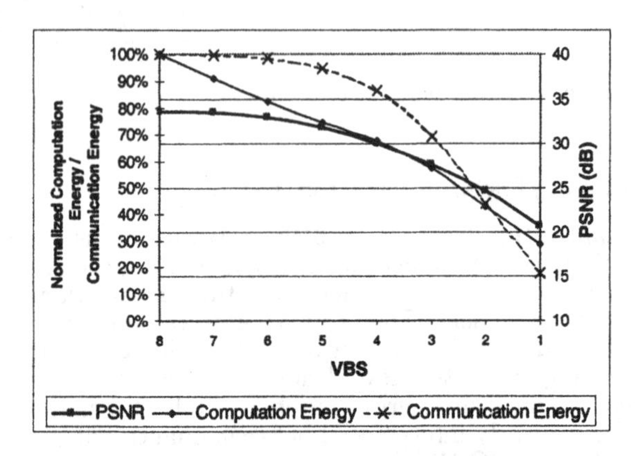

Fig. 9. Effect of varying the virtual block size on image quality, computation energy, and communication energy (number of bits transmitted)

Knowing the results of varying JPEG image compression parameters permits an intelligent tradeoff between energy, latency, and image quality in a multimedia radio. In the following section, we present a methodology which uses knowledge of image compression parameter effects to lower the energy consumption of a multimedia radio.

4. Selecting Image Compression Parameters for Optimal Energy/Latency/Image Quality Tradeoffs

As discussed in sections 2 and 3, varying image compression parameters can significantly affect the quality of image transmitted, the Quality of Service (latency) of transmission, and the energy required by a multimedia radio. In this section, we propose a methodology to select the optimal image compression parameters to effect the desired energy/latency/image quality tradeoff.

The objective of our proposed methodology is to minimize the overall (computation as well as communication) energy consumption, while meeting the specified QoS (latency) and QoMD (image quality) requirements. The complete methodology, shown in Figure 10, consists of a precomputation stage, which is performed off-line, and a table lookup step, performed by the multimedia radio. The off-line precomputation stage calculates an optimal parameters table consisting of the VBS and quantization level parameters to be used for each possible latency and image quality combination, such that the total energy is minimized. The NAOS in the multimedia radio uses the lookup table to determine the correct parameters for the current latency and image quality requirements. This approach ensures that the majority of the work done to determine the optimal parameters is performed off-line, thereby ensuring that the reconfiguration of the multimedia radio is fast and power efficient.

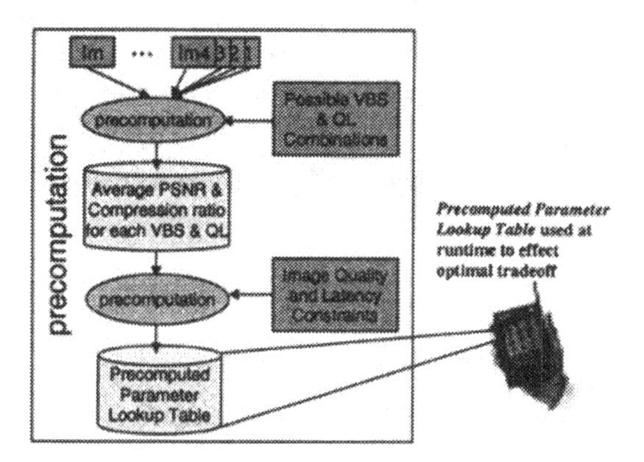

Fig. 10. Methodology for determining the optimal parameters for image compression

The precomputation of the optimal parameters table is performed in two steps. In the first step, the image quality (PSNR) and latency (number of bits to transmit as determined by the compression ratio) is precomputed. Since these values vary from image to image, an average over a large number of images is used. The result of step 1 is a table of PSNRs and compression ratios referenced by VBS and quantization level, as shown in Figure 11.

	VBS	8	7	6	...
QL					
...		...	...	...	
19		35, 1.74	34.8, 1.72	34, 1.69	...
...		...	...	...	
25		34.1, 1.47	34, 1.46	33.4, 1.44	...
26		34, 1.46	33.9, 1.45	33.3, 1.43	...
...		...	...	...	

Fig. 11. Example PSNR and compression ratio table referenced by VBS and quantization level

The goal of the second step of precomputation is to generate a lookup table which, given the image quality and latency constraints, minimizes the total energy consumed in compressing and wirelessly transmitting an image. We present below two algorithms which can be used in implementing the second step. Both algorithms rely on the table generated in the first step to determine the VBS and quantization level to use.

The *VBS-first* algorithm starts by finding the smallest VBS which can yield the required image quality (PSNR), then choosing the largest quantization level which still meets the image quality requirement. If latency constraints are not met (as determined by the compression ratio obtained), then the next larger VBS is chosen and the process repeats. The advantage of using the *VBS-first* algorithm to select the compression parameters is that the resulting image compression (performed by the multimedia radio) is very energy efficient.

However, choosing the smallest VBS possible may not always be sufficient for lowering the overall energy consumption. For a constant image quality (PSNR=32dB), Figure 12 shows the effects of varying the VBS on the computation and communication energy, normalized for when the VBS is 8. As the VBS decreases, the computation energy decreases, as expected. To maintain the same PSNR, however, the quantization level must be decreased, as shown at the top of the graph, leading to an increase in compression ratio (and number of bits transmitted), and hence the communication energy required. This shows that lowering the VBS can actually cause an increase in communication energy.

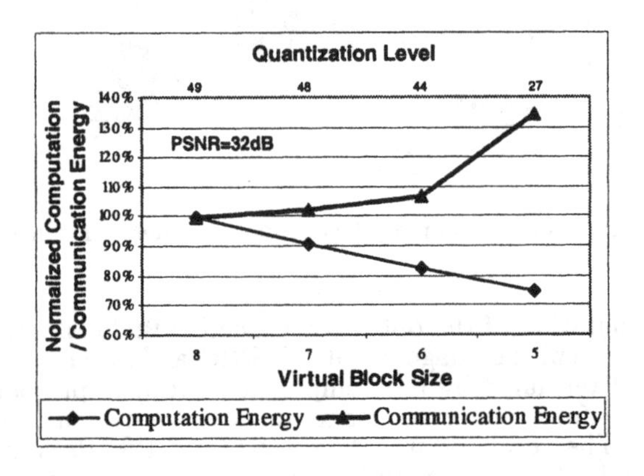

Fig. 12. Difference in Computation and Communication Energy with a constant PSNR and decreasing VBS

An alternative algorithm (*simultaneous*), aims to minimize the total computation and communication energy consumed by a multimedia radio to transmit an image by simultaneously determining the VBS and quantization level parameters. To determine both parameters at the same time, the algorithm first searches through all possible VBSs and determines the quantization levels needed to meet the image quality constraint for each VBS. For each VBS and corresponding quantization level, the overall energy can be computed. The computation energy is determined by the VBS chosen, while the communication energy is determined by using the compression ratio found in the PSNR and compression ratio table referenced by VBS and quantization level. In this way the algorithm can compare the results of all possible VBSs and choose the one which minimizes overall energy consumption while still meeting the constraints of the communication.

For example, given the PSNR and compression ratio table found in Figure 11, a VBS of 7 and quantization level of 25 would be chosen by *simultaneous*, while *VBS-first* would chose a VBS of 6 and quantization level of 19. This is because the necessary increase in quantization level from VBS=7 to VBS=6, and corresponding compression ratio (communication energy) increase, cancel out the decrease in computation energy.

Figure 13 compares the total energy consumption of the multimedia radio when the image compression parameters are selected by the *VBS-first* and *simultaneous* approaches. The top line, which often consumes more energy, is the overall energy consumption when using the parameters chosen by *VBS-first*. The bottom line represents the overall energy consumption when using the *simultaneous* algorithm. As shown, the total energy consumed by the multimedia radio can be significantly lower when the *simultaneous* algorithm is used as opposed to the *VBS-first* algorithm. For example, to compress and transmit an image with PSNR=37dB, using the *simultaneous* algorithm leads to an energy consumption that is 1.97 times smaller than the energy consumption when *VBS-first* is used. The lower energy consumption achieved by the *simultaneous* algorithm is because it considers both computation and communication energy when effecting tradeoffs.

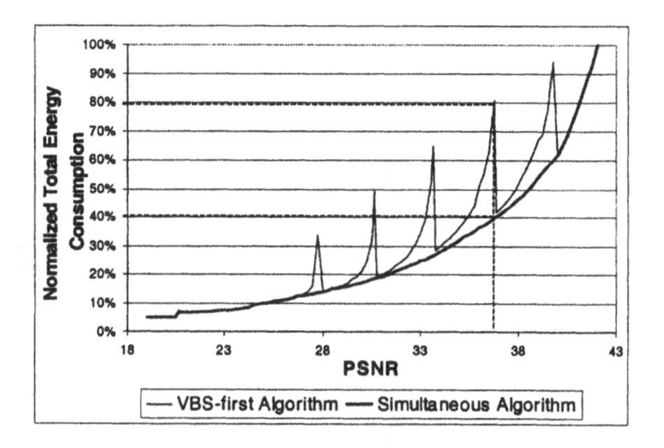

Fig. 13. A comparison of total energy consumption in a multimedia radio using image compression parameters from *VBS-first* and *simultaneous*

Once the optimal parameters table has been computed in step 2, the table can be stored in the multimedia radio for on-line use to determine the optimal compression parameters for the desired energy/latency/image quality tradeoff. The NAOS in the radio determines the image quality (QoMD) and latency (QoS) requirements and performs a table lookup to determine the optimal parameters. It then configures the adaptive image/video coder appropriately. This ensure the multimedia radio meets the QoMD and QoS requirements of the multimedia communication, while minimizing the energy consumed by the radio.

5. Conclusion

In this paper, we have discussed some of the major bottlenecks facing mobile multimedia communication. These bottlenecks include bandwidth, energy consumption, and channel noise. While upcoming wireless standards address the bandwidth limitation; energy consumption and channel noise are still significant bottlenecks to mobile multimedia communication. We proposed that by adapting to current communication conditions and requirements, we can overcome the bottlenecks to mobile multimedia communication.

An important component of multimedia communication is the source coder. The source coder affects not only the quality of multimedia data and bandwidth required, but also the quality of service provided and energy consumed by a multimedia radio. We proposed a new radio architecture which, in addition to traditional radio components, includes a Network Aware Operating System and adaptive image/video coder to adapt the source coding algorithms and parameters to the conditions and constraints of the current communication.

To better understand the effects of source coding on energy consumption, latency in transmission, and quality of multimedia data transmitted, we presented the results of varying parameters of the JPEG image compression algorithm. Using this information, we presented a methodology for choosing image compression parameters which allows us to efficiently tradeoff energy, latency, and image quality.

6. References

[1] E. Dahlman, P. Beming, J. Knutsson, F. Ovesjö, M. Persson, and C. Roobol, "WCDMA - The Radio Interface for Future Mobile Multimedia Communications", *IEEE Transactions on Vehicular Technology*, vol. 47, pp. 1105–1118, November 1998.

[2] R. Prasad and T. Ojanpera, "A survey on CDMA: evolution towards wideband CDMA", in *1998 IEEE International Symposium on Spread Spectrum Techniques and Applications - Proceedings*, pp. 323–31, 1998.

[3] N. R. Prasad, "GSM evolution twoards third generation UMTS/IMT2000", in *1999 IEEE International Conference on Personal Wireless Communications*, pp. 50–4, 1999.

[4] Qualcomm HDR Standard, http://www.qualcomm.com/hdr.

[5] The Official Bluetooth Website, http://www.bluetooth.com.

[6] IEEE Wireless Standards, http://standards.ieee.org/wireless.

[7] F. Hendessi, A. U. Sheikh, and R. M. Hafez, "Co-Channel and Adjacent Channel Interference in Wireless Cellular Communications", *Wireless Personal Communications*, vol. 12, pp. 239–253, March 2000.

[8] S. Kallel, S. Bakhtiyari, and R. Link, "An Adaptive Hybrid ARQ Scheme", *Wireless Personal Communications*, vol. 12, pp. 297–311, March 2000.

[9] P. Cherriman and L. Hanzo, "Error-rate Based Power-controlled Multimode H.263-Assisted Video Telephony", *IEEE Transactions on Vehicular Technology*, vol. 48, pp. 1726–38, September 1999.

[10] M. Goel, S. Appadwedula, N. R. Shanbhag, K. Ramchandran, and D. L. Jones, "A Low-power Multimedia Communication System for Indoor Wireless Applications", in *1999 IEEE Workshop on Signal Processing Systems. SiPS 99*, pp. 473–82, October 1999.

[11] G. K. Wallace, "The JPEG still picture compression standard", in *IEEE Transactions on Consumer Electronics*, vol. 38, February 1992.

[12] Independent JPEG Group, version 6a: http://www.ijg.org.

[13] O. K. Al-Shaykh, I. Moccagatta, and H. Chen, "JPEG-2000: A new still image compression standard", in *Conference Record of Thirty-Second Asilomar Conference on Signals Systems and Computers*, vol. 1, pp. 99–103, 1998.

[14] JPEG2000, http://www.jpeg.org/JPEG2000.htm.

[15] J. Bracamonte, M. Ansorge, and F. Pellandini, "VLSI systems for image compression. A power-consumption/image-resolution trade-off approach", in *Proceedings of the SPIE - The International Society for Optical Engineering*, vol. 2952, pp. 591–6, October 1996.

An Efficient Network Protocol for Virtual Worlds Browsing

R. Bernardini and G. M. Cortelazzo

Dipartimento di Elettronica ed Informatica
Università di Padova
Via Gradenigo 6A, 35100 Padova

Abstract: Three-dimensional virtual environments are gaining popularity in such applications as e-commerce, virtual visits of cultural heritage and video-games. The way virtual environments are currently accessed over the Internet is plagued by two major problems: the first one is due to the fact that the user must spend a considerable amount of time to download the 3D data which will be rendered on the user's local machine (whose computational power can be inadequate for a smooth navigation); the second problem is that the control of the copyright over 3D models can be very difficult if anyone can download the 3D data. In this paper we propose an alternative approach for the access to virtual worlds which is based on the idea of performing the rendering at the server's side and sending the resulting images to the client. The proposed protocol solves all the problems related to the downloading the 3D object description.

1. Introduction

Three-dimensional virtual environments, possibly with animated "virtual humans" or "avatars" [1], are gaining popularity in such applications as video-games, e-commerce, medical and surgery training, virtual fashion exhibits and virtual visits of cultural heritage.

VRML, an acronym for "Virtual Reality Modeling Language," is a language purposely written for the Internet exchange of 3D virtual worlds. It is the reference standard for this application "de facto" incorporated also in MPEG4 standard for handling 3D data. The way VRML and its current variations and/or extensions operate is plagued by two major problems for the practical use of 3D worlds via Internet.

The first drawback is related to the fact that the interactive inspection of the virtual worlds is entirely handled at client-side. This means that the user, in order to interact with the virtual world, must first download the 3D data, an operation of unpredictable duration as it depends both on the size of the 3D data (which a priori can be any) and on the transmission channel bandwidth (rather unpredictable, especially for home users connected via phone lines).

Once the data are downloaded, the interactive inspection of the 3D world entirely relies on the computational power of the user's machine, which is often inadequate for a smooth navigation of articulated 3D environments.

Another important problem for the diffusion of 3D worlds over the Internet is copyright control. The construction of tridimensional models is extremely labor-intensive and one can expect that the author of a model will want to control its diffusion. Unfortunately, as long as virtual object browsing will imply downloading of the 3D object, copyright control will not be possible.

In this paper we propose an alternative approach for the access to virtual worlds over the Internet which is based on the idea of performing the rendering at the server's side and sending the resulting images to the client. The task of decoding and visualizing the received images is well within the capabilities of home computers. The client may communicate the server its capability and possibly negotiate the visualization tasks it may take care of. The proposed protocol solves all the problems related to the downloading the 3D object description.

2. A Split-browser Structure

The recent development of 3D imaging technologies, e.g. the availability of range cameras and semiautomatic 3D modeling tools, makes feasible the construction of 3D models of real objects and suggests their use for interactive applications such as virtual visits.

The only way currently available protocols allow one to view a 3D model, is by locally downloading a file (e.g. in VRML format) and having the client rendering it. This approach, however, has several drawbacks

- Descriptions of 3D models can be quite large. As an example, the full, uncompressed model of "Madonna con bambino" by Giovanni Pisano (Fig. 1) is approximately 100 Mbytes. For home users, often connected to Internet via phone lines, the long loading times needed by files of such a size are a major nuisance. Interactive updating of virtual world descriptions (e.g. in games) by straightforward downloading schemes is clearly impossible.
- Rendering with good quality a VRML file requires a lot of memory and computational power, often not available to home users. Interacting with virtual objects and environments can be quite frustrating if the rendering is too slow.
- Since at the moment there are several file formats for 3D models, the end user must install a viewer for each format.
- 3D model construction is a labor-intensive task, both if the object is synthesized by CAD software, and if the 3D data are taken from the real world via range cameras. Because of this, it is reasonable to assume that one may want to keep the copyright of the 3D model. If the electronic description of the model can be downloaded via Internet, copyright control becomes very challenging.

In order to solve these problems, let us first analyze how a VRML browser works. In Fig. 2a one can see the internal structure of a generic VRML browser: the end user interacts with a Graphical User Interface (GUI) which, by means

Fig. 1. 3D View of the complete model of "Madonna con Bambino" by Giovanni Pisano (Arena Chapel, Padova).

of events, controls a virtual world and a graphical engine (GE) whose task is to produce 2D views of the virtual world.

The solution proposed in this work splits the browser in two: the part with the GE is moved to the server and the part with the GUI remains at the client (see Fig. 2b). The internal events generated by the GUI in Fig. 2a are now replaced by events transmitted by the client to the server along the network. At the server's side the GE responds to such events by sending back to the client the updated views as images. This approach, which essentially turns the interactive inspection of a 3D model into an image transmission task, has the following advantages

- The amount of data transmitted from the server to the client is much smaller than sending the full 3D model. For instance, a JPEG image the view shown in Fig. 1 is approximately 5K, i.e., more than four orders of magnitude smaller than the 3D model. By using specific compression techniques (see Section 3.), one can expect even greater gains. In the case of dynamic virtual worlds, this approach make feasible updating the view in order to keep up with changes.
- Visualization at the client side can be very simple since it does not necessarily require strong computational capabilities anymore.
- With the proposed solution the client will receive a sequence of images independently on the format used for the 3D data (VRML models, triangle meshes, point clouds, image-based models or proprietary formats), therefore the user will not need a viewer for every 3D data format.
- The copyright control of the 3D data is preserved, since the end user does not receive the whole model, but only 2D views of it.

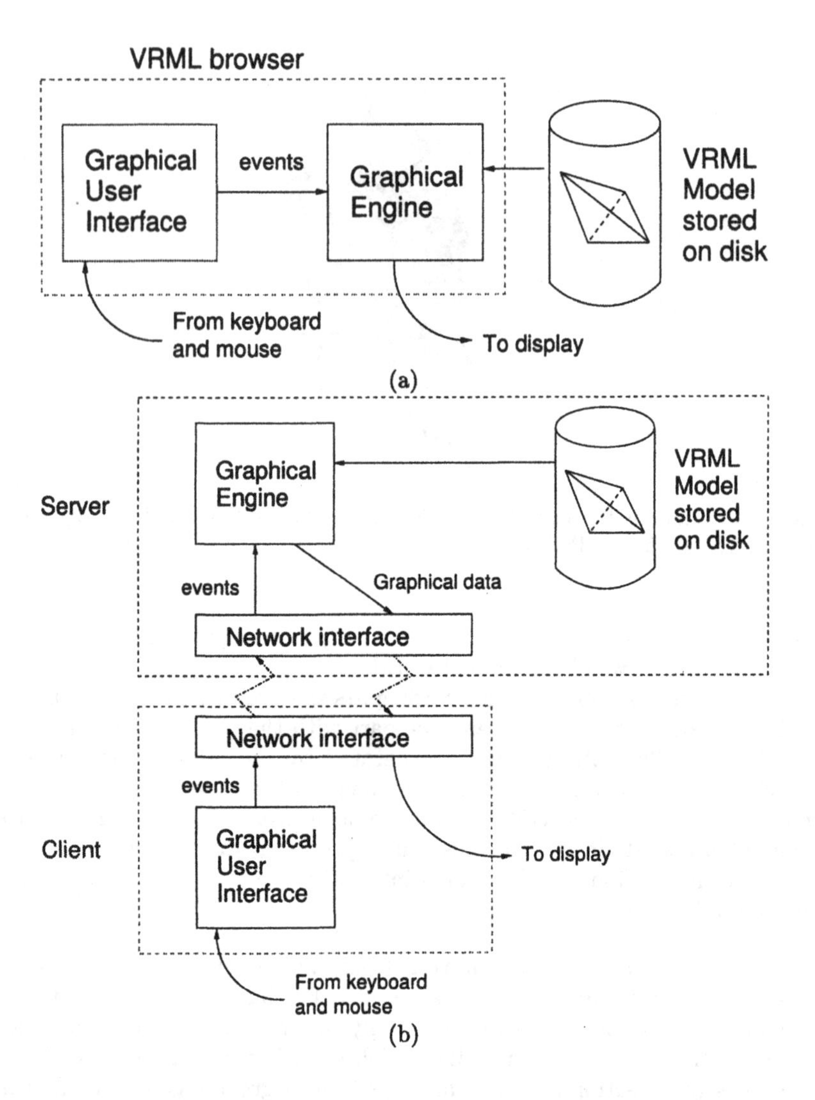

Fig. 2. The Split-browser approach: (a) The internal structure of a generic VRML browser. (b) In the split-browser approach the graphical engine remotely runs at the server and sends the computed views to the client.

In the following we will exemplify this protocol's concept by referring to a network game like, for example, a chess game or an action video-game. This situation, with a virtual world which is accessed by several users, is shown in Fig. 3. Corresponding to each user, there is a "local" application running at the user's computer and a "remote" application running at the server. Each user interacts with a local GUI which generates some events that are converted into packets and transmitted to the server via the network.
Suppose, for example, that in a chess game User 1 drags with the mouse a pawn

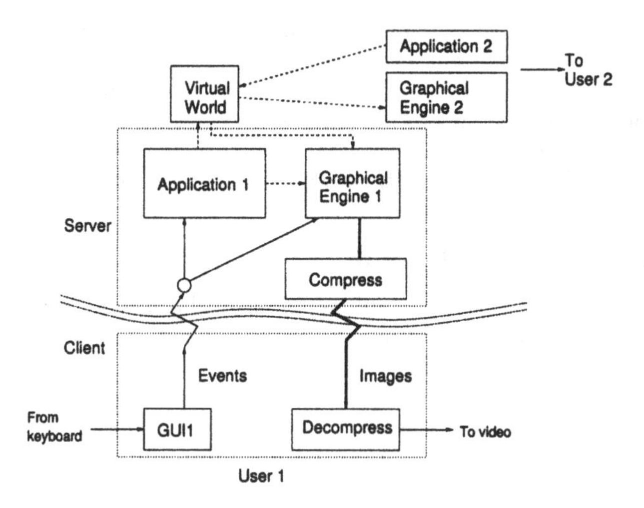

Fig. 3. Example of an application with a virtual world accessed by several users.

from b2 to b3. In response to this, a command like "Move the pawn in b2 to b3" will be sent to the remote application which will update the virtual world description. Since the virtual world is changed, the GE of User 1 and User 2 will recompute the corresponding views which will be sent to the local applications. The image module of each local application will interpret the received data and will update the image shown on the screen.
Suppose now that User 1 resizes his window or decides to change his point of view. Such a request will be sent directly to the GE (since it does not change the virtual world). In response to this request, the GE will recompute the check-board view corresponding to the new parameters and will send back to User 1 the resulting image.

From this exemplification we can deduce several facts.

1. In a general application there will be two types of commands going from the local (client-side) to the remote application (server-side): graphical commands (such as "the new window size is..."), directly sent to the GE and application specific commands (ASC) (such as "move the pawn") sent to the application itself. It is clear that the protocol that we are developing will not directly concern ASC, but it will only consider them as data to be transmitted to the server and it will create a mechanism for embedding ASC together with the stream of graphical commands.

2. The GE at the server-side will update the views in response to two types of events: change of the graphical status (e.g. window resize) and virtual world change (e.g. pawn's moves).

3. Most of the data carried by the downstream will be image data (although it is not possible to exclude other types of data).

Although the proposed solution is conceptually very simple, several questions must be answered in order to make it suitable for applications. Two fundamental issues concern about the client/server protocol and compression schemes suited to make more efficient the downloading of the updated views. Minor (but important) issues concern the kind of "intelligence" to equip the client (or server) with in order to make the interaction with the virtual world as smooth as possible. In the following sections we are going to address such questions.

3. A Compression Scheme

Although a JPEG compressed rendered view can be several times smaller than the original 3D model, it is well worth developing compression schemes tailored to this particular application.

Since the data-stream from server to client is just a sequence of images, one could simply use MPEG2. However, the sequence of images is not a generic one, but it is obtained by rendering a virtual world description. It is reasonable to suppose that an "ad hoc" MPEG4 compliant compression scheme could achieve better performance than the general-purpose MPEG2.

3.1 The Sprite Model

Let us return for a moment to the example of the chess game. If one opponent moves a pawn, it would be nice if the GE could send the update information as "Delete the image of the pawn in b2 and redraw it in b3." This is possible if the view is not sent as a single image, but as a set of layered images which will be composed at the client side. This brings us to the *sprite model* introduced in this section.

Within the sprite model, each image is considered as a set of several independent layered images (with non-rectangular support) called *sprites*. Each sprite lives on a *layer* which is a replica of the image plane shown to the user. The layers are linearly ordered and sprites on nearer layers may cover sprites on farther ones. We put no limit (at least theoretically) to the number of layers. It will prove useful to consider a farthest layer or background (that is, no other layer can be behind it) and a closest layer or foreground (that is, no other layer can be in front of it). One can imagine the layers indexed by real numbers in the interval $[0, 1]$, where 0 indexes the background and 1 the foreground.

A sprite is completely described by its support and by the RGB values associated to points of its support. The support can be described as a polygon (for simple supports, like the support of the sprite relative to the check-board) or as a union of rectangular support together with a mask, see Fig. 4 for an example. Alternatively, instead of using a mask one could use an RGB image with a transparency component, also known as alpha-channel.

It is worth observing that it is quite simple to compose several sprites into a single image and that the computational complexity of such a task is well within the capabilities of current personal computers. Moreover, even if the client has very poor computational power, one can still use the sprite model by using, in the extreme case, a single sprite for the whole image.

Another advantage of the sprite model is that makes easier the updating task on the server side too. Indeed, with the sprite model the server does not

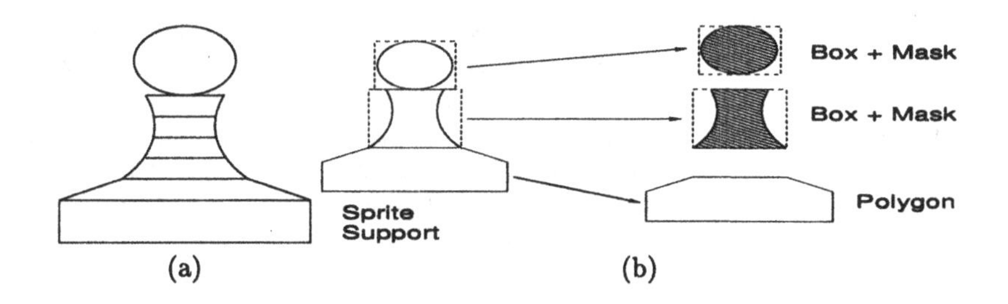

Fig. 4. Example of sprite support description. (a) Sprite representing a pawn. (b) The support of the sprite in (a) is described as the union of three areas.

have to redraw the whole image, but only the sprites which need to be updated, with a considerable computational saving.

3.2 Sprite Compression

The sprite model allows one to update the view in an efficient way when the virtual world changes. Within the previous chess game example, the server could send the following two commands to the client: "Delete sprite `pawn-1`," "Draw sprite `pawn-1` at position..." However, since the position of the pawn with respect to the observer is changed, the sprite image must be updated, possibly in an efficient way.

The most probable change in the virtual world is the movement of a single object. If the object is planar, the old view and the new one are related by a projective transform [2, 3] and one needs only to send to the client the projective transformation corresponding to the movement.

If the object is not planar, it can be approximated by a polyhedron, and one could code the new sprite image by associating to each planar component a projective transformation. For better quality, whenever necessary, one could send together with the transform parameters a "prediction error" component to be added to the transformed image. In some sense, the proposed compression scheme is similar to MPEG2, but with "generalized" motion vectors (GMV).

If a new sprite is created, the sprite image will be transmitted in a compressed format (e.g., JPEG). The same solution will be used if no transform can efficiently predict the new sprite image.

Observe that sometimes it could be more convenient to code the sprite image by giving the generalized motion vectors relative to a sprite image preceding the previous one. For example, suppose the user turns around a virtual object while observing it. After a complete revolution around the object, it could happen that the new view is very close to the first one and it could be more convenient to send the GMV relative to such an image. It is clear that this requires that both server and client keep a cache of sprite images. Since the server cannot know in advance how much space the client reserves to the sprite cache, the client will have to declare, during the negotiate phase, its caching capacity and to notify the server when a sprite is going to be deleted from the cache.

Another possible bandwidth provision is to lower image quality when the user is moving fast in the virtual world. Since the views will rapidly change, the user will not be able to notice the lower image quality. Higher image quality could be restored when the user slows down. Suppose, for example, that the user in a video-game asks to run towards an object. The client, upon this command could be programmed in order to ask the GE to lower the image quality before sending to the remote application the "run" order. When the user requires a stop, the client will ask to the GE to restore the image quality.

A possible interesting alternative is to use a multiresolution scheme. In order to illustrate this possibility, suppose that the user is playing a car-race game with two other users. A possible view is shown in Fig. 5a. The image in Fig. 5a can be conveniently coded with three sprites: one sprite for each car and one sprite for the background. The most natural way to update the image is shown by the pseudo-code of Fig. 5b which uses three instructions, namely `TRANSFORM` which sends to the client the projective transformation parameters, `LO_RES_ERR` which sends the client a low resolution version of the prediction error and `HI_RES_ERR` which sends the prediction error residue. The code of Fig. 5b sends first the complete description of the new background, followed by the complete description of each car. If one wanted to save bandwidth by trading it with image quality, it should include to the code of Fig. 5b a further specific request about this to the GE.

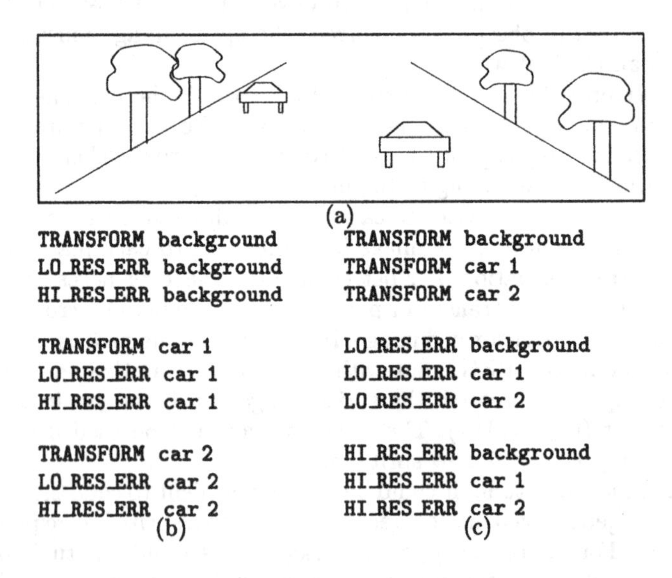

Fig. 5. (a) Example of a scene with a background (road and trees) and two moving cars. (b) and (c) two ways of coding the updates of the scene in (a).

However, bandwidth/image quality trading can be automatic if one uses the code of Fig. 5c, with the provision that the instructions list of Fig. 5c can be interrupted and restarted at anytime by a new view update request.

In this way, if the time between subsequent updates is short, the client may receive, for example, only the data relative to the LO_RES_ERR instructions, or even the data relative to the TRANSFORM .

In other words, if the virtual world changes rapidly, the user automatically receives a low-quality version of the image, that is, the trading bandwidth/image quality happens automatically.

We can outline our proposed compression scheme as follows

- The whole image is obtained by overlapping several independent images called *sprites*.
- Each sprite belongs to a *layer* and "hides" the sprites which belong to farther layers. Each layer is indexed by a real number belonging to $[0, 1]$. Layer 0 corresponds to the background.
- The first sprite image is compressed and its updates are coded by means of *generalized motion vectors* and a prediction error.
- A multiresolution scheme is enforced in order to trade bandwidth with image quality.

3.3 Intelligent Clients

Further bandwidth can be saved if we accept to do some computation at the client side. It is worth observing that a first example of this bandwidth/client-computation tradeoff is the provision of coding the whole image as a single sprite if the client has very poor computational capabilities. Another example of this tradeoff is the idea of coding a new sprite image as an old one distorted by means of some transformation. Such an approach allows us to save bandwidth by asking the client to carry out the necessary image warping.

This tradeoff can be exploited even further if the client has the capability of reconstructing an intermediate views. The use of the essential matrix [4] or of the trilinear tensor [5] appear to be quite interesting in this context, since the reference images are taken with a virtual camera with completely known parameters. For example, if the client declares in the negotiation phase of having the intermediate view reconstruction capability, the server can update a sprite by only sending the corresponding trilinear tensor (27 numbers) instead of the whole image, assuming that the reference images are still in the client cache.

4. A Protocol Between Client and Server

This section outlines a possible protocol between client and server. For sake of convenience, we will split the whole protocol in several sub-protocols (see Fig. 6). The four most important sub-protocols are the graphical sub-protocol, the negotiation sub-protocol, the compression sub-protocol (presented in Section 3.) and the application sub-protocol.

The graphical sub-protocol contains commands which directly act on the GE. In this sub-protocol the user controls one or more virtual cameras. Each

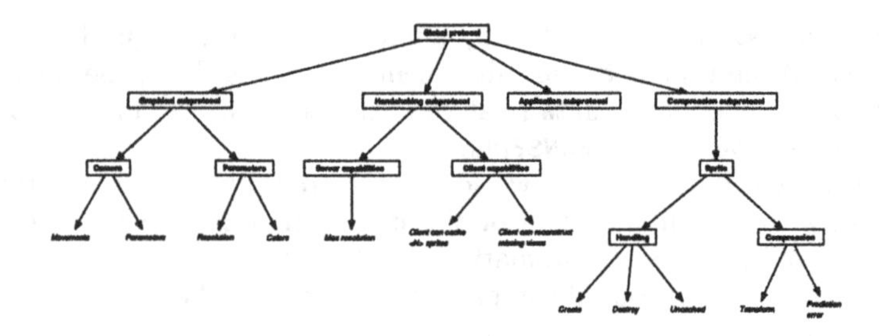

Fig. 6. client-server protocol organization

camera has some geometrical parameters (orientation, position, focal length) and some graphical parameters (resolution, number of colors, image quality). The user can create virtual cameras, change their parameters or destroy them. An "iterative" form of the primitives acting on the the camera position and/or orientation (e.g. "Every 2 seconds move the camera 1 unit to the left. Repeat 5 times") would help towards better interactivity.

The negotiation sub-protocol allows the client and the server to reach an agreement about image quality, compression services, and so on... For example, with this sub-protocol the server may declare the resolution range of the image it generates and the details of the image compression schemes it implements; on the other side, the client may declare its capability of caching or reconstructing missing views (see Section 3.).

The only aspect of the application sub-protocol that concerns this work is the need for a mechanism which embeds ASC within the discussed protocol.

From a general point of view, each command exchanged between client and server should have an unique progressive ID number for reference purposes. As an example of the usefulness of such an uniqueness of ID instruction numbering, consider the case of a client willing to delete a sprite image from the cache. In such a case, according to Section 3., the client must send a message to the server "I'm going to delete sprite image number N." However, it could happen that the server has already sent a compressed image which refers to the deleted sprite. In order to avoid this race condition, the server should acknowledge the receipt of the command and the client should wait for the server's response before deleting the sprite. Since the two channels (from server to client and from client to server) are independent, the command ID uniqueness is necessary in order to avoid confusion.

Table 1 shows a list of primitives for the proposed protocols. The primitives are detailed in Appendix 6..

5. Conclusions

The construction of articulated and photo-realistic 3D virtual worlds is at reach of current computer graphics and 3D imaging technology. The philosophy currently adopted for the diffusion of 3D virtual worlds over the Internet via the

Table 1. A list of primitives for the proposed protocol. Supscripts refer to the notes in Appendix 6..

Protocol	Class	Command	Examples	
Graphical	Camera	Movements	TRANSLATE[1]	$ID_{\text{camera}}, \Delta_x, \Delta_y, \Delta_z$
			ROTATE[2]	$ID_{\text{camera}}, \phi, \theta, \alpha$
			FOCAL[3]	ID_{camera}, f
		Parameters	RESOLUTION[4]	$ID_{\text{camera}}, N_x, N_y$
			COLORS[5]	ID_{camera}, N_c
			QUALITY[6]	ID_{camera}, Q
		Handling	CREATE[7]	ID_{camera}
			DESTROY[8]	ID_{camera}
Negotiation	Server Capabilities	Resolution	MAXRES[9]	N_x, N_y
		Colors	MAXCOLOR[10]	N_c
	Client Capabilities	Sprite caching	SPCACHE[11]	max_size
		Missing views	MISSINGV[12]	B
Compression	Sprite	Handling	SPCREATE[13]	ID_{sp}
			SPDESTROY[14]	ID_{sp}
			UNCACHED[15]	ID_{sp}, N
		Support	POLYGON[16]	$ID_{sp}, N, x_1, y_1, \ldots, x_N, y_N$
			MASK[17]	$ID_{sp}, x_{ll}, y_{ll}, x_{ur}, y_{ur}, M$
			EXTMASK[18]	$ID_{sp}, N_b, x_{ll}, y_{ll}, x_{ur}, y_{ur}, M$
		Compression	TRANSF[19]	$ID_{sp}, N, \mathbf{p}$
			ERR[20]	$ID_{sp}, N, R, data$

VRML and/or current related extensions has a fundamental drawback: it calls for downloading at client side the 3D virtual world description.

This work addresses this crucial issue and proposes an alternate scheme which essentially transform the interactive inspection of a virtual world into an image transmission task from server to client with a light communication feedback from client to server.

The advantages of the proposed protocol are several. It allows for a design of the communication system so that prescribed QoS demands about visualization can be met. This is not even conceptually feasible with the current approach where visualization is bound to happen after the downloading of data of unknown entity.

The client may be partially or even completely relieved from the visualization tasks, making feasible the 3D world navigation also with client with limited resources.

The issue of the copyright control over the 3D data is completely solved as the 3D data are never delivered to the client.

6. Notes

1. Translate the virtual camera of $\Delta_x, \Delta_y, \Delta_z$.
2. Rotate the virtual camera of angles ϕ, θ, α.
3. Set the camera focal length equal to f.

4. Set the image resolution to $N_x \times N_y$.
5. Set the number of colors equal to N_c.
6. Set the image quality equal to Q (a quality parameter whose precise meaning is still to be determined).
7. Create a new camera.
8. Destroy the camera $ID_{\mathbf{camera}}$.
9. The maximum resolution that the server can handle is $N_x \times N_y$.
10. The server can handle up to N_c colors.
11. The client can cache up to max_size sprites.
12. If Boolean parameter B is true, the client can reconstruct missing views.
13. Create a new sprite.
14. Destroy a sprite. Observe that this is different from *deleting* a sprite, since the former removes every information about the sprite from the client's memory, while the latter merely removes the sprite image from the screen.
15. Image number N of sprite $ID_{\mathbf{sp}}$ is going to be deleted from the cache.
16. Add to the support of sprite $ID_{\mathbf{sp}}$ the polygonal region with vertices $x_1, y_1, \ldots, x_N, y_N$.
17. Add to the support of sprite $ID_{\mathbf{sp}}$ the region with bounding box $x_{ll}, y_{ll}, x_{ur}, y_{ur}$ and mask M.
18. Command EXTMASK is like MASK, but it uses an N_b-bit mask (used for sprite partitioning).
19. Apply projective transformation with parameter $\mathbf{p}$ to the polygonal region with vertices $x_1, y_1, \ldots, x_N, y_N$.
20. Command ERR introduces prediction error of resolution R for image N of sprite $ID_{\mathbf{sp}}$.

7. References

[1] D. Thalmann, "Virtual humans in virtual environments: a new view of multimedia applications," in *International Workshop on Synthetic-Natural Hybrid Coding and Three Dimensional Imaging* (N. Sarris and M. G. Strintzis, eds.), (Rhodes, Greece), pp. 3–7, Sept. 1997.

[2] R. Szeliski, "Video mosaic for virtual environments," *IEEE Computer Graphics and Applications*, pp. 22–30, Mar. 1996.

[3] R. W. Picard and S. Mann, "Video orbits of the projective group: A simple approach to featureless estimation of parameters," *IEEE Transaction on Image Processing*, vol. 6, pp. 1281–1295, Sept. 1997.

[4] H. C. Longuet-Higgins, "A computer algorithm for reconstructing a scene from two projections," *Nature*, pp. 133–135, 1981.

[5] A. Shashua, "Algebraic functions for recognition," *IEEE Transaction on Pattern Analysis Mach. Intell.*, pp. 779–789, 1995.

Part 2

Software Radio for Multimedia Communications

Software Radio is Walking into Implementation Stage

H.Che[1], M.Hajian[2], L.P.Ligthart[3] and R.Prasad[4]

[1] Delft University of Technology (author for correspondence: YilunChe@ieee.org)
[2] Delft University of Technology
[3] Senior Member,IEEE, Delft University of Technology
[4] Senior Member, IEEE, Aalborg University

Abstract: This paper starts with a brief review of the origin, development and the certainty of the appearing of software radio; after that, we describe the different development stages. We indicate that now software radio is walking into the Implementation stage and predict the trend of the development.

1. Introduction

Software radio was first proposed in 1991 to predict the transition in radio technology from digital radio with 80 percent hardware and 20 percent software to radio with 80 percent software and 20 percent hardware. Since then, there has been increasing importance attached to it in universities and industries. We divide the development process of software radio into several stages: Definition stage, Feasibility & Related Technologies Study stage, and Implementation stage. During the definition stage, the conceptions of ideal software radio and practical software radio have emerged. Generally people refer to practical software radio when software radio is mentioned at the moment. In the Feasibility & Related Technologies Study stage, we classify technologies into three categories: rudimental technologies, soft handset & soft base station technologies, and network technologies. In the Implementation stage, we anticipate that this stage will go in a spiral way. A project related to software radio implementation is introduced.

2. Origin and Development of Software Radio

Just like most of wireless technologies, software radio (SR) is originally for military applications. The military application of SR techniques was proposed in 1970s for higher reliability in high frequency. SR has more flexibility than traditional analog radios, which makes it desirable in civil applications, [1]. At the '70s and '80s, the "software radio" was actually digital radio. In 1991 Joseph Mitola III proposed the conception of "Software Radio" to predict the transition in radio technology from digital radio with 80 percent hardware and 20 percent software to radio with 80 percent software and 20 percent hardware and published the first paper on SR [2]. Three years later, this conception of SR got increased attention. In May 1995, the functions, performance, software and hardware of SR were investigated, [3]. Recently, Analog-to-Digital and Digital-to-Analog (A/D/A) conversion becomes more accurate with higher resolution, higher sampling rate and better linearity; Silicon is evolving into lower voltage, larger memory, smaller geometry size; DSP (Digital Signal Processing) power continues to decrease, [4, 5]. The development of the deterministic technologies makes it improved multiband and multimode personal communications. This facilitates the realization of software radio, although there are still some limitations in these technologies to achieve the full benefits of SR. SR was defined in its Definition Stage. Some projects about feasibility and concepts of different compositions of SR have been performed, which enables the initialization of the Implementation Stage.

3. Why Software Radio (SR) is needed

SR is the necessity and inevitable outcome for future wireless communications due to the requirements of multiband, multimode and multistandard personal communication systems. SR's reconfigurability of parameters and DSP reprogramming by software downloding make it "intelligent" or adaptive in different applications and different environments. The reasons are: [2, 4]

1. **Versatility, Small Volume.** With the swift improvement in wireless personal communications, wireless people are confronted with a new task to realize multiple functions (indicated later) on a single device. SR is a platform to implement such a task of versatility, simple hardware architecture and small volume.
2. **Global Standards.** Although in 1994 a lot of work had been done related to standardization, it is still difficult by now to implement global accepted standards. The standards in use for wireless communication systems are all designated for certain areas. This leads to inconvenience for subscribers when roaming internationally. SR is expected to solve this problem using reconfigurability technologies.
3. **Efficient Frequency Spectrum.** The frequency spectrum is a limited source. Due to the advancement of wireless communications (e.g., more services are needed, more users), commercial spectrum becomes more and more precious. It is anticipated that SR can provide a solution for efficient use of spectrum by, for example, adaptive spectrum management.
4. **Compatibility.** SR is reprogrammable and reconfigurable. That is to say, the incompatibility between different kinds of equipments may not prevent people from communicating with each other by SR.
5. **Personalizing Services.** With reconfigurability, reprogrammability and software downloading, SR can also offer more extensive services and enhanced personalized services, which will bring even stronger functions and more practical facilities of communication to users.

4. Definition Stage

Since 1991, some work has been done to study the role and function of SR in wireless communications. Till 1999, conception of SR had been given [4]. We designate this period as Definition Stage. In this stage, a new task was set up and people know "what to do" for SR. The conception of SR with radio terminal implemented almost completely by DSP arises from the requirement of multiband, multimode and multistandard wireless communications. To realize SR, A/D/A conversion is expected to be "as close as possible to the antenna [2]". Below we describe the concepts of ideal SR and practical SR respectively. Because of the limitation of the relative technologies, e.g., A/D/A conversion, ideal SR can not be realized within a short period of time.

4.1 Ideal Software Radio

Ideal SR has full benefits of SR. Ideal SR can reconfigure parameters and algorithms automatically according to the instantaneous situations of different environments to optimize signals with an architecture of only three sections: antenna section, DSP section and Information section, as shown in Figure 1. The performance (e.g., speed, power) of the DSPs is an essential factor to decide when ideal SR can be realized.

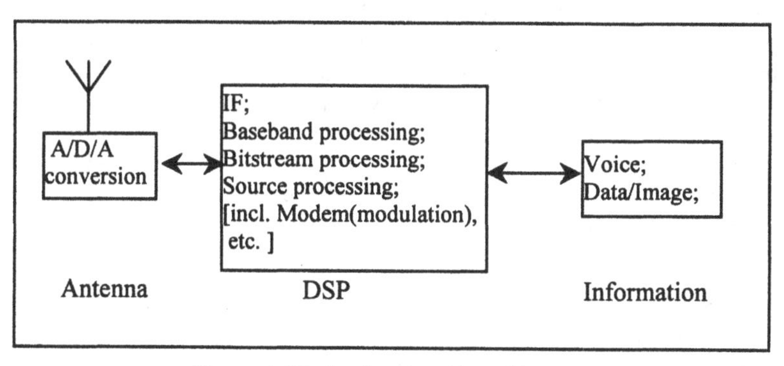

Figure 1. Ideal software radio architecture

4.2 Practical Software Radio

Practical SR sets a beginning and path to achieve ideal SR in the future. It allows A/D/A conversion in RF circuit, which is not together with antenna; and allows some signal processing in RF circuit instead of in DSPs. Reconfigurability and reprogrammability can be realized by wireless or wired download, not necessary to be real-time. The architecture is shown in Figure 2. This architecture is more practical at present. The practical significance makes this definition popular. When SR is mentioned at the moment, generally it refers to the practical SR. Thus, in the following two stages (Feasibility & Related Technologies Study stage and Implementation stage), we refer to Practical software radio.

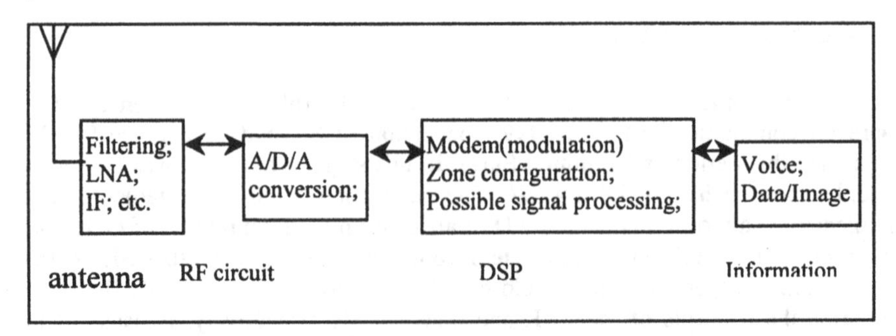

Figure 2. Practical software radio architecture

5. Stage of Feasibility & Related Technologies Study

SR performs more functions than traditional radio. It is adaptive or "intelligent" compared with traditional radios. For example, before transmitting (in order to optimize transmission), SRs check the channel properties and possible transmission routes by channel estimation and channel modeling, set up modulation scheme by adaptive modulation algorithm, use smart antenna for space multiplexing, designate transmission power by adaptive power control, etc. And before receiving (in order to optimize the received signal), SRs check the receiving power, discern the modulation method of the received signal, cancel or decrease the multipath interference adaptively by corresponding algorithms, etc. When SR was defined, these additional functions were not readily available or not well developed. SR needs support from technology and thus stimulates the development of a certain technologies. What's more, "software radio engineering is fraught with pitfalls", [3]. Therefore, the feasibility and related technologies of SR need further studies.
The work that has been done so far about feasibility & related technologies includes:

- From 1994 to 1995 a number of papers related to radio standardization were published, e.g., [6].
- In 1995, some articles had been written about reviewing state of art of some SR technologies, e.g., ADC [7], DSP [8], smart antenna [1].
- Development of enabling technologies, e.g., Signal digitization, silicon technology, of SR has been summarized in some articles, [4,5].
- The European ACTS project (1994 – 1998) including more than 200 individual projects has explored SR feasibility in second generation and third generation wireless communication systems, for example, the feasibility of flexible, modular, multiband and multimode radio systems, software portability and hardware reconfigurability, [http://www.infowin.org/ACTS/].
- IST (Information Society Technologies) started a project "Re-configurable low power radio architecture for software defined radio (SDR) for third generation mobile terminals". The project is to study and define the impact of the SR concept on the RF architecture and to define the RF architecture

that is able to provide the best solution for Re-configurable radio, and to validate the feasibility of the selected radio architecture, [http://www.cordis.lu/ist/projects/99-11243.htm].

- The German Integrated Broadband Mobile System project studied the dynamic download concept, which set a path to software downloading, [http://www.comnets.rwth-aachen.de/project/ATMmobil/IBMS.htm].
- Wide-band RF hardware circuits have been improved.
- The European ESPRIT project has developed software libraries for multiple hardware platform, [4]. This will help to take advantage of reconfigurability.

Due to the limit of available spectrum and the increasing demand for radio spectrum, the efficient use of radio spectrum has to be considered. This results in the development of adaptive technologies. According to the variance of channel, the adaptive technologies change the corresponding parameters in order to improve the spectrum efficiency, [9].

From 1994 on, a lot of work with respect to feasibility & related technologies study of SR has been done, but it is still far from mature. So this stage is not over yet. Following is the generalization of different SR technologies, which are divided into rudimental or basic technologies, soft handset and soft base station technologies, and network technologies. Among these, soft handset and soft base station technologies are dominant and have been the focus of exploration of SR.

5.1 Rudimental Technologies

ADC & DSP

Just as mentioned above, SR brings the need to develop relative technologies. In this process, there are some limitations to impede or delay the related technology development, and impede or delay the advancement of SR. Most of the limitations are concerned with ADC and DSP. ADC and DSP play a rudimental role in SR. By now, ADC and DSP technologies are both not developed well enough to meet the requirements of the technologies needed by SR, especially for base station applications. Although it is already be noticed, no efficient solution has been accepted over the world yet, [10]. For example, SR requires fast, cheap ADC with high resolution, which are difficult to be met at the same time. When the speed goes up, the resolution goes down, vice verse. The work is needed to improve the product of the speed and the resolution. But it is not optimistic by now. And the price of ADC is still a consideration. As for DSP, a base station in third generation W-CDMA system requires processing speed to be at least hundreds of thousands of MIPS (Million Instructions Per Second), maybe much more. But up to now, the highest performance of DSP is 8800MIPS, the processing speed of TMS320C6000 of TI (Texas Instruments). Although TI made a great progress in it, we can see it is still far from the requirement of SR.

Standardization
Although from 1994 to 1995 much effort had been done to set up communication standards and radio standards, no mature and generally accepted open architecture standard for high-performance, real-time SR exists at the moment.

When we asked Joseph Mitola why the current SR work is "labor-intensive", he said, "because it takes dozens of people a lot of time to put together a reasonable amendment to a standard", "If we had computer-based semantics (e.g., model-based definitions) we could do things more efficiently".

To take the full advantage of SR, to simplify and accelerate the process of SR development, a uniform standard is necessary for different software to run, for different hardware platforms to be based on, for different networks to be connected with. Of course, it is a long-term objective and needs a lot of inputs from different parties.

5.2 Soft Handset & Soft Base Station Technologies

Soft handset and soft base station have similar architectures except that base station is fixed, larger and consumes more power, and handset is and is going to be even mobile, smaller and consumes less and less power. Therefore, some of their respective technologies are also similar. The related technologies of soft handset and soft base station are listed below.

Smart Antenna
Smart antenna is one of the essential SR technologies and is developed rather early among the SR technologies. The use of smart antenna is to increase the system capacity and to cover larger area in mobile personal communications. Smart antenna makes a leap in the development of antennas technology. It demands new architecture, and traditional analog architecture for second generation can not provide solutions any more to realize smart antenna. Many people in Europe and America are working on smart antennas, but there is still much work to do before the benefits are fully realized. It is anticipated that smart antenna will not be integrated at the early time of third generation timeframe, [4].

Reconfigurable Radio Interface
Reconfigurable radio interface belongs to the technology of handset, not of base station. To achieve the versatility, capacity, multiband, multimode, SR systems require new architecture concepts. One important concept is reconfigurable radio interface with adaptability, which is designed to meet different configuration conditions in different environments, to increase flexibility, to increase the using efficiency of resources. Reconfigurable radio interface is regarded as an essential component of soft handsets. And it is expected that reconfigurable radio interface by software downloading will be feasible within the timeframe of Third Generation (3G) systems.

Real Time Software Download
Real time software download is a technology for soft handset. If SR wants to reach its full benefits of reconfigurability, real time software download at the handset

interface is indispensable. In [4], download mechanisms have been classified into static download, pseudo static download and dynamic download. These three steps, which become more and more complicated and have stronger and stronger function, may be the evolution steps of software download and indicate a valuable solution to achieve software download. Real time software download-and-run is still a challenge at the moment.

Adaptive Modulation

Modulation is a basic technology for transceivers. Adaptive Modulation is to switch modulation formats adaptively to acquire optimal throughput when the received SNR level varies with time over a wide range. It can increase system capacity [11] significantly. Higher modulation format can bring higher data rate, but at the same time, a higher modulation format will also result in a higher Frame Error Rate (FER). In [12], adaptive modulation schemes are classified into slow- and fast-adaptive modulation based on modulation parameter adaptation speed.

Adaptive modulation (mode switching) is not easy to implement. For example, to determine modulation format, the feedback of precise SNR from the receiver is necessary, but it is difficult to get; sometimes the interference varies so fast that the feedback can not keep up with it, which makes it difficult to get in-time feedback.

Adaptive Zone Configuration

Adaptive coverage zone configuration is an important technology for soft base station. At the moment the zone configuration in use is uniform and unchanged for all environments and all traffic. That is to say, the service area of every base station (BS) is equal in size and in shape no matter how much the traffic is different in each area (small zone/cell). The disadvantage is that there may be a waste of power in low traffic zone; and in high traffic zone, communication quality may decrease heavily due to interference. Adaptive zone configuration is a solution to the disadvantage mentioned above. The concept is to assign zone or cell dynamically according to the traffic distribution. Higher traffic density corresponds to smaller zones; lower traffic density is corresponding to larger zones. Not much work has been done related to adaptive zone configuration to date.

Adaptive Power Control

Power control is to decrease or dispel the near-far effect that refers to base station receiving larger power from a closer mobile than a farther mobile. Near-far effect will make mobile communication system less efficient. Adaptive power control is to optimize the transmission power. It makes the power reaching the BS from different mobile station (MS) equal by regulating the transmission power of every MS. It is an essential condition to achieve the benefit of CDMA.

5.3 Network Aspect

Network aspect is in its infancy as a SR technology. People's concentration is extending to network from handset and base station since SR is going to deploy downloadable software. But of course, the network aspect refers to data link too, just as the general meaning of network in wireless communication systems. With the development of network application in mobile communication systems, it is estimated that network functions demand to be divided into functions related to radio and functions unrelated to radio. And the network aspect in SR refers to the functions related to radio. The ACTS (1994-1998) project has explored the network aspect in wireless applications. In this project functions have been divided into control part and transport part.

Network aspect needs further study and there are difficulties, e.g., no effective solution yet for the network evolution from second generation to third generation.

6. Implementation Stage

Because international roaming is not a problem in GSM, European third generation – Universal Mobile Telecommunication System (UMTS) will be the application market for SR. While in North America, Personal Communication System (PCS) can not perform seamless international roaming as GSM, so there is a requirement of SR at present, that is, even in second generation PCS. Therefore, UMTS and PCS are the drive for SR.

The introduction of third generation (3G) system is inevitable. The reasons are, firstly, second-generation will not be able to provide enough capacity for the more and more subscribers in different environment; and secondly, second-generation can not provide enough services that the users require. 3G systems are expected to provide broadband multimode, multistandard, multimedia systems. SR implementation will have a profound effect on the multiband, multimode, multistandard communication market.

Implementation stage refers to the implementation of SR as a whole instead of in a certain part or in a certain technology. Consideration of the relationship among different technologies and parts in the whole SR system is required. For example, in order to optimize the performance of the system, how they co-operate and compensate the disadvantage of each other. In the past almost nine years, SR has experienced the first (definition) stage and the second (feasibility & related technologies study) stage. Although the second stage has not been completed yet so far, actually some technologies are far from mature enough to support the full potential of SR, conditions enable SR to walk into third (implementation) stage.

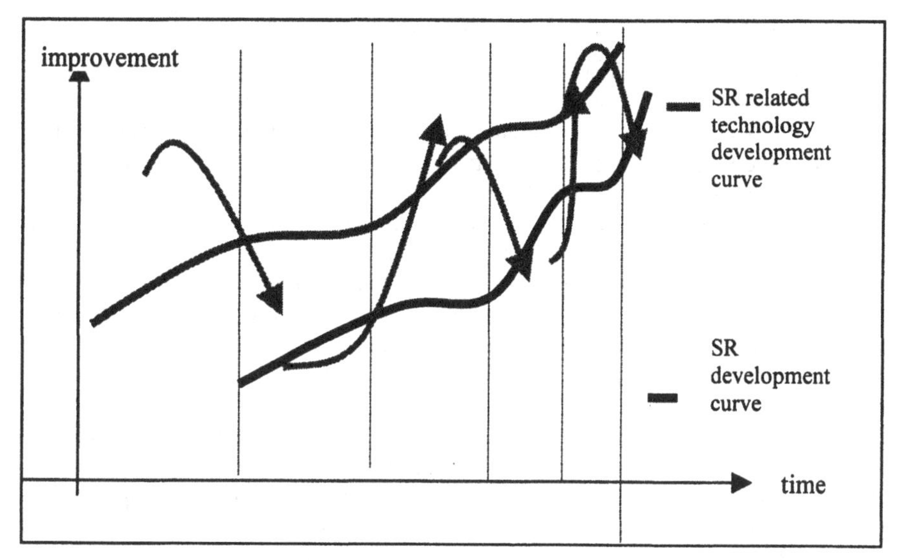

Figure 3. Relationship between SR and its related technology developments

Implementation stage will advance in a spiral way, shown in Figure 3. With the applications in military and commercial market, SR has caused and will continue to stimulate the development of the relative technologies, and further influences the industry distribution. When they develop forward to some extent, technologies and industry markets will facilitate the SR improvement. That pushes SR to advance. The advancement of SR drives the development of technology and industry again. This process will run recursively. That is to say, SR and its related technologies interact and stimulate the development of each other. This will lead to the improvement for both parties. With the improvement, the interaction speed will increase and the period of each development step will become shorter.

6.1 Enabling Conditions for Implementation Stage of Software Radio

- The well-developed Definition Stage gives an explicit idea of what need to be done for SR in the short term and in the long term. This stage also demonstrates the difference from traditional analog and digital radio, and indicates an path to realize ideal SR gradually;
- The rapid developing Stage of Feasibility & Related Technologies Study lays a foundation for the Implementation Stage. Although some technologies still need further study to reach their full potential, almost all the necessary concepts and technologies for the basic SR architecture have been studied and proved to be feasible.
- The study of enabling technologies for SR gives an optimistic estimation for the aspects required by SR of different technologies.
- The industry architecture has the trend to change towards the direction beneficial to SR development.

6.2 A Software Radio Testbed (SRTB) Project at IRCTR

Due to the enabled conditions of SR implementation, a SR implementation project has been introduced in IRCTR (International Research Center for Telecommunications-Transmission and Radar) of Delft University of Technology in co-operation with Aalborg Univeristy. The project is entitled " Realization of Real Time Adaptive Software-Definable Radio (ASDR) Wide Band CDMA Testbed to be used for Third Generation Mobile Communications". It is beneficial to judge the architecture and technologies in use for advanced mobile communication systems by employing efficient tools. To implement the flexibility of SR a testbed is an optimal option to utilize latest result of hardware- and software developments. Of course, however, some techniques need study, for example, adaptive modulation, due to the evolvement into a new generation in wireless communication.

Although a lot of work has been done related to adaptive modulation, most of them are for TDMA system. Adaptive modulation in TDMA system and that in CDMA system are quite different. Adaptive modulation becomes less important in third generation wireless system since system flexibility mainly realized by changing spread spectrum (SS) parameters, not modulation parameters any more. The adaptation parameters are changed from symbol rate, modulation level, and coding rate in TDMA system to processing gain, channel activation and coding rate [13] in CDMA system. Confronted with challenges from OFDM modulation, adaptive modulation in CDMA system needs to consider to combat the drawback of CDMA, for example, increasing user capacity in single cell or multi-cell environment, improving performance in multi-cell environments with more frequencies, increasing data rate, simplifying algorithm operation, etc.

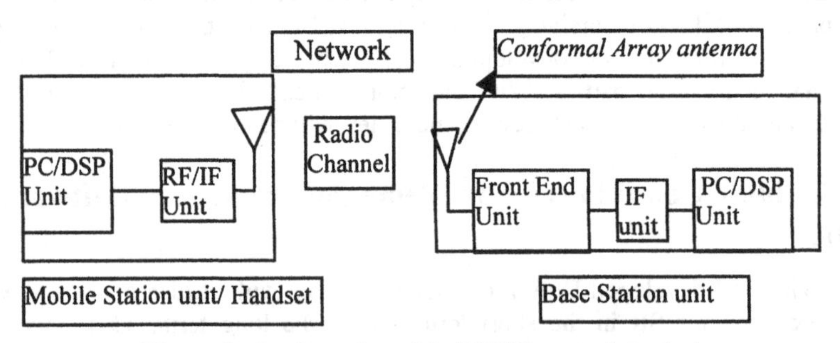

Figure 4. Configuration of the IRCTR research testbed

This Testbed considers the combined effects of a range of techniques key to SR, aiming at providing a powerful tool to develop and assess advanced wireless communication techniques (adaptive algorithms) which could be used in outdoor, real-time environment for the European third generation wireless communications – UMTS. Figure 4 is the configuration of the SR testbed in the project. Base station, handset and network are considered as a whole to implement the functionality of SR in real time. IRCTR and its partners have much interest in realizing a research demonstrator testbed.

The final goal of the project is to:

- implement a SR using the adaptive techniques in a real time and in a fully operational network;
- demonstrate the performance and cost efficiency of deploying adaptive techniques in a testbed that can be used within the infrastructure of third generation mobile communications system. This will be done in two steps: first in the research demonstrator, and then test in real network;
- enhance the performance of the mobile communications systems for the third generation;

Figure 5 gives the classification of the adaptive technologies in the SR testbed. These techniques are the key issues to the personal wireless communications systems that require high reliability and high capacity. Here traffic performance and adaptive spectrum management are explored to optimize the efficiency of other algorithms and to improve the system capacity. Downloadable software is the source of reconfigurability. The availability of downloadable software determines the SR flexibility.

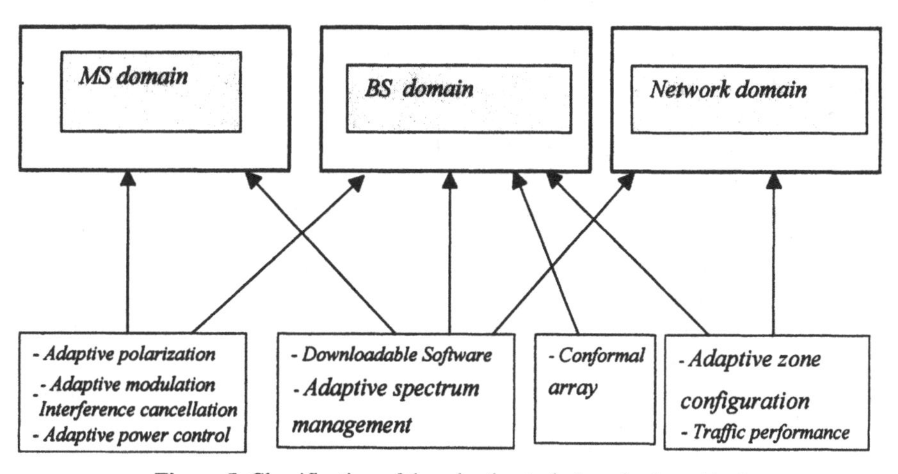

Figure 5. Classification of the adaptive technique in the testbed

At present there are almost no BS and MS in the world, which uses so many adaptive techniques in real time and in real operational network. The result of this project will accelerate the development of the future 3G wireless communication system.

7. Conclusions

Software radio (SR) came to the world with the rapid development of Wireless communications inevitably. Since the appearance of its conception, increasing attention has been paid. Correspondingly, a lot of work has been done to make SR go to the regular orbit of development gradually. We indicate the process of SR development is experiencing several stages: first stage – Definition Stage almost passed successfully, and its second stage – Feasibility & Related Technologies Study Stage is advancing smoothly and optimistically. We indicate that the situation allows SR to walk into its third stage – Implementation Stage. According to this, we discussed the SR implementation project " Realization of Real Time Adaptive Software-Definable Radio (ASDR) Wide Band CDMA Test Bed to be used for Third Generation Mobile Communications" at IRCTR.

8. Acknowledgement

The authors would like to express sincere thanks to Zoran Zvonar (Analog Devices, USA) for his help.

9. References

[1] Joseph Kennedy "Direction Finding and 'Smart Antennas' Using Software Radio Architectures" IEEE Communications Mag. May 1995

[2] Joe Mitola "Software Radios" IEEE Communications Magazine .May 1995

[3] Joe Mitola" The Software Radio Architecture" IEEE Communication Mag. May 1995

[4] W. H. W. Tuttlebee "Software-Defined Radio:Facets of a Developing Technology" IEEE Personal Communications. April 1999

[5] Dimitrios Efstathiou "Recent Developments in Enabling Technologies for Software Defined Radio" IEEE Communications Magazine August 1999

[6] Donovan Nak "Coordinating Global Standards and Market Demands"IEEE Communications Mag. January 1994

[7] Jeffery A. Wepman "Analog-to-Digital Converters and Their Applications in Radio Receivers" IEEE Communications Magazine . May 1995

[8] Rupert Baines "The DSP Bottleneck" IEEE Communications Mag. May 1995

[9] Mohamed-Slim Alouini "An Adaptive Modulation Scheme For Simultaneous Voice And Data Transmission Over Fading Channels" 1998 IEEE VTC

[10] Apostolis K. Salkintzis "ADC and DSP Challenges in the Development of Software Radio Base Station" IEEE Personal Communications August 1999

[11] R. van Nobelen "Towards Higher Data Rates for IS-136" IEEE VTC 1998

[12] Norihiko Morinaga"New Concepts and technologies for Achieving Highly Reliable and High-Capacity Multimedia Wireless Communications Systems"IEEE Comm. Mag. Jan. 1997

[13] Rajamani Ganesh, <<Wireless Multimedia Network Technologies>> 2000

Capacity Considerations on the Uplink of a Multi-user DMT OFDMA System Impaired by Time Misalignments and Frequency Offsets

Andrea M. Tonello, Nicola Laurenti, Silvano Pupolin

DEI - Department of Electronics and Informatics - University of Padova
Via Gradenigo 6/A - 35131 Padova - Italy
tonello@dei.unipd.it - nil@dei.unipd.it - pupolin@dei.unipd.it

Abstract: The uplink of an asynchronous DMT/OFDM based communication system is studied and capacity implications are derived. Multiple users share a Gaussian channel through DMT OFDMA modulation and multiplexing. We consider a bank of single user detectors, at the output of which multiple access interference arises whenever time misalignments and frequency offsets exist among users. Consequently, we determine an inner bound to both the capacity of a given user link, and the region of achievable information rates for joint reliable communications (i.e., capacity region). Such capacity inner bounds are random variables and function of several system parameters. The associated complementary cumulative distribution functions are defined, and are evaluated for several system scenarios characterized by different tone assignment schemes and strategies of power allotment to sub-carriers. As a result, a pragmatic approach for enlarging the capacity region is devised. The approach is based on the appropriate insertion of a guard time, the partition of tones to users, and the choice of the power profile to be assigned to the sets of tones.

1. Introduction

Several advantages of multi-carrier modulation and in particular of DMT-OFDM (discrete multi-tone/orthogonal frequency division multiplexing) for wireless communications are well recognized. These include robustness against frequency selective fading, frequency diversity, simple implementation with FFT (fast Fourier transform) based joint modulation and multiplexing [1], [2]. However, a major limit is its sensitivity to time misalignments and frequency offsets that can arise in an asynchronous multi-user uplink scenario. Some investigation of the effect of the time asynchronism can be found in [3]. In [4] and [5] a general analysis of the effects of time and frequency asynchronism in both Gaussian and multi-path fading channels is presented. In particular, in [4], the detrimental effect due to multiple access interference is measured in terms of signal energy over noise plus interference ratio, while in [5] symbol error rate performance is also investigated.

In this paper we address the problem of defining the *achievable information rates* for reliable communications (i.e. with arbitrarily small probability of error) of a set of time and frequency asynchronous users that share a Gaussian channel with an orthogonal frequency division multiplexing scheme. In particular, demodulation is accomplished with a bank of parallel single user detectors, each perfectly time and frequency synchronized to a given user [4], [5].

Part of this work was supported by the Italian National Research Council (CNR) under "Progetto Multimedialitá ".
A.M. Tonello is on leave from Lucent Technologies - Bell Labs, Whippany Laboratory, NJ, USA.

Considering N_U users, the determination of the *capacity region*, which is the closure of the set of all achievable rate N_U-tuples, of such a scenario, is in general a formidable task [6],[7]. We determine an *inner bound*, i.e. a subset of the capacity region that can be achieved when deploying the proposed receiver. A trivial *outer bound* is found by considering the perfectly time and frequency synchronous system.

We emphasize that in this scenario the capacity region is determined as a function of several system parameters. Once we have constrained the total number of sub-carriers, and the total power per user, the capacity region still depends on the number of users, the number of sub-carriers per user, the sub-carrier allocation scheme to users, the power allotment to sub-carriers, the deployment of time and frequency guards [4]. Finally, there is a dependency on the time and frequency offsets of each user, which are random variables.

To proceed into the analysis we take the approach of treating the maximum rate for reliable communications of a given user link (link capacity) as a random variable. We fix a maximum time and frequency offset within the system, and then we compute the complementary cumulative distribution functions (ccdf) of the link capacity. These curves show the probability that a user link has a larger capacity than the corresponding abscissa, over all system realizations. The dependency on the tone allocation and power allotment scheme is illustrated by numerically computing, through Monte Carlo simulations, the ccdf for several assignment schemes in a fully loaded system. Among these, a promising dynamic tone allocation scheme yields increased capacity, compared to the conventional block and interleaved tone allocations.

We further consider the joint ccdf of a set of users, from which it possible to determine, for instance, the N_U-tuples of achievable rates with a given probability.

Finally, from the comparison of the capacities achieved with the aforementioned schemes, system design guidelines are devised.

This paper is organized as follows. In Section 1 we revise the OFDM based communication system and the proposed detector [4]. In Section 2 the resulting multiple user interference channel model is described. Sections 3 and 4 discuss the system capacity problem. The evaluation of the ccdfs for several system scenarios is carried out in Section 5. Finally, the conclusions follow.

2. Asynchronous Multi-user Communication System

We consider a communication system where N_U users deploy discrete multi-tone modulation (DMT) and share a Gaussian channel with an orthogonal frequency division multiple access scheme (OFDMA). The overall bandwidth W is subdivided into N equally spaced sub-carriers (tones) among which K_u are assigned to the u-th user. The transmitter of user u is shown in Figure 1. Its discrete time complex transmitted signal x_n^u, $n=-\infty,\ldots,+\infty$, is serially to parallel converted into blocks of length K_u, $\underline{x}^{u,i}=[x_0^{u,i},\ldots,x_{K_u-1}^{u,i}]^T$, which are mapped into blocks of length N, $\underline{c}^{u,i}=\underline{T}_u\,\underline{x}^{u,i}$, by the excitation matrix $\underline{T}_u$ of size N by K_u. The excitation matrix is determined by the tone assignment scheme. Basically, it permutes the elements of $\underline{x}^{u,i}$ into the elements of $\underline{c}^{u,i}$ whose indices correspond to the tones assigned to user u, and inserts zeros in correspondence of the other tones.

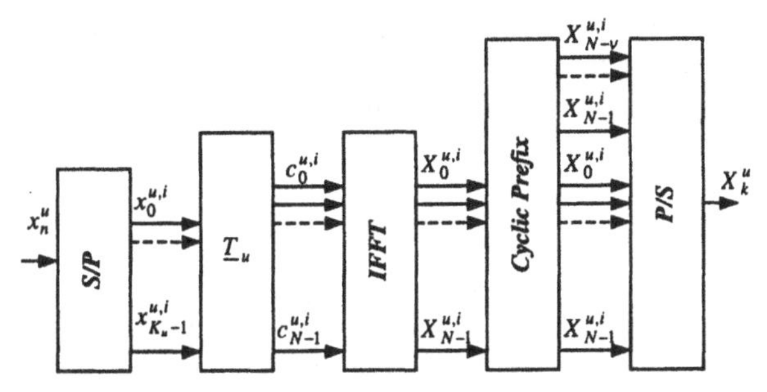

Figure 1. Baseband DMT-OFDMA transmitter of user u.

DMT modulation is implemented through a N-point FFT, yielding

$$X_k^{u,i} = \sum_{n=0}^{N-1} c_n^{u,i} e^{j\frac{2\pi}{N}nk} \tag{1}$$

We consider the insertion of a cyclic prefix (guard time) of ν symbols. As shown in [4], the guard time has the beneficial effect of reducing the MAI interference detrimental effects in the presence of asynchronous users, at the expense of some bandwidth loss.

Finally, the modulated symbol stream is transmitted at rate $W=F_c=1/T_c=(N+\nu)/T$. After RF conversion, and channel propagation, the signals of all users are superimposed and received at the base station.

In a wireless up-link scenario the users are asynchronous and are received with a time and frequency offset relatively to, say, user u'. The time offsets originate from different transmission starting epochs and/or different propagation delays. The frequency offsets are due to a mismatch among the local oscillators and/or carrier frequency Doppler shifts arising from movements. At the base station the overall signal is RF down converted, and sampled at rate F_c. We assume ideal sampling, so that the overall time and frequency offset effect, in the presence of additive white Gaussian noise, can be represented with the discrete time equivalent channel model in Figure 2.

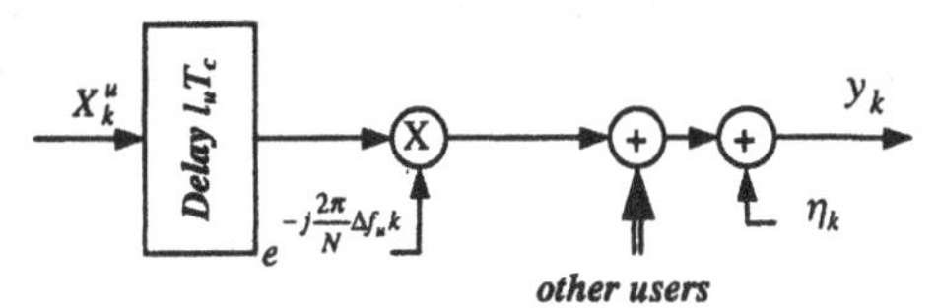

Figure 2. Baseband equivalent channel model for user u.

In Figure 2, $l_u T_c$ represents the relative time delay of the *u-th* user with respect to the u'*-th* user, while $\Delta f_u = (f_{u'} - f_u)NT_c$ is the constant normalized frequency offset between the *u-th* transmitter oscillator at frequency f_u and the local oscillator at frequency $f_{u'}$.

In order to reconstruct the transmitted information symbol stream of all users, we consider a bank of N_U single user detectors identical to the one shown in Figure 3 [4]. Demodulation for the u′*-th* user is accomplished by first acquiring time and frequency synchronization with user u' (that is equivalent to setting $l_{u'}=0$ and

$\Delta f_{u'} = 0$). Then, blocks of $N+\nu$ samples are extracted. A window of N samples is set starting from the middle of the cyclic prefix. Finally, a N-point FFT is applied. From the FFT output we extract blocks of $K_{u'}$ decision variables belonging to the u'-*th* user.

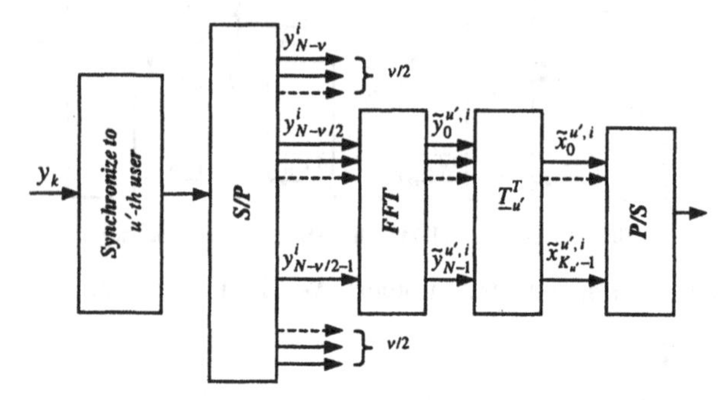

Figure 3. Baseband DMT-OFDMA receiver synchronized to user u'.

The k-*th* sample of the i-*th* received block can be written as

$$y^i_k = \sum_{u=0}^{N_U-1} e^{-j\frac{2\pi}{N}\Delta f_u k} X^{u,i,l_u}_k + \eta^i_k \tag{2}$$

where $i=-\infty,\ldots,\infty$, $k=0,\ldots,N-1$, and η^i_k is the thermal noise contribution, while X^{u,i,l_u}_k is the k-*th* element of the N-point window shifted by l_u

$$\ldots, X^{u,i-1}_{N-1}, X^{u,i}_{N-\nu}, \ldots, \left[X^{u,i}_{N-\frac{\nu}{2}}, \ldots, X^{u,i}_{N-1}, X^{u,i}_0, \ldots, X^{u,i}_{N-\frac{\nu}{2}-1} \right], \ldots, X^{u,i}_{N-1}, \ldots \tag{3}$$

$$\underset{-|l_u|}{\leftarrow} \quad \underset{+|l_u|}{\rightarrow}$$

From this point on, we consider the sets of tones that are assigned to distinct users to be disjoint. Thus, let $\Gamma_{u'} = \{n^{u',i}_0, \ldots, n^{u',i}_{K_{u'}-1}\} \subset \{0, \ldots, N-1\}$ be the set of $K_{u'}$ tone indices univocally assigned to user u'. Then, the set of decision variables for the i-*th* block of user u' is [4]

$$\tilde{y}^{u',i}_n = \frac{1}{N}\sum_{k=0}^{N-1} y^i_k e^{-j\frac{2\pi}{N}nk} = c^{u',i}_n + \sum_{u=0,u\neq u'}^{N_U-1} \tilde{c}^{u,i}_n + w^{u',i}_n \tag{4}$$

where $n \in \Gamma_{u'}$ and

$$\tilde{c}^{u,i}_n = \frac{1}{N}\sum_{k=0}^{N-1} X^{u,i,l_u}_k e^{-j\frac{2\pi}{N}(n+\Delta f_u)k} \tag{5}$$

$$w^{u',i}_n = \frac{1}{N}\sum_{k=0}^{N-1} \eta^i_k e^{-j\frac{2\pi}{N}nk} \tag{6}$$

Thus, each sub-channel output is the sum of the symbol transmitted on that sub-carrier $c^{u',i}_n$, a noisy term $w^{u',i}_n$, and a multiple access interference (MAI) term

$z_n^{u',i} = \sum_{u \neq u'} \tilde{c}_n^{u,i}$. The MAI term differs from zero whenever at least one of the other users has a time delay $|l_u| > \nu/2$ and/or a frequency offset $\Delta f_u \neq 0$.

3. Multiple-access Interference Channel Representation

In the presence of time and frequency asynchronous users the link of user u' can be modeled as shown in Figure 4, where the indices $n_l \in \Gamma_{u'}$, $l=0,\ldots,K_{u'}-1$.

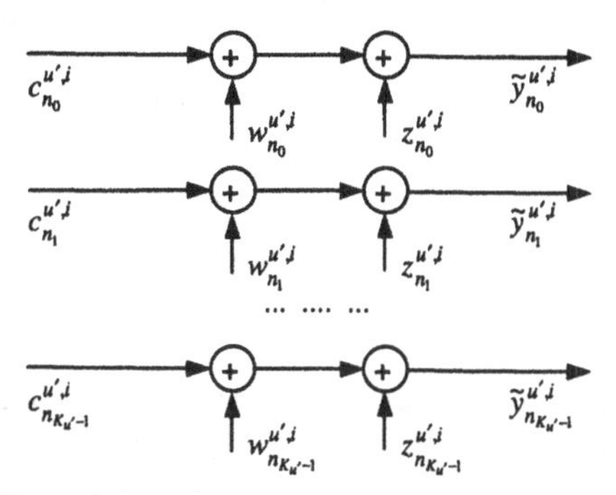

Figure 4. MAI channel model for the link of user u'.

The *u'-th* user transmits its information through $K_{u'}$ parallel channels that experience additive multiple user interference and thermal noise. The thermal noise vector $\underline{w}^{u',i} = [w_{n_0}^{u',i},\ldots,w_{n_{K_{u'}-1}}^{u',i}]^T$ is Gaussian distributed with zero mean and covariance $\underline{K}_{ww}^{u'} = E[\underline{w}^{u',i}(\underline{w}^{u',i})^H] = N_0 \underline{I}$, where $\underline{I}$ is the $K_{u'}$ by $K_{u'}$ identity matrix. In general the MAI is correlated in time (along index i) and in frequency (along the sub-carriers). However, we make the assumption of temporal uncorrelation, based on the fact that this is true for time offsets $|l_u| \leq \nu/2$, and is a valid approximation for small time offsets exceeding the guard interval.

Now, for a capacity evaluation standpoint, we constrain to consider input signals with Gaussian distribution that have zero mean and covariance $\underline{K}_{cc}^{u'} = E[\underline{c}^{u',i}(\underline{c}^{u',i})^H]$. Inputs signals of distinct users are considered independent. It follows that the overall MAI is Gaussian distributed since each user generates a MAI contribution that is Gaussian. The mean of the MAI is zero and the covariance is $\underline{K}_{zz}^{u'} = E[\underline{z}^{u',i}(\underline{z}^{u',i})^H]$. In conclusion, the link of user u' can be modeled as $K_{u'}$ additive Gaussian noise channels that are correlated. The time index i can be omitted and we can write in vector notation

$$\underline{\tilde{Y}}^{u'} = \underline{C}^{u'} + \underline{Z}^{u'} + \underline{W}^{u'} \qquad (7) \qquad\qquad \underline{Z}^{u'} = \sum_{u=0,\, u \neq u'}^{N_U-1} \underline{A}(u,u') \underline{C}^{u} \qquad (8)$$

where for instance $\underline{C}^{u'} = [c_{n_0}^{u'},\ldots,c_{n_{K_{u'}-1}}^{u'}]^T$, and $\underline{A}(u,u')$ is the interference matrix of user u on user u' . The interference matrix is a function of the system parameters and in particular of the relative time and frequency offset of user u with respect to user u'.

If we assume that the input covariance of each user is equal to $\underline{K}^{u}_{cc} = diag(P^{u}_{n_0},...,P^{u}_{n_{K_u-1}})$, and that all users have a time delay $|l_u| \leq v/2$ relatively to user u', then the interference covariance seen by user u' computed over carrier indices n and $n+p$ (belonging to $\Gamma_{u'}$) can be easily evaluated, yielding

$$K^{u'}_{zz}(n,n+p) = \frac{1}{N^2}\sum_{\substack{u=0\\u\neq u'}}^{N_U-1}\sum_{c\in\Gamma_u} P^{u}_{c} \frac{2-e^{j2\pi\Delta f_u}-e^{-j2\pi\Delta f_u}}{1-e^{j\frac{2\pi}{N}(n+p+\Delta f_u-c)}-e^{j\frac{2\pi}{N}(-n-\Delta f_u+c)}+e^{j\frac{2\pi}{N}p}} \tag{9}$$

The result in (9) will be used in the next section for the link capacity evaluation. Note that there is no dependency on the time offset since we have considered delays satisfying $|l_u| \leq v/2$. It has to be said that more general expressions can be calculated to include delays $|l_u| > v/2$.

4. Capacity Evaluation

The presence of asynchronous users (interferers) induces some degradation on the performance of the single user receiver. The problem we are dealing with is to quantify such degradation as a function of the system parameters. In [4] we have used the signal energy over noise plus interference ratio as a figure for the performance degradation. In this paper we aim to evaluate the capacity of the *u'-th* link, and more in general the capacity region where jointly reliable communications are possible.

Within the model in Section 2, the capacity of the *u'-th* user link is evaluated by maximizing the mutual input-output information $I(\underline{\tilde{Y}}^{u'};\underline{C}^{u'}|\underline{C}^{0},...,\underline{C}^{u'-1},\underline{C}^{u'+1}...)$, over all possible input covariance matrices $\underline{K}^{u'}_{cc}$ subject to the trace constraint $trace(\underline{K}^{u'}_{cc})=\sum_{n\in\Gamma_{u'}} P^{u'}_{n} \leq P$. Since we consider Gaussian inputs, the capacity in bit/s/Hz is obtained as [6]

$$C(u'|\underline{K}^{u'}_{zz}) = \frac{1}{K_{u'}}\max_{\underline{K}^{u'}_{cc}}\left\{\log_2\left(\frac{|\underline{K}^{u'}_{cc}+\underline{K}^{u'}_{ww}+\underline{K}^{u'}_{zz}|}{|\underline{K}^{u'}_{ww}+\underline{K}^{u'}_{zz}|}\right)\right\} \tag{10}$$

where $|\underline{A}|$ is the determinant of the matrix $\underline{A}$, and $K_{u'}$ is the number of tones assigned to user u' [(1)]. Note that (10) is conditioned on the covariance of the interference, which in turn is a function of the input covariance of the interferers, their time/frequency offsets, and their set of assigned tones. For a particular system realization these parameters can be considered constant.

All information rates $R_{u'} \leq C(u'|\underline{K}^{u'}_{zz})$ allow communications with arbitrarily small probability of error over the *u'-th* user link. Further, we do not consider the case where cooperation among decoders of distinct users exists. In this case a way to proceed would be to augment the dimensions of the model in (7)-(8).

To get insight, we constrain the input covariance to be diagonal for all users. With this hypothesis, knowing the frequency offset, and the set of assigned tones of each

[(1)]: The standard capacity formula for real vectors is *(1/2)log₂[.]*. Since we are dealing with Gaussian complex vectors it can be shown that the factor (1/2) vanishes out.

user, (9) gives the covariance matrix of the interference (when $|l_u| \leq v/2$ for all u). Thus, (10) can be computed, yielding an achievable rate for a given user link. In this case, suppose to assign an equal number of tones to all users, and to distribute equally the power among tones, then the individual link capacities can in general differ. This is due to the fact that each of the links sees on its set of tones a different MAI contribution. In this scenario, the conditional capacity region, which is the closure of the set of all rate vectors $\underline{R}=[R_0,...,R_{N_U-1}]$ that satisfy $R_u \leq C(u|\underline{K}^u_{zz})$ for $u=0,...,N_U-1$, can be thought to be an hyper-parallelepiped. This capacity region (inner bound) is contained into the capacity region (outer bound) defined by all rate vectors whose components satisfy $R_u \leq (1/K_u)\log_2 \prod_{n \in \Gamma_u} (1+P_n^u / N_0)$ for $u=0,...,N_U-1$. This outer bound is certainly achieved by the synchronous system where the MAI detrimental effect disappears.

5. Capacity as a Random Variable

In general, even if we constrain the input covariance matrices to be the same for all users, simple meaningful closed expressions that define the conditional capacity region (i.e. conditioned on the time and frequency offsets) cannot be found. This is because the interference covariance in (9) still has a dependency on how we assign the tones to the users. Furthermore, the time and frequency offsets are random parameters. Thus, to proceed we follow the approach of treating the link capacity (i.e. rate that grants reliable communications over a user link) a random variable. We consider the time and frequency offsets of all users independent and equally distributed. Then, we fix the tone assignment scheme together with the power allotment to sub-carriers, and we constrain the input covariance matrices of all users to be diagonal.

Our aim is now to compute the complementary cumulative distribution function (ccdf) of the link capacity. This is defined as the probability that the capacity of a given user is greater than a fixed value under all possible system realizations where all users have a time and frequency offset independently distributed

$$ccdf_{u'}(\overline{C}) = P[C(u') > \overline{C}] \tag{11}$$

If the tone assignment scheme of each user satisfies symmetry rules, the power allotment is also the same, and the system is fully loaded, then the ccdfs in (11) are the same for all users. Thus, rate $R_{u'}=\overline{C}$ is achievable by the u'-th link with probability given by (11), and the same applies to the other links, individually considered.

Following the same approach it is possible to define the joint link capacities ccdf

$$ccdf_U(\overline{C}_0,...,\overline{C}_{N_U-1})=P[C(u=0)>\overline{C}_0,...,C(u=N_U-1)>\overline{C}_{N_U-1}] \tag{12}$$

Now, the capacity region takes on a probabilistic significance. It can be interpreted as the region determined by all rate vectors for which jointly reliable communications are possible with a given probability. This, capacity region is outer bounded by the capacity region that is determined in the absence of MAI (achieved with probability 1 in the synchronous case).

Finally, we emphasize that (11) and (12) still depend upon the tone allocation strategy, and the power allotment to tones.

6. System Scenarios and Capacity Performance Comparison

Evaluation of the ccdfs (11) and (12) was carried out numerically through Monte Carlo simulations. As a result it was possible to evaluate the capacity performance of several systems that differ on the tone allocation and power allotment strategy.

6.1 Single link capacity complementary cumulative distribution functions

We start considering the ccdf of the single user link (i.e., Equation 11). The system under investigation is a fully loaded system scenario characterized by the following parameters: fixed number of overall carriers $N=256$, $N_U=16$ users, $K_u=16$ tones per user. A guard time is inserted such that the MAI effect due to time misalignments is null. Further, the bandwidth penalty is also ignored assuming that N is much larger than the cyclic prefix. Obviously, this assumption depends upon the cell size. However, we could always think of having some degree of centralized control that confines the time offset within the guard time length [8]. Vice versa the frequency offset detrimental effect cannot be completely canceled out by the insertion of frequency guards [4]. Thus, we have considered users with frequency offsets that are independent and uniformly distributed in $(-\Delta f_{\max}, +\Delta f_{\max})$.

In Figures 5 and 6 the users are allocated with an *interleaved tone assignment* scheme [4]. In other words, the tones of distinct users are regularly interleaved across the overall set of N tones. All sub-carriers of all users have the same power. The signal-to-thermal-noise ratio is set to 10 dB and 30 dB and several maximum frequency offsets are considered. We plot a vertical line in correspondence of the capacity that is achieved by the synchronous system. This represents a capacity upper bound. For SNR=10 dB the upper bound is 3.46 bit/s/Hz, while for SNR=30 dB is 9.97 bit/s/Hz.

Now, looking for instance at Figure 5, we set a percentage, say 90%, then we read the minimum capacity we can provide to a user link with that probability over all system realizations. Note that, since the system is fully loaded, the tones are regularly assigned, and the power allotment to sub-carriers is the same, all N_U links have the same ccdf. The same applies in the other schemes described in this section. From Figure 5 it can be seen that the capacity lower bound is about 0.5 bit/s/Hz from the upper bound for 90% of the system realizations when Δf_{max}=0.1 and the SNR=10 dB. The capacity loss dramatically increases to about 2.5 bit/s/Hz for Δf_{max}=0.5. As the SNR increases (Figure 6), the capacity performance is dominated by the MAI, such that the difference between the lower and the upper bound is more pronounced.

In Figures 7 and 8 we consider the same system with, however, a *block tone assignment* where disjoint blocks of K_u contiguous tones are assigned to each user. Fixed the same outage probability (i.e. 10%), the block tone allocation scheme exhibits a strong capacity improvement over the interleaved. For instance, at 30 dB SNR the block scheme provides a minimum capacity ranging from about 8.4 to about 9.1 bit/s/Hz for Δf_{max} ranging from 0.5 to 0.1. On the other hand the interleaved scheme yields capacities ranging from 1 to 4.5 bit/s/Hz.

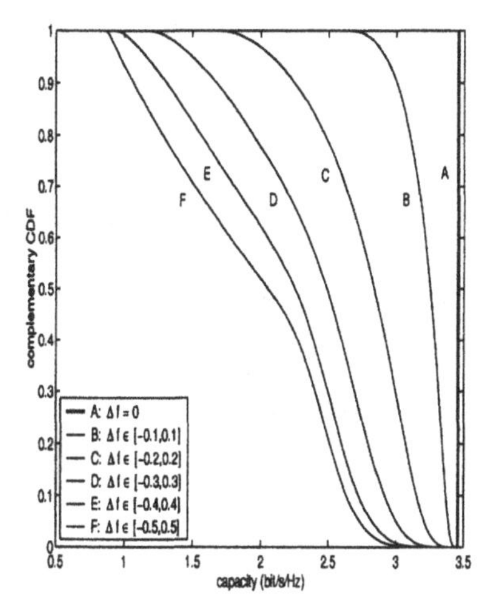

Figure 5. Ccdf of the capacity of a user link for several maximum frequency offsets. Fully loaded system with N=256, N_U=16, K_u=16. **Interleaved tone assignment**. Equal power tones. SNR=10 dB.

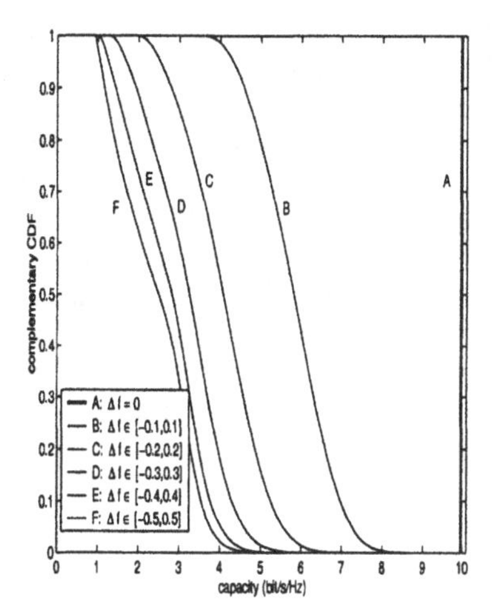

Figure 6. Ccdf of the capacity of a user link for several maximum frequency offsets. Fully loaded system with N=256, N_U=16, K_u=16. **Interleaved tone assignment**. Equal power tones. SNR=30 dB.

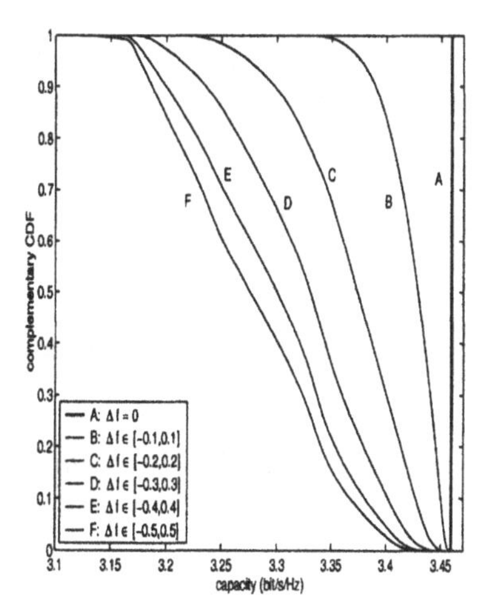

Figure 7. Ccdf of the capacity of a user link for several maximum frequency offsets. Fully loaded system with N=256, N_U=16, K_u=16. **Block tone assignment**. Equal power tones. SNR=10 dB.

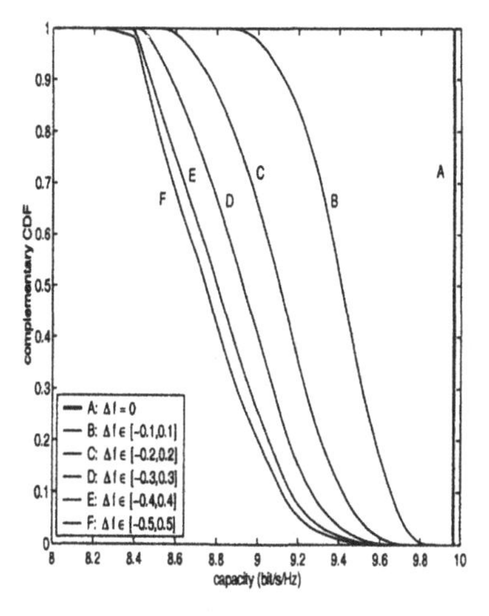

Figure 8. Ccdf of the capacity of a user link for several maximum frequency offsets. Fully loaded system with N=256, N_U=16, K_u=16. **Block tone assignment**. Equal power tones. SNR=30 dB.

At this point, a first approach to improve further the single link capacity can be to differentiate the power transmitted on the sub-carriers. We pursued this idea in a simplified manner. In Figure 9, the tones are block allocated. Then, the power of the first and last tone of each user block is set to zero. Basically, this corresponds to inserting frequency guards [4]. Unfortunately, this helps only for high Δf_{max} , as shown by comparison of Figure 9 with Figure 8. This is due to the fact that the diminished MAI level does not always compensate the capacity loss due to no transmission on two sub-carriers.

In a second experiment (Figure 10), the total available power is distributed in the following manner. The first and the last tone of each block have half power, while the power of two middle tones is increased by half. As Figure 10 confirms, this strategy shows some improvement over Figure 8.

It is clear that although we pursued an intuitive and simple method to allocate the power to sub-carriers, the results in Figure 10 confirm that distributing the power uniformly among the tones is not optimal. A systematic application of the water-filling method [6] should lead to more significant improvements.

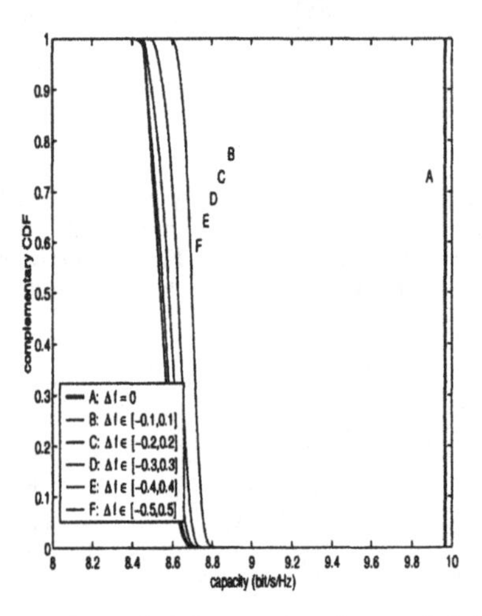

Figure 9. Ccdf of the capacity of a user link for several maximum frequency offsets. Fully loaded system with N=256, N_U=16, K_u=16. **Block tone assignment.** Power zero on first and last tone in each block. SNR=30 dB.

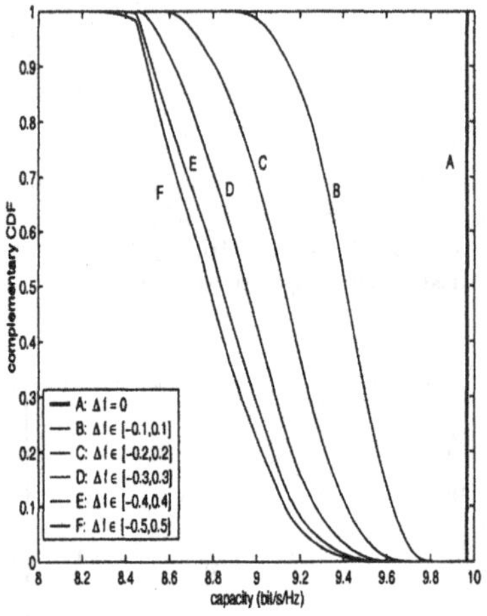

Figure 10. Ccdf of the capacity of a user link for several maximum frequency offsets. Fully loaded system with N=256, N_U=16, K_u=16. **Block tone assignment**. Half power on first and last tone, 3/2 power on two mid-tones. SNR=30 dB.

A second proposed method to improve capacity is based on *adaptively allocating* the tones to users. The main idea is that if we know for instance what the frequency offsets of the users are, we can assign block of tones in a way such that each block has adjacent blocks ordered with increasing frequency offset. This can be simply accomplished by imagining of allocating the N_U users in a circular queue of N_U cells. The user with the lowest offset is assigned to a cell. Then, pick other two users with the remaining lowest offset and allocate them one to the right and one to the left adjacent cells. Proceed until the cells are filled. This adaptive method of allocating tones in order of increasing frequency offset can be generalized to include time offsets exceeding the guard time.

The ccdf of the user link capacity with this adaptive scheme is plotted in Figures 11 and 12. A deep improvement is found over the schemes that we have considered so far. For instance at 30 dB SNR the lower bound in capacity granted with probability 90% loses less than 1 bit/s/Hz over the upper bound for all Δf_{max} from 0.5 to 0.1. Clearly, the practicality of such an approach needs to be investigated.

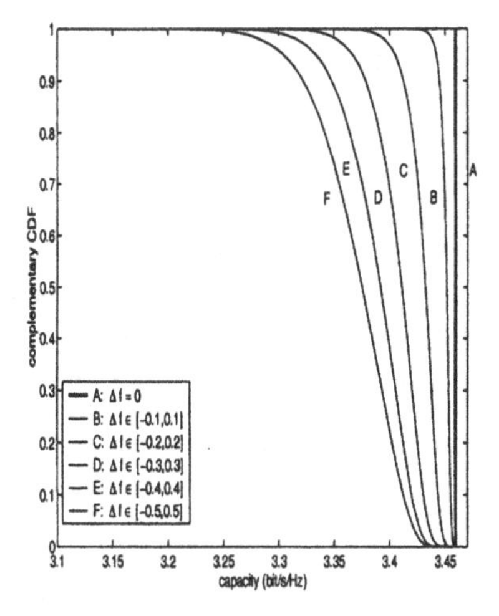

Figure 11. Ccdf of the capacity of a user link for several maximum frequency offsets. Fully loaded system with N=256, N_U=16, K_u=16. **Adaptive block tone assignment**. Equal power tones. SNR=10 dB.

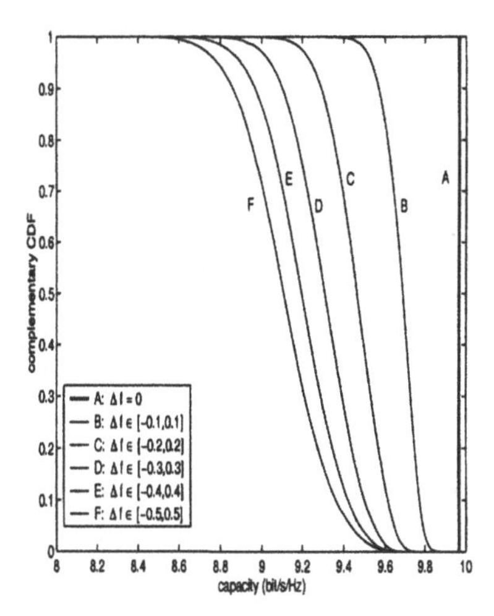

Figure 12. Ccdf of the capacity of a user link for several maximum frequency offsets. Fully loaded system with N=256, N_U=16, K_u=16. **Adaptive block tone assignment**. Equal power tones. SNR=30 dB.

6.2 Joint link capacity complementary cumulative distribution functions

In this section we report some results obtained by numerically computing (12). We considered a fully loaded system with $N=64$, where 4 users have $K_u=16$ tones block assigned with equal power, and experience independent frequency offsets with uniform distribution in (-0.25,+0.25). Figure 13 shows the joint ccdf of two users out

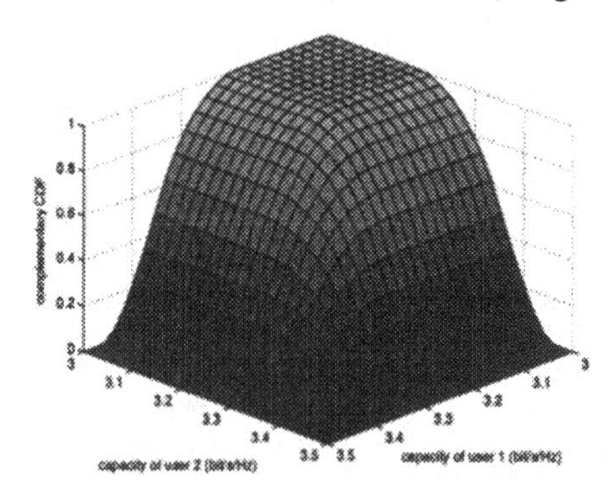

Figure 13. Joint ccdf of user 1 and 2 out of 4 users in a fully loaded system. **Block tone assignment**, N=64, N_u=4, K=16. Equal power tones. SNR=10 dB. $\Delta f_u \in U$[-0.25,0.25].

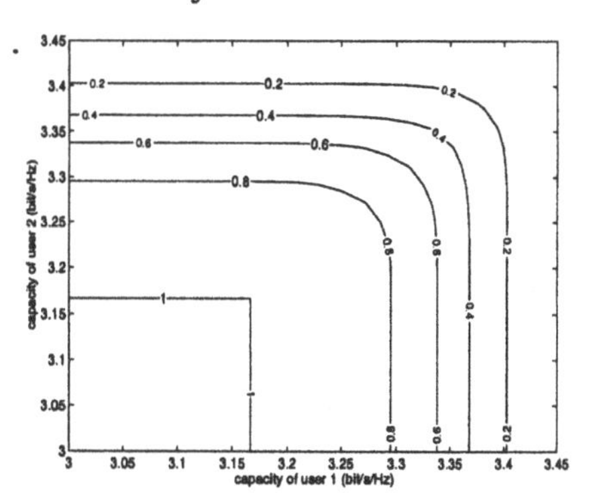

Figure 14. Contour lines of joint ccdf of user 1 and 2 out of 4 users in a fully loaded system. **Block tone assignment**, N=64, N_u=4, K=16. Equal power tones. SNR=10 dB. $\Delta f_u \in U$[-0.25,0.25].

of four. Figure 14 is the contour plot of the surface in Figure 13, each curve is the set of rate pairs for which joint reliable communications are possible with a given probability over all system realizations. The closure determined by such curves can be thought to be the set of all achievable rate pairs with probability at least equal to the corresponding contour label. The capacity region that has probability one is a square, and is the one that would be determined in the case all users experienced the maximum frequency offset in all realizations.

7. Conclusions

In this paper we have studied the problem of defining the achievable information rates for reliable communications of a set of time and frequency asynchronous users that share a Gaussian channel through a DMT OFDMA access scheme. Demodulation is accomplished with a bank of single user FFT based detectors, at the output of which multiple access interference arises. We have shown that each user link can be modeled with a multiple input multiple output correlated Gaussian interference channel. Consequently, we have determined an inner bound to both the capacity of a given user link, and the region of achievable information rates for joint reliable communications (i.e., capacity region). These bounds are conditioned on a system realization (i.e., time and frequency offsets), and depend on the tone allocation scheme and the power allotment to carriers. Thus, by treating these bounds as a random variable we have numerically computed their ccdfs for several fully loaded system scenarios characterized by different tone assignment schemes, and power allotment to sub-carriers.

Based on the results, the following conclusions and guidelines for maximizing capacity in a multi-user DMT OFDMA system are derived. The insertion of an appropriate cyclic prefix completely eliminates the MAI due to time misalignments. For large cell size some degree of centralized synchronization is required to reduce the prefix length, thus save bandwidth. Multiplexing the users with a block tone allocation yields improved capacity, compared to interleaving the users tones, in the presence of the irreducible frequency offset detrimental effect. Other capacity improvements can be obtained with a dynamic tone allocation that adaptively allocates the tones on a per user time/frequency offset base. Finally, the power of a user should be distributed in an ad hoc fashion among the user tones, following for example a water-filling approach.

8. References

[1] L. Cimini Jr. "Analysis and simulation of a digital mobile channel using orthogonal frequency division multiplexing", *IEEE Trans. Comm., July 1985*, pp. 665-675.

[2] H. Sari, G. Karam, "Orthogonal frequency-division multiple access and its application to CATV Networks", *European Trans. on Telecom.*, December 1998, pp. 507-516.

[3] S. Kaiser, W. Krzymien, "Performance effects of the uplink asynchronism is a spread spectrum multi-carrier multiple access system", *European Trans. on Telecom.*, July-August 1999, pp. 399-406.

[4] A. Tonello, N. Laurenti, S. Pupolin, "On the effect of time and frequency offsets in the uplink of an asynchronous multi-user DMT OFDMA system", *Proc. of International Conference on Telecom.* 2000, Acapulco, Mexico, May 22-25, 2000, pp. 614-618.

[5] A. Tonello, N. Laurenti, S. Pupolin "Analysis of the uplink of an asynchronous DMT OFDMA system impaired by time offsets, frequency offsets, and multi-path fading", *to appear on Proc. of VTC 2000 Fall*, Boston, USA, September 24-28, 2000.

[6] T. Cover, J. Thomas, "Information theory", *John Wiley & Sons*, 1991.

[7] A. Carleial, "Interference channels", *IEEE Trans. Info. Theory*, January 1978, pp. 60-70.

[8] J. van de Beek, P.O. Borjesson, et al. "A time and frequency synchronization scheme for multiuser OFDM", *IEEE JSAC*, November 1999, pp. 1900-1914.

A Software Radio Platform for New Generations of Wireless Communication Systems

Christian Bonnet, Giuseppe Caire, Alain Enout, Pierre A. Humblet, Giuseppe Montalbano, Alessandro Nordio, and Domnique Nussbaum[*]

Institut Eurécom[**]
B.P. 193, 06904 Sophia–Antipolis CEDEX, France
Tel: +33 4 93 00 26 08, Fax: +33 4 93 00 26 27
E-mail: `firstname.name@eurecom.fr`

Abstract: A major concern of today's and near future mobile communication systems is represented by the need of providing universal seamless connection to the users. Software Defined Radio (SDR) systems lend themselves to handle several different standards and types of service, and this motivates the already large and still increasing interest in this research area. Eurécom and EPFL have started a joint project whose objective is to study and implement a real-time SDR communication platform to validate advanced algorithms for wireless communications. Due to various practical design issues, the platform implements the essential physical layer features of the time-division duplex (TDD) mode of the UMTS standard proposal (air-interface and signal processing), although frequency-division duplex (FDD) mode and even other standards could also be implemented. The platform is a real-time PC based system that can handle wide-band radio resources. It provides hardware, DSP software, and link level software functionality. In this paper we address the major issues related to the design of a real-time SDR system. We also describe the general architecture, the signal processing techniques adopted to implement the transmitter and receiver SDR front-end, the current platform set-up, and we provide DSP performance measurements. Finally, we consider the future perspectives for the platform evolution.

1. Introduction

The presence of several different wireless communication standards and the wide variety of services provided by mobile communication operators poses the problem of providing universal seamless connection to customers with different service requirements or simply needing to access different wireless networks supporting different standards. Software defined radio (SDR) terminals able to reconfigure themselves to handle several different standards and different services represent a solution to this problem (notice that SDR is a very broad term involving several levels in the protocol stack, see e.g. [1], [2], [3] and references therein). Also motivated by the intensive world wide activity around the

* Authors appear in alphabetical order.

** Eurécom's research is partially supported by its industrial partners: Ascom, Cégétel, France Télécom, Hitachi, IBM France, Motorola, Swisscom, Texas Instruments, and Thomson CSF third generation mobile communication systems, Eurécom and EPFL (École Polytechnique Fédérale de Lausanne) have started a joint project with the objective of designing and implementing a real-time software radio communication platform to validate advanced mobile communication signal processing algorithms. The platform is characterized by the following major features:

- Flexibility, achievable by a software driven system
- Duplex communication
- Multiple antennas transmit and receive signal processing

Flexibility remains a key word for a software defined system. In our case it serves on several purposes. For instance, the receiver can be properly configured (and calibrated) to perform propagation channel measurements and transmitter characterization, while the whole system can be programmed to implement different signal processing algorithms for both single user and multiuser systems under different operating conditions. *Duplex communication* is also necessary to allow higher layer protocol services, and to analyze more complex system aspects, such as multi-access and power control, and optimize down-link signal processing from up-link measurements. The platform will allow *multiple antennas signal processing*, or more generally, joint spatio-temporal signal processing. This is a very promising ensemble of techniques to significantly increase the capacity of wireless communication systems.

In the sequel we first provide a description of the general platform architecture. Then we consider end-to-end SDR signal processing solutions for efficient DSP algorithm implementation, followed by a description of the current platform set-up and a detailed DSP performance measurements. Finally we address the near future perspectives for the platform evolution.

2. Platform Architecture

A single antenna architecture for both the Mobile Terminal (MT) and the Base Transceiver Station (BTS) has been chosen for a first implementation. This architecture can be easily upgraded without requiring any substantial new design for the essential hardware and software components. The BTS and the MT are based on a similar hardware. The hardware is highly partitioned in order to maintain the maximum modularity and flexibility, allowing the use of different standard cards and components. The platform consists of a signal processing subsystem and a radio subsystem. The signal processing subsystem comprises

- A reconfigurable data acquisition system based on Field Programmable Gate Array (FPGA) and PCI bus technology
- A PCI bus-based DSP system employing a combination of embedded DSP's and workstations
- Data management software (data routing, framing, synchronization)
- Signal processing software (digital transceiver algorithms, multiple-access protocols, error coding/decoding)

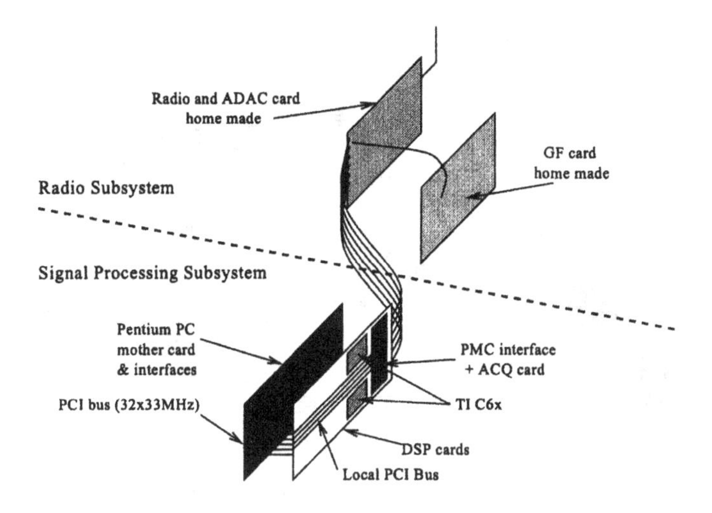

Fig. 1. Mobile terminal architecture

These elements may be replicated in a parallel fashion to implement a multiple-antenna systems (both at the BTS and MT). Figure 1 shows the current hardware setup:

- A radio card capable to handle a 5 MHz bandwidth radio signals
- An A/D, D/A conversion card (ADAC)
- A data acquisition card (ACQ) that is connected as a *mezzanine* (PMC) via a local PCI bus
- A 2-processors DSP card based on Texas Instruments TMS320C6x (in the sequel denoted as TI C6x) technology supporting signal processing at chip level
- A Pentium PC card (main CPU) that supports the processing at symbol level and the higher layer protocols

The target architecture of the BTS (figure 2) will include the following elements:

- 8 radio frequency cards
- A clock and frequency generation card (GF card)
- 8 data acquisition cards (ACQ) with 8 ADAC's.
- 8 DSP cards (the 2-processors DSP cards might be replaced by 4-processors DSP cards as shown in figure 2)

It is important to notice that the same architecture can be adopted to develop different real-time and non-real-time applications by properly defining different operating modes and developing the related software. The operating modes will include real-time signal processing pre-detection for UMTS like applications, real-time processing for narrow-band signals (e.g. GSM, EDGE), non-real-time off-line processing to test high complexity algorithms or collecting measured data, and hardware simulation.

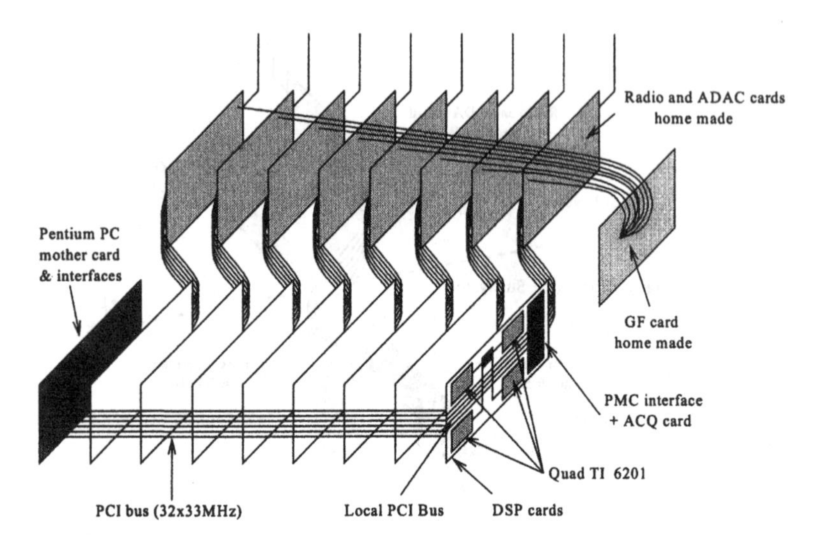

Fig. 2. Base transceiver station target architecture

Software plays an essential role at each stage of the digital processing chain: at the acquisition level via the use of programmable FPGAs, and at signal processing level via the use of DSP's. Once the application and the corresponding operating mode are defined the dedicated software can be down-loaded and the platform completely reconfigured.

3. End-to-End Signal Processing

This section gives an overview of the signal processing algorithms which have been designed and coded on the DSP to implement in real-time the essential UMTS TDD physical layer. We start the analysis considering the transmitted data flow (for example a video stream) already coded and mapped in the QPSK or BPSK alphabet.

3.1 Transmitter Front-End

Let $a[k]$ denote a sequence of chips and let $\psi(t)$ denote the pulse-shaping filter, band-limited over $[-W/2, W/2]$, and T_c the chip interval. The corresponding continuous-time complex base-band equivalent linearly modulated signal is given by

$$x(t) = \sum_k a[k]\psi(t - kT_c) \tag{1}$$

For Direct-Sequence CDMA system [4], using a spreading gain N, $a[k] = b[\lfloor k/N \rfloor]c[k]$, where $b[m]$ is the m-th modulation symbol (QPSK in the UMTS case) and $c[k]$ is the k-th chip and N is the spreading gain (this generalizes trivially to systems with several spreading layers, like IS-95). Notice that (1) can also represent the sum of the chip sequences associated with different users.

In general, for digital transmitters, the signal $x(t)$ is the output of a D/A converter which takes as input the discrete-time signal

$$x[n] = x(n/f_s)$$

with sampling frequency $f_s \geq W$. In classical I-Q modulators, the continuous baseband components $\mathrm{Re}\{x(t)\}$ and $\mathrm{Im}\{x(t)\}$ are generated by low-pass filtering the output of two separate D/A converters, and the IF signal

$$y(t) = \mathrm{Re}\{x(t)\exp(\mathrm{j}2\pi f_{\mathrm{IF}}t)\} \tag{2}$$

is then produced by mixing $\mathrm{Re}\{x(t)\}$ and $\mathrm{Im}\{x(t)\}$ with IF carrier signals in phase and quadrature and by summing the modulated real signals [4]. This approach requires two D/A converters, two low-pass filters, two analog mixers and one adder. Thus it looks quite costly from the hardware point of view.

Another approach consists of producing a sampled version of the IF modulated signal $y(t)$ by using a sampling rate f_s greater than $2f_{\mathrm{IF}}$. The continuous-time signal is obtained by bandpass filtering the output of a single D/A converter. This solution (for the receiver front-end) is described in [5] and [6]. However, since intermediate frequencies usually range between tens of MHz up to 100 MHz, this approach is extremely computationally intensive as it requires the generation of signal samples and the multiplication by the carrier signal at extremely large sampling frequency. We may state that with the current DSP technology this approach is infeasible on a SDR system.

In the following, we propose an efficient transmitter front-end architecture that allows a working sampling frequency of the order of the baseband signal bandwidth (and not of the order of the IF carrier), does not require explicit multiplication by the carrier signal, and only needs a single D/A converter and analog filter centered at f_{IF}.

IF-Sampling and Up-Conversion. We choose the sampling rate f_s according to the expression

$$f_s = \frac{f_{\mathrm{IF}}}{\ell \pm 1/4} \qquad \text{for a positive integer } \ell \tag{3}$$

Then, we generate the discrete-time real signal

$$x'[n] = \mathrm{Re}\{x[n]e^{\mathrm{j}2\pi(f_{\mathrm{IF}}/f_s)n}\} = \mathrm{Re}\{\mathrm{j}^{\pm n}x[n]\} \tag{4}$$

In this way, the periodic spectrum of $x'[n]$ has a replica centered at f_{IF} (see figure 3). After ideal D/A conversion, a pass-band filter centered at f_{IF} removes the other replicas, generating the desired IF modulated signal. The discrete-time modulation by $f_s/4$ and taking the real part of the modulated signal as expressed in (4) requires an almost negligible computational cost since the whole processing reduces to changing alternately the signs of $x[n]$. In order to avoid aliasing when taking the real, the sampling rate must satisfy also the condition $f_s \geq 2W$.

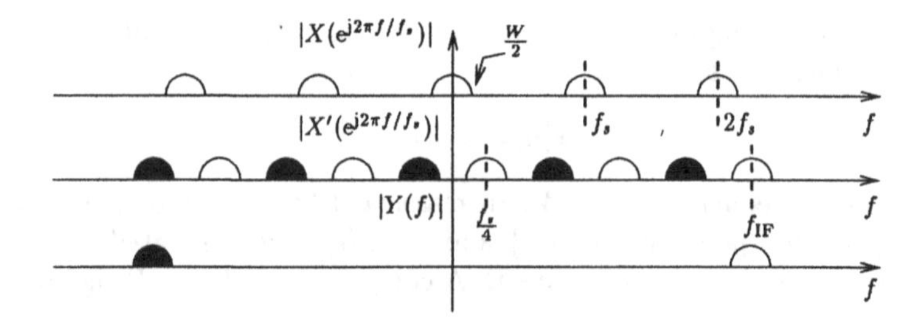

Fig. 3. Spectrum of $x[n]$, $x'[n]$, and $y(t)$ with the integer $\ell = 2$ and the sign (+) chosen in (3)

D/A Conversion. In the above description we assumed an ideal D/A converter with flat frequency response. Actual D/A converters exhibit a low-pass frequency response (approximately) of the form $\mathrm{sinc}(f/f_s)$ that does not extend to IF. A way to extend the D/A converter response so as to reduce the attenuation at IF consists of clocking the converter at rate $f_d = L_{D/A} f_s$, where $L_{D/A}$ is a positive integer such that $f_d \gg f_{IF}$, and up-sampling $x'[n]$ by the factor $L_{D/A}$. In our setting we chose $L_d = 8$. Notice that this approach has the undesired effect of reducing the average signal energy per sample by a factor $L_{D/A}$. Hence for high $L_{D/A}$ the IF analog signal can be very weak after D/A conversion, therefore it may require to be strongly amplified to be transmitted giving rise to significant non-linear distortion. As an alternative, one may think of pre-compensating the linear distortion of the D/A converter by introducing a pass-band FIR filter between the up-sampler and the D/A converter. The filter must be designed in order to enhance the spectrum replica at IF while attenuating the other replicas. In this way the analog signal at IF would require lower amplification gains in the IF-RF conversion stage reducing the signal distortion due to the non-linearities of the power amplifiers. A low-complexity implementation is addressed in [7].

3.2 Receiver Front-End

IF-Sampling and Down-Conversion. Once the RF signal incoming from the antenna has been down-converted to IF, it is sampled by an A/D converter at a rate f_s (see figure 4). Let $r_{IF}(t)$ denote the received IF analog signal at the input of the A/D converter. By choosing $f_s \geq 2W$ according to (3), because of the periodicity of the discrete-time signal spectrum, the resulting real sampled signal $r[n] = r_{IF}(n/f_s)$ is pass-band with a spectrum replica centered at $f_s/4$. We shall remark that although f_{IF} and f_s at the receiver can be different from f_{IF} and f_s at the transmitter, for the sake of simplicity we use the same notation. Notice that here in order to avoid signal re-sampling we suppose the rate f_s to be a multiple integer of the chip rate (i.e. $f_s = N_c f_c$ where in our implementation we set $N_c = 4$).

A base-band version of the received signal can be obtained by multiplying $r[n]$ by $(-j)^n$ followed by low pass filtering. We shall show that both channel

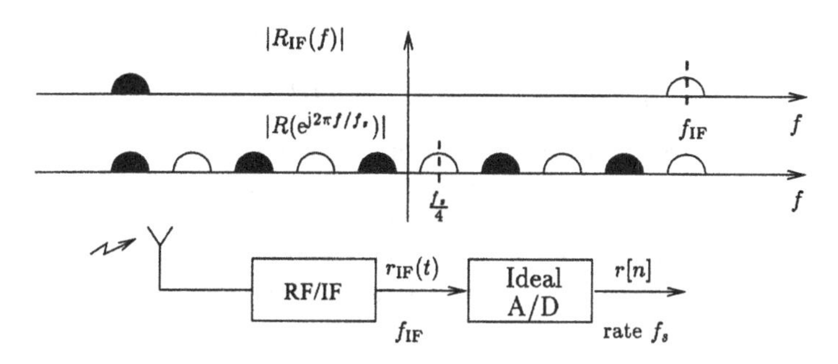

Fig. 4. Receiver front-end

estimation and the data detection process can also be performed in pass-band with the same complexity and avoiding explicit demodulation.

Slot-Timing Acquisition. For the acquisition of the slot-timing we superpose to the transmitted data a primary synchronization sequence (see [8]) at the beginning of every slot. The signal at the output of the A/D converter is filtered by a correlator matched to the primary synchronization sequence. The magnitude square of the filtered signal is averaged with an exponential window over several noise realizations (i.e., over several slots) and over the observation window. Then the slot-timing is detected as the time instant, with a resolution of one chip period, corresponding to the maximum of the averaged square output of the correlator. We remark that there is no need for higher resolution since the residual synchronization errors are accounted for and compensated by the channel estimation algorithm.

Channel Estimation. Here we consider the training-sequence based multiuser channel estimation procedure for block-synchronous CDMA described in the UMTS/TDD standard proposal.

In this scheme users are roughly synchronized to a common time-reference and transmit their training sequences at the same time (user timing errors are included as effect of the channel and taken automatically into account by the estimation procedure). The maximum allowed channel length (including possible timing errors) is Q chips and the training sequence sent by each user is built from a common base sequence $\mathbf{a} = [a[0], a[1], \ldots, a[M-1]]^T$ of length M chips, adding a cyclic extension of Q chips. This solution allows joint least-square (LS) estimation of all users' channels if $M \geq QU$, where U is the number of interfering users [9, 10]. In our set-up we assume $M = 192$ and $Q = 64$ (corresponding to a user's training sequence of length 256 chips) according to [8]. Under these assumptions we can write the received signal sampled at frequency $f_s = N_c f_c = 4 f_c$ during the M chips spanned by the base sequence as

$$\boldsymbol{w} = \bar{\mathbf{A}}\boldsymbol{g} + \boldsymbol{\nu}$$

where $\boldsymbol{w} = [w[0], w[1], \ldots, w[MN_c - 1]]^T$ is the received signal,

$$\boldsymbol{g} = [\boldsymbol{g}_1^T, \ldots, \boldsymbol{g}_u^T, \ldots, \boldsymbol{g}_U^T]^T$$

is a vector containing the channel impulse responses of the U users,

$$\boldsymbol{g}_u = [g_u[0], g_u[1], \ldots, g_u[QN_c - 1]]^T$$

is the u-th user's channel filter vector and $\boldsymbol{\nu}$ is a vector of interference plus noise samples, assumed white. The $MN_c \times MN_c$ matrix $\bar{\mathbf{A}}$ is defined as $\bar{\mathbf{A}} = \mathbf{A} \otimes \mathbf{I}_{N_c}$ where $(\otimes)$ denotes the Kronecker product and $\mathbf{A}$ is a circulant matrix containing all the possible cyclic shifts (by columns) of the base sequence $\mathbf{a}$. The matrix $\bar{\mathbf{A}}$ is also circulant and it is unitary similar [11] to the diagonal matrix diag($\bar{\boldsymbol{\alpha}}$), where

$$\bar{\boldsymbol{\alpha}} = [\underbrace{\boldsymbol{\alpha}^T, \ldots, \boldsymbol{\alpha}^T}_{N_c \text{ times}}]^T$$

and where $\boldsymbol{\alpha}$ is the discrete Fourier transform of $\mathbf{a}$. After some algebra, it is possible to show that the Least Squares estimation of the overall channel impulse response $\boldsymbol{g}$ is given by

$$\hat{\boldsymbol{g}} = \text{IDFT}\left\{\frac{\text{DFT}\{\boldsymbol{w}\}}{\bar{\boldsymbol{\alpha}}}\right\} \tag{5}$$

where DFT and IDFT denote direct and inverse discrete Fourier transforms and the ratio of two vectors should be interpreted as the element-by-element division.

This approach can be applied to both base-band and pass-band signals. In our case since the received signal is real and pass-band, its spectrum shows both the left and right-side replicas. As it is shown in figure 5 for $N_c = 4$, the left-side replica occupies the first and the second quarter of the DFT spectrum, while the right-side replica occupies the third and the fourth one.

The receiver can also use the a priori information that the signal bandwidth is limited to W. Then it can limit the computation to the range $[f_s/4 - W/2, f_s/4 + W/2]$, setting to 0 the rest of the channel estimates spectrum. Notice that this operation in the frequency domain corresponds to a low-pass filtering in the time-domain, moreover it reduces the computational cost since only a part of the MN_c products (by the element-wise inverses of $\bar{\boldsymbol{\alpha}}$ in (5)) needs to be computed. Finally the IDFT produces the estimated pass-band channels complex responses.

Matched Filter Synthesis and Data Detection. Given the channel estimates, we are then interested in synthesizing a Matched Filter (MF) matched to the channel (including also the pulse shaping filter)-spreading sequence cascade associated with the user of interest. The overall discrete-time matched channel-spreading sequence cascade is given by

$$f[k] = \sum_{i=0}^{N_c N - 1} s[i] g[k - i] \tag{6}$$

where $s[i]$ denotes the up-sampled version of the chip rate spreading sequence $s[n]$, defined as $s[i] = s[n]$ if $i = 4n$, $s[i] = 0$ otherwise and $g[i]$ denotes the channel estimate associated with the user of interest.

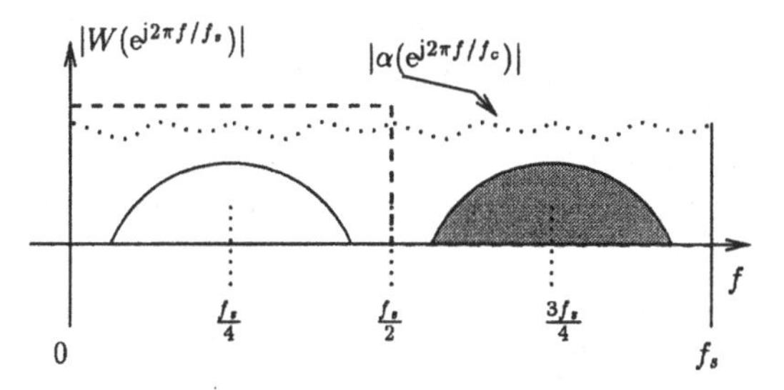

Fig. 5. Pass-band channel estimation

The data symbols are detected by filtering the received signal with the MF, and sampling the filter output with the right timing at symbol rate. In this way the demodulation by $f_s/4$ is automatically achieved and the symbol rate sequence is base-band. Hence the MF output at symbol rate can be written as

$$\hat{b}[k] = (-\mathrm{j})^{NN_ck} \sum_m r[m]f^*[m + NN_ck] = \sum_m r[m]f^*[m + NN_ck] \quad (7)$$

where the last equality holds in our setting with $N_c = 4$.

Carrier Synchronization and Decoding. The carrier synchronization is done at symbol rate with a classical decision directed algorithm [4]. The algorithm then takes a decision on the symbols and recovers the data flow (in our example a video stream).

4. Validation of the Existing Platform

The platform described in this paper has been validated by the transmission and the reception of of two user's real-time flows in an indoor environment. Two H263 video streams are transmitted in parallel and decoded in real time. For this we use the following parameters:

- Spreading factor $N = 16$
- Peak bit rate per user equal to 397.44 kbps
- Symmetric TDD slot arrangement (transmission occurs every 2 slots).
- Two synchronous users per slot
- RF band: 5 MHz at 2.1 GHz

5. End-to-end Processing DSP Performance

In this section we provide performance figures for the transmitter and receiver front-end algorithms previously described with the above set-up. The performance are evaluated in terms of DSP clock cycles. The algorithms have been implemented on a TI-TMS320C6201 DSP. The code is hand-optimized in parallel assembly [12].

The transmitter front-end processing includes the following operations:
Spreading and Scrambling. The spreading and scrambling routine takes as input the user information bit streams. Then it spreads, scrambles and eventually maps it to QPSK symbols. User's symbols are then amplified in order to assign the corresponding power and then summed up together. The routine also creates the slot structure filling each TX slot with the midamble, the primary synchronization sequence and the user's symbols. This process takes for each slot a fixed amount of 7700 DSP cycles plus 6300 cycles per user corresponding to $(38.5 + 31.5 \times U)\,\mu$s.

Pulse-shaping, oversampling by 4 and modulation by $f_s/4$. All these operations are implemented in a single assembler routine. The routine up-samples by a factor 4 the chip sequence previously generated, then passes it through a root-raised cosine filter with roll-off factor equal to 0.22, truncated over a symmetric window of 12 chip periods, and modulates the output of the filter by $f_s/4$, taking only the real part of the modulated signal. The oversampled version of the pulse-shaping filter is implemented as a polyphase filter bank with 4 phases. This processing requires 6 DSP cycles per output sample. Processing an entire slot of 2560 chips requires $6 \times 2560 \times 4 = 61440$ cycles to which one must add the cycles required by the prolog and the epilog needed for the routine pipeline [12]. Finally about 61500 cycles are needed to process a slot of 2560 chips corresponding to about $308\,\mu$s per slot.

The receiver front-end processing includes the following operations:

Primary synchronization code correlation. A routine is designed to compute the real correlation between the primary code and the samples (from the ACQ card) for the initial synchronization. This convolution is done at 4 times the chip rate and exploits the hierarchical properties of the primary synchronization code. The primary code only contains -1 and $+1$ and all users share the same code. Hence, the convolution is performed by using ADD and SUB instructions [12]. The routine also takes the square magnitude of the output. The routine loop kernel takes 33 cycles to produce an output sample at chip rate (about $422\,\mu$s per slot). To acquire the slot-timing we also perform both noise and temporal averaging. Noise averaging is done by accumulating several slots and by averaging with an exponential window with a properly chosen forgetting factor. Temporal averaging is performed over a time interval equal to two slots by an exponential window with a forgetting factor. A single routine performs all these operations in 3 cycles per output sample at chip rate ($38\,\mu$s per slot) and returns the slot-timing estimate.

Joint channel estimation. Up to 3 users' channels each one with a duration of 64 chips, can be estimated with the current set-up. The training sequence (the midamble) period is 192 chips that, accounting for the oversampling factor of 4, corresponds to $192 \times 4 = 256 \times 3$ samples. Therefore a joint LS pass-band channel estimate can be obtained by one mixed radix FFT (radix 4 and radix 3) and one mixed radix inverse FFT (IFFT). The samples from the output of the ACQ card, corresponding to a real pass-band signal, are sent to the channel estimator that performs one real FFT. The LS channel estimate is produced in the FFT domain by multiplying the corresponding samples of the FFT of the received signal with the inverse of the FFT of the basic midamble period, which has been precomputed. A pass-band filtering is also performed to reject the image spectrum of the input real signal, by setting to zero the undesired samples. An IFFT produces a channel pass-band estimate. All the processing described above requires about 15600 cycles per slot (about 78 μs).

Channel analyzer. This routine analyzes the channel estimates, computing the channel energy, the channel length and the channel position that serve to the slot timing tracking operation. The routine also cleans the estimates from the round-off noise, clips the significant portion of each channel response. This process requires about 5.5 μs per slot.

Matched filter synthesis. The pass-band channel response of 64 chips is convolved with each user's spreading sequence of 16 chips. This operation generates the user's symbol pass-band matched filter and can be performed in about 5000 cycles (25 μs) per user per slot.

Pass-band matched filtering and data detection. Once both slot-timing and channel have been estimated, and the users' matched filters built, the pass-band signal at the output of the DAQ card is sent directly to each user's matched filter and the output is down-sampled at symbol rate (note that this operation automatically involves a demodulation to baseband avoiding the need of explicit demodulation). This processing requires 44000 cycles (220 μs) per user per slot.

Table 1. Transmitter: time required by DSP routines for 3 users

Operation	μs per slot
Spreading and Scrambling	133
Pulse-shaping, oversampling and modulation	308
Total	441

Table 2. Receiver: time required by DSP routines for 3 users

Operation	μs per slot
Joint channel estimation	78.0
Channel analyzer	16.5
Matched filter synthesis	25.0
Matched filtering and data detection	660.0
Total	779.5

6. Conclusions and Future Perspectives

In this paper we presented the major features of a real-time SDR platform implementing the essential physical layer of UMTS TDD. This first prototype demonstrated the viability of SDR systems based upon the DSP technology to provide universal seamless connection to wireless communication users.

For the next platform upgrade we envision the implementation of a multiple antenna system, more sophisticated signal processing algorithms (e.g. multi user detection and iterative decoding), and higher layer protocol stacks (e.g. MAC layer). We also envision to improve the design of the radio subsystem in terms of both flexibility and sensitivity. Along with these activities the platform will be opened to both academic and industrial partners to activate collaborations on specific research topics.

References

[1] S. Srikanteswara, J. H. Reed, P. Athanas, and R. Boyle, "A soft radio architecture for reconfigurable platforms," *IEEE Communications Magazine*, February 2000.

[2] "Special issue on software radio," *IEEE JSAC*, vol. 4, April 1999.

[3] "Software radio," *IEEE Personal Communications*, vol. 4, August 1999.

[4] J. G. Proakis, *Digital Communications*. NY: McGraw Hill, 2nd ed., 1989.

[5] J. Mitola, "The software radio architecture," *IEEE Communications Magazine*, pp. 26–38, May 1995.

[6] J. Razavilar, F. Rashid-Farrokhi, and K. J. R. Liu, "Software radio architecture with smart antennas: A tutorial on algorithms and complexity," *IEEE JSAC*, vol. 17, pp. 662–676, April 1999.

[7] G. Caire, P. A. Humblet, G. Montalbano, and A. Nordio, "Transmission and reception front-end algorithms for software radio." submitted to IEEE JSAC Wireless Communications Series, July 2000.

[8] 3GPP-TSG-RAN-WG1, "TS-25.2xx series," tech. rep., January 2000.

[9] B. Steiner and P. Jung, "Optimum and suboptimum channel estimation for the uplink of cdma mobile radio systems with joint detection," *European Transaction on Communications*, vol. 5, pp. 39–49, Jan.-Feb. 1994.

[10] G. Caire and U. Mitra, "Structured multiuser channel estimation for block-synchronous DS/CDMA." Submitted to IEEE Transaction on Communications., July 1999.

[11] G. H. Golub and C. F. V. Loan, *Matrix Computation*. The John Hopkins University Press, 1996.

[12] Texas Instruments, *TMS320C62/C67x Programmer's Guide*, Febraury 1998.

Adaptive Access Scheme Selection in Software-Based Wireless Multimedia Communications

Shinsuke Hara, Daisuke Kitazawa and Hiroyuki Yomo

Graduate School of Engineering, Osaka University
2-1, Yamada-Oka, Suita, Osaka, 565-0871 Japan

Abstract: "Software Radio" has drawn much attention as a means to support multimedia services in wireless communications. Software radio gives us a possibility that we can adaptively change not only modulation/demodulation and channel coding schemes but also multiple access scheme. Here, one question naturally arises; "In software radio, can just a single access scheme such as TDMA or CDMA effectively support multimedia services?" or "If we can select even an access scheme, is it more beneficial?"

In this paper, we discuss the traffic performance enhancement when we have a freedom in selecting an appropriate access scheme out of TDMA and CDMA. With typical multimedia service model supposed to be provided in third or later generation mobile communications systems, we evaluate the traffic performance in terms of average delay time by computer simulation.

1. Introduction

Looking at the history of mobile communications systems, just a single access scheme has been successfully being standardized in each generation system, such as TDMA (time division multiple access) scheme in PDC and PHS (Japan), IS-54 (USA), GSM and DECT (Europe) and CDMA (code division multiple access) scheme in IS-95 (USA) and IMT-2000. Considering fourth or later generation mobile communications systems where wide range of multimedia services should be provided, one question naturally arises: "Can just a single access successfully support it?" or rather say, "If we can adaptively select even an access scheme, is it beneficial for us?"

Recently, "Software Radio" has drawn much attention as a means to support multimedia services in wireless communications[1]-[3]. Software radio is a technique which replaces hardware-based devices/components by software-controlled digital signal processors with enough memories. Mobile terminal can get its best-suited transmission format from base station, by downloading software programs on modulation/demodulation and encoding/decoding scheme, and furthermore multiple access scheme. Therefore, software radio has a possibility to adaptively change even appropriate access scheme according to user's QoS (quality of service) such as requested transmission rate, tolerable bit error

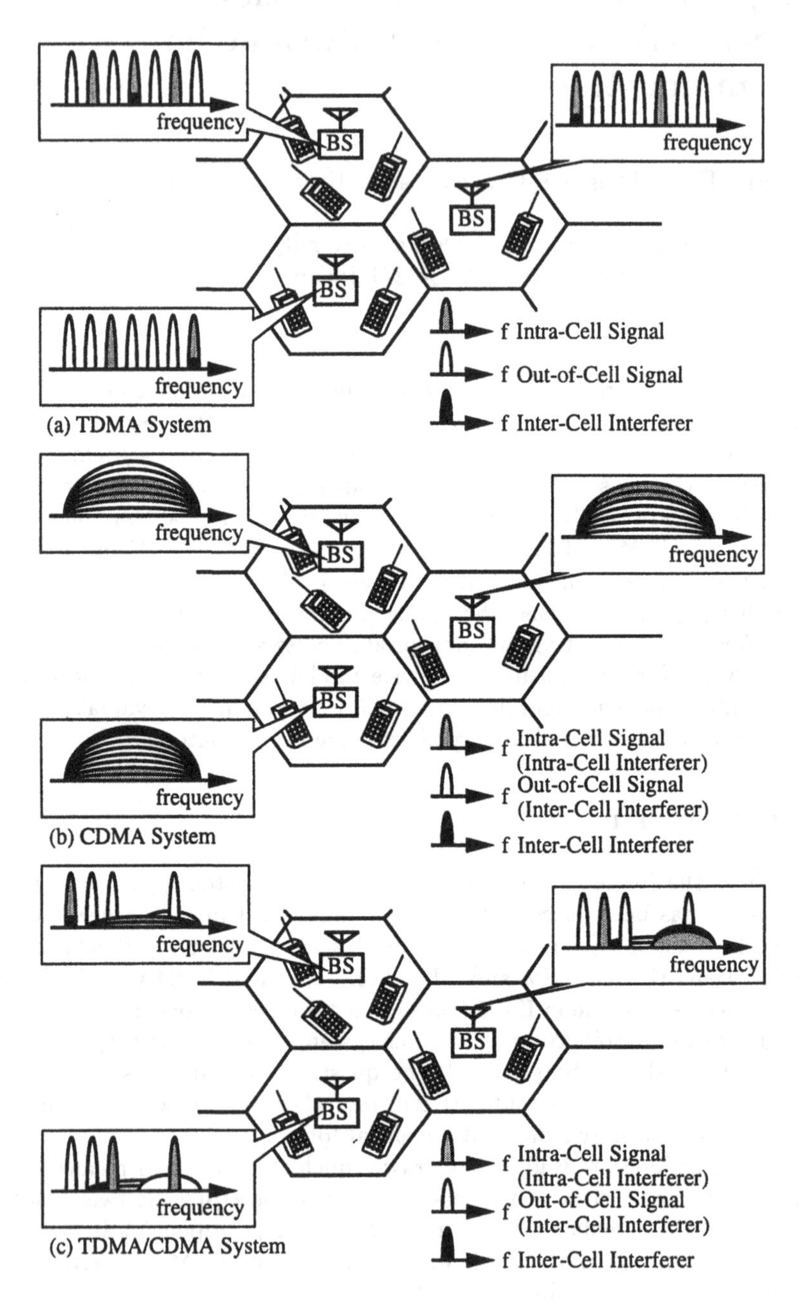

Fig. 1. System model

rate/delay and various factors to maximize system capacity such as channel condition and available frequency bandwidth.

In this paper, we discuss benefit from access scheme selectability in software-based multimedia communications. Here, in order to focus much attention only on role of multiple access scheme, we do not take into account channel coding effect. Therefore, we define service QoS as not *BER* (bit error rate) but *target SNIR* (signal to noise plus interference ratio). We assume three different communications systems, such as a (hardware-based) TDMA system, a (hardware-based) CDMA system and a software-based TDMA/CDMA system where one appropriate access is selectable, and with a multimedia service model containing seven typical applications, we discuss their attainable traffic performance by computer simulation.

2. System Model

Taking into account both uplink and downlink traffic flow, we apply FDD (frequency division duplex) to all the three systems. In the FDD system, which is currently used in GSM, PDC and so on, just a half of total frequency bandwidth given for the system is allocated for downlink, whereas the remainder is allocated for uplink.

2.1 Hardware-Based TDMA System

Fig.1 (a) shows the TDMA system, where all the services are supported in TDMA scheme. Here, we assume the FCA (fixed channel assignment) as a channel assignment algorithm. In order to provide various transmission rate services, the appropriate number of time slots can be assigned to a user according to the required transmission rate and channel condition.

2.2 Hardware-Based CDMA System

Fig.1 (a) shows the CDMA system, where all the services are supported in DS (direct sequence)-CDMA scheme. The processing gain is fixed (1023), and for various transmission rate services, multiple code assignment is employed, where the appropriate number of spreading codes can be assigned to a user also according to the required transmission rate and channel condition.

2.3 Software-Based TDMA/CDMA System

Fig.1 (c) shows the TDMA/CDMA system, where TDMA and CDMA schemes are adaptively selectable according to user's QoS and channel condition. The processing gain is changeable, so different gains can be employed in different cells, but the same gain is employed within a cell. Namely, the TDMA and CDMA signals *coexist* in the frequency band allocated to the system, but these two signals are never overlapped each other within a cell. Here, we assume that the lower frequency band is allocated for the TDMA scheme and the higher for the CDMA scheme. The system employs the FCA algorithm for the TDMA channel assignment.

3. Adaptive Selection of Multiple Access Scheme

Since we must in advance determine a criterion for selecting either of the two access schemes in the TDMA/CDMA system, we adopt a selection algorithm proposed in [4]. Namely, we select either TDMA or CDMA scheme for a user according to the user's QoS and channel condition. Here, we categorize multimedia services provided in mobile radio communications system into two types: Low-Quality service (which is permitted relatively low *SNIR*) and High-Quality service (which requires relatively high *SNIR*). The selection method is based on the following idea: for the case of providing Low-Quality service, the CDMA scheme has a higher priority since the CDMA signal is vulnerable to interference but the required *SNIR* can be admitted to be lower. On the other hand, for the case of High-Quality service, the TDMA scheme has a higher priority since higher *SNIR* is required and the TDMA signal can attain higher *SNIR* because the interference power among users is relatively small.

In order to determine whether one access scheme (carrier frequency position and time slot(s) for the TDMA scheme or carrier frequency position and spreading code(s) for the CDMA scheme) can be used for one service (denotes subscript i), the following condition is examined:

$$SNIR_i \leq SNIR_{act} = \frac{S(i)}{N + \sum_{\substack{j=1 \\ j \neq i}}^{J} I(j)U_j + I(i)(U_i - 1)} \tag{1}$$

where

$SNIR_i$: Target *SNIR* for service i
$SNIR_{act}$: actual *SNIR* for service i
$S(i)$: received desired signal power for service i
N : noise power (single sided AWGN)
$I(j)$: received interference power from service j
U_j : the number of users for service j
J : the number of services

If one of the two access schemes cannot be used, then the condition for the other scheme is examined. In addition, it is possible that different multiple access schemes are selected in downlink and uplink for a call. Fig.2 summarizes the selection procedure.

This is a very simple strategy, however, it is already examined that the strategy can improve traffic performance when taking into account only uplink traffic flow[4]-[7].

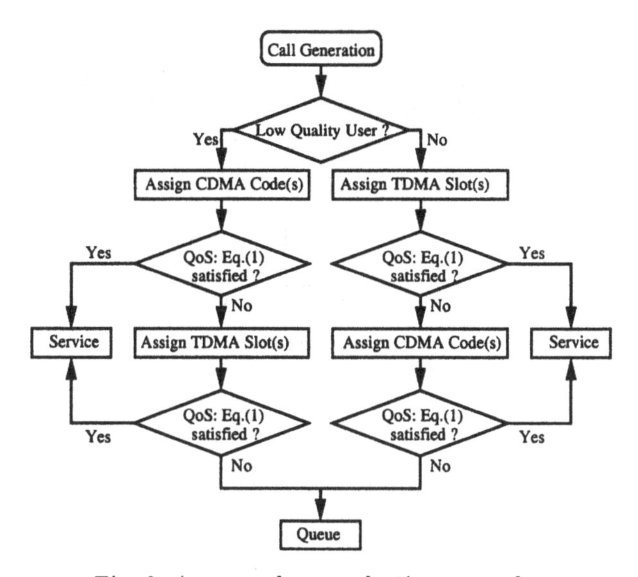

Fig. 2. Access scheme selection procedure

Table 1. Service model (circuit-switched type)

Service Name	Transmission Rate [kbps]		Duty Factor	Target *SNIR* [dB]	Mean Holding Time [sec]	Illustrative Application
	Uplink	Downlink				
LQ-Voice	8	8	0.4	4.3	120	LQ Telephone
HQ-Voice	32	32	0.4	4.3	120	HQ Telephone
LQ-Video	32	32	1.0	8.4	120	LQ TV-tel
HQ-Video	256	256	1.0	8.4	120	HQ TV-tel
BD-Data	16	16	0.4	0.4	120	Chat

4. Multimedia Service Model

We classify multimedia services into two types. Tables 1 and 2 show the circuit-switched type and the packet-switched type, respectively. In these tables, "Duty Factor" means the average time interval of actual signal transmission during one holding time. In addition, the target *SNIR* of 4.3 [dB] corresponds to the BER (bit error rate) of 10^{-2}, whereas *SNIR* of 8.4 [dB] the BER of 10^{-4} without FEC (forward error correction).

Table 2. Service model (packet-switched type)

Service Name	Link	Transmission Rate [kbps]	Duty Factor	Target *SNIR* [dB]	File Size		Illustrative Application
					ζ	σ	
DL-Packet	Uplink	8	0.03	4.3	11.61	2.26	DataBase Access Image Download WWW
	Downlink	384, 256, 128 64, 32, 16, 8	0.97	8.4			
UL-Packet	Uplink	384, 256, 128 64, 32, 16, 8	0.97	8.4	11.61	2.26	Image Upload
	Downlink	8	0.03	4.3			

4.1 Circuit-Switched Type Services

Table 1 shows the circuit-switched type services: LQ (low quality)-Voice, HQ (high quality)-Voice, LQ (low quality)-Video, HQ (high quality)-Video, and BD (bidirectional)-Data. In these services, the characteristics in terms of transmission rate are all symmetric, since the amounts of the information born in downlink and uplink are considered to be almost equal. However, the transmission rates widely vary according to services.

For voice-like services including "BD-Data", the duty factor is approximately 3/8. On the other hand, for video services, the duty factor equals 1 because such services have a great deal of information.

For the circuit-switched type services, the CDMA scheme has a higher priority for "LQ-Voice" and "HQ-Voice", whereas the TDMA scheme for the remainder, regardless of link transmission rates.

4.2 Packet-Switched Type Services

Table 2 shows the packet-switched type services: DL (down load)-Packet and UL (up load)-Packet. These services mainly provide data transfer application, and have an asymmetric feature in downlink/uplink . For "DL-Packet", the traffic in downlink is overwhelmingly heavier than in uplink, and ACK (acknowledgement) is sometimes transmitted in uplink, and vice versa for "UL-Packet". It is assumed that ACK can be transmitted without errors in the link where the transmission rate and target *SNIR* equal 8 [kbps] and 4.3 [dB], respectively. Hence in Table 2, the duty factors of "DL-Packet" uplink and of "UL-Packet" downlink indicate the ratio that is occupied by ACK in the total amount of the information born in both links, which is assumed to 0.03.

On the other hand, for the downlink of "DL-Packet" and the uplink of "UL-Packet", the transmission rates are determined by *best effort*. That is, the link must employ as a higher transmission rate as possible, and its maximum rate is assumed to 384 [kbps] here.

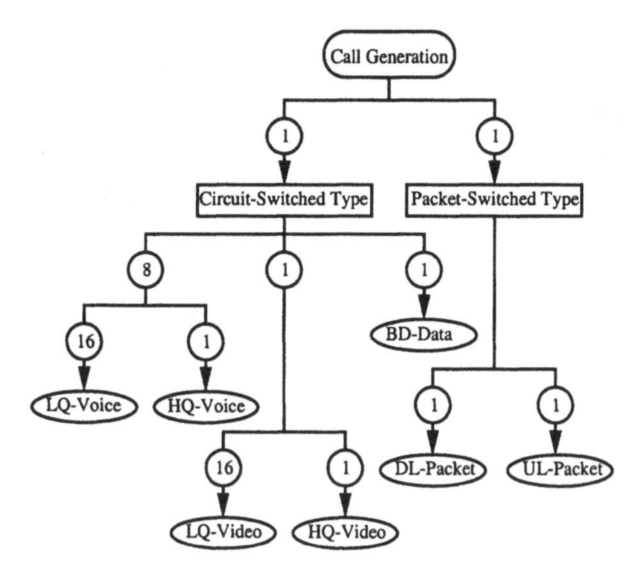

Fig. 3. Service generation model

Paying attention to the amount of born information, especially file size in once downloading or uploading, it has been examined that variation of file size in wired communications such as LAN (local area network) and Internet traffic follows a log-normal distribution[8]. The log-normal distribution of file size X [Bytes] is written as:

$$p(X) = \frac{1}{\sqrt{2\pi}\sigma X \ln 2} \exp\left[-\frac{(\log_2 X - \zeta)^2}{2\sigma^2}\right] \tag{2}$$

where ζ and σ in Table 2 show the parameters in Eq.(2).

For the packet-switched services, the CDMA scheme has a higher priority for the uplink of "DL-Packet" and the downlink of "UL-Packet", whereas the TDMA scheme for the downlink of "DL-Packet" and the uplink of "UL-Packet".

4.3 Average Ratio of Each Service

The number of users employing these seven services is still unknown and may not be equal. However, the average ratio of the users employing each service should be determined for numerical demonstration in advance. Fig. 3 shows the service generation model employed in our computer simulation.

The values in the figure correspond to the ratios of services in each branch, for example, the ratio of "LQ-Voice" to "HQ-Voice" equals 16 to 1.

Table 3. Common simulation parameters

Call Generation	Poisson Process
The Number of Cells	19 (Omni)
Mobile Terminal Distribution	Uniform
Mod./Demod. Scheme	QPSK Coherent Detection
Baseband Filter	Root Nyquist ($\alpha = 0.5$)
Power Decay Factor	3.5
Shadowing	Log-Normal Distribution ($\sigma = 6.0$ [dB])
Fading	Rayleigh Distribution
Total Bandwidth	12.8 [MHz]
Base Transmission Rate	8 [kbps]
Tx. Power margin on Uplink	5 [dB]
Tx. Power Margin on Downlink	10 [dB]
TDMA Scheme	
Carrier Separation	320 [kHz]
Channel Assignment Scheme	FCA
Frequency Reuse Factor (FCA)	3
The Number of Time Slots per Carrier	48
Channel Frequency Selectivity	Non-Selective
CDMA Scheme	
Processing Gain	31, 63, 127, 255 511, 1023
The Number of Fingers onRake Receiver	3
Channel Frequency Selectivity	Selective

5. Numerical Results and Discussions

The cellular systems discussed here is composed of multiple (19) omni-cells. In each cell, users are distributed uniformly and communicate with a centered base station. If there is no radio channel (no time slot in TDMA scheme or no spreading code in CDMA scheme) or the received signal cannot satisfy Eq.(1) due to interference from other users or other factors, the user cannot be in service. Such a user waits in a FIFO (first-in first-out) queue until the user becomes to be served. We evaluate the system performance in terms of average delay (waiting) time for each service. Here, the average delay time means *time from call generation to channel assignment*. Also, note that we define "service time" as *time interval when QoS is kept satisfied.* In other words, the forced termination rate due to interference during a call is exactly zero in our computer simulation. Fig.2 also shows the computer simulation flow.

For both the TDMA and the CDMA signals, TPC (transmission power control) is perfectly employed, and it is applied only for uplink. Furthermore, as a radio channel model on CDMA RAKE receiver, we assume the ITU-R "Outdoor to Indoor and Pedestrian Test Environment B model" [9], where there are three paths in the multipath delay profile, and its average received power ratio is 1.0 : 0.8 : 0.3. RAKE combining is performed in both downlink and uplink. The simulation parameters are summarized in Table 3.

Figs.4 (a)-(g) show the numerical results on the traffic performance: the average delay time versus the call generation rate. Here, there are three curves in these figures: the (hardware-based) TDMA system and the (hardware-based) CDMA system, and the (software-based) TDMA/CDMA system. Note that in the figures, the curve of a better system should go lower.

In each cell, once a HQ-Video call is generated, which requires high transmission rate and high target *SNIR* in both downlink and uplink, the TDMA system needs to support it with a lot of time slots whereas the CDMA system with a lot of spreading codes. Furthermore, in the CDMA system, in order to suppress multiple access interference in the cell (to satisfy the high target *SNIR*), it also limits the number of multiple access users. Therefore, as a result, the TDMA performs better than the CDMA system for almost all the services, although the CDMA system can enjoy path diversity effect by RAKE combining.

The TDMA/CDMA system can flexibly select an appropriate access scheme for each call according to the required QoS and channel condition, basically the CDMA schemes for low quality users whereas the TDMA schemes for high quality users. Therefore, for all the seven services provided, the TDMA/CDMA system performs best among the three systems in wide range of the call generation rate.

6. Conclusions

In order to discuss role of multiple access scheme in wireless multimedia communications, we have evaluated the traffic performance in terms of average delay time among three systems, namely, the hardware-based TDMA system, the hardware-based CDMA systems, and the software-based TDMA/CDMA system where we can adaptively select an appropriate access scheme out of TDMA and CDMA. We have shown that once a high rate/high quality-service is included in multimedia services provided, the TDMA system has an advantage over the TDMA system, because it guarantees a perfect orthogonality among users within each cell. Furthermore, we have clarified that the TDMA/CDMA system outperforms the conventional hardware-based systems, because it can flexibly select an appropriate access scheme for each call according to the required QoS and channel condition. It is clear from our numerical demonstration that not fixed-multiple access scheme but its selectability has a large impact on system capacity enhancement.

It is not only by "software radio" that can give access scheme selectability to mobile users. For example, hardware-based dual mode (TDMA and CDMA) transmitter/receiver does have the same functionality. However, it is true that "software radio" can easily realize such an environment by its attractive characteristics of *Downloadability* and *reconfigurability*.

Further investigations include theoretical analysis on traffic performance, more sophisticated multiple access scheme selection, effect of adaptive modulation and channel coding schemes, and so on. Software Radio, which has several aspects to be analyzed from software to hardware, must be a key technology to realize *"an ultimate goal of wireless multimedia communications."*

7. Acknowledgements

This work was supported in part by the Mazda Foundation's Research Grant.

8. References

[1] Special Issues on "Software radios," *IEEE Commun. Mag.*, Vol. 33, No. 5, May 1995.

[2] Special Issues on "Globalization of software radio," *IEEE Commun. Mag.*, Vol. 37, No. 2, Feb. 1999.

[3] W. H. W. Tuttlebee, "Software-defined radio: facets of a developing technology," *IEEE Personal Communications*, Vol. 6, No. 2, pp. 38-44, Apr. 1999.

[4] H. Yomo, S. Hara, "Flexible multiple access scheme assignment in software radio environments," in *Proc. WPMC'98*, pp.416-421, Nov. 1998.

[5] H. Yomo and S. Hara, "Impact of access scheme selectability in software-based wireless multimedia communications," in *Proc. PIMRC '99*, pp. 6-10, Sept. 1999.

[6] S. Hara, D. Kitazawa and H. Yomo, "Impact of access scheme selectability on traffic performance in software-based wireless multimedia communications system," in *Proc. of VTC 1999-Fall*, pp.2805-2809, Sept. 1999.

[7] D. Kitazawa, H. Yomo and S. Hara, "Flexible downlink/uplink frequency resource allocation for software-based wireless multimedia communications," in *Proc. of VTC 2000-Spring*, pp.670-674, May 2000.

[8] M. Nabe, et al., "Analysis and modeling of WWW traffic for designing Internet access networks," (in Japanese) *IEICE Trans.* B-I, Vol. J80-B-I, No. 6, pp. 428-437, June 1997.

[9] *"Guidelines for evaluation of radio transmission technologies for IMT-2000/FPLMTS,"* ITU-R COMMUNICATION STUDY GROUP, June 1996.

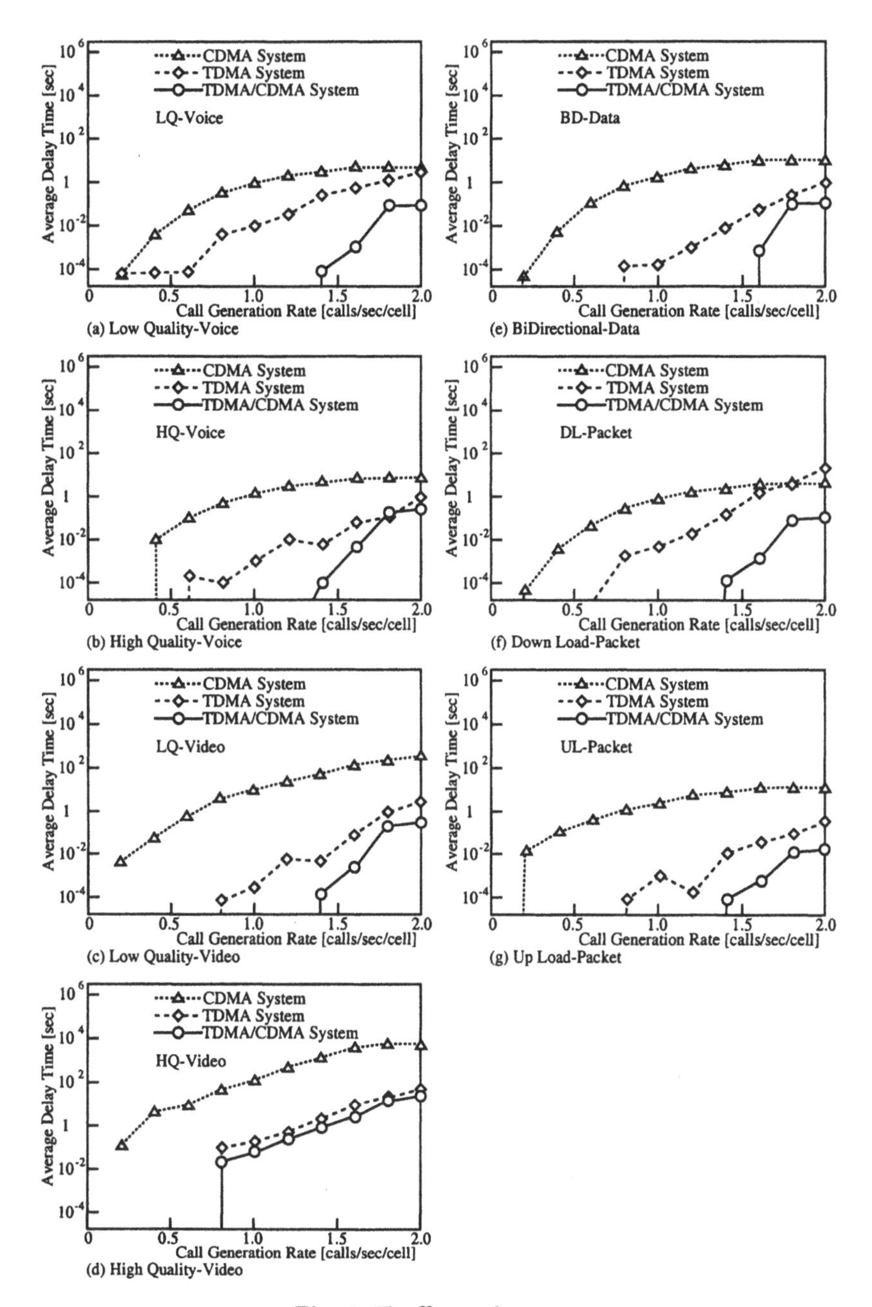

Fig. 4. Traffic performance

Distribution of Intelligence and Radio Link Configurability in Wireless Video-based Surveillance Networks

Claudio Sacchi, Gianluca Gera, and Carlo S. Regazzoni

University of Genoa, Department of Biophysical and Electronic Engineering (DIBE), Via Opera Pia 11/A, I-16145 Genoa (Italy)

Abstract: This work[1] considers the distributed decomposition of the image processing tasks in wireless video-based surveillance networks. We considered the use of a Spread Spectrum-based wireless radio link, dynamically configurable by means of software radio technology, for remote transmission of the visual information from the peripheral sensor level to the central operator level. The considered video-based surveillance application is related to an image processing system for people counting. The distribution of the image processing tasks is performed by embedding at the sensor level some low-level image processing modules. The results obtained by the distributed architecture in terms of counting precision, are then compared with the ones provided by a centralised architecture performing the overall image processing tasks at the operator level, considering in both cases different configurations of the radio link in terms of spreading factor.

1. Introduction

Historically, advanced video-based (AVS) surveillance systems can be classified into three generations. The First Generation AVS systems (1960-1980) [1] were based on analog or digital CCTV transmission [2] of the visual information from the guarded sites to a control desk, where it is displayed to the human operator by means of TV monitors, without any kind of processing. In such systems, both the recognition of interesting parts of the monitored scene and the understanding of the meaning of the scene were demanded to the human operator.

Second Generation AVS systems (1980-2000) [1] introduced PC-based digital image processing architectures in order to driving the attention of the human operator in complying with video-surveillance tasks. The elaboration capabilities of the 2nd generation systems are oriented to information filtering, recognition of interesting parts of the monitored scene, and low-level understanding functions. Moreover, 2nd generation AVS systems can exploit the most recent

[1] This work has been partially supported by the Italian University and Scientific Research Ministry (MURST) within the framework of the National Interest Scientific Research Programme.

developments concerning image compression coding and digital transmission techniques, in order to consider the actual implementation of *remote* video-based surveillance systems, where the visual information acquired at the sensor level is remotely processed at the operator level. The problems to be faced in such kind of applications mainly concern with the limited bandwidth resources available for continuous transmission of multimedia information from the guarded sites to the network head-end [2] (i.e., upstream transmission).

From the market point of view, the current reality is focused on the commercialisation of 2nd generation AVS systems both for local and remote applications. Considering this favourable perspectives, the current research on AVS is already working on the new generation systems: the so-called 3rd generation ones (2000 - ?) [1]. A basic concept of 3rd generation AVS systems is the *distribution of the intelligence* across the computer network. 1st and 2nd generation systems are both characterised by *centralised* intelligence (i.e., image processing tasks entirely managed at the operator level). This kind of AVS system concept was constrained by the limited processing capabilities of IC architectures, which avoided for many years the effective transferring of complex image processing functions from the operator level to the sensor level.

The distribution of intelligence would have required the installation of a PC-based image processing architecture close to the video-camera, as pointed out in [3], where the feasibility study of a distributed AVS system for detection of abandoned objects in unattended railway station is proposed.

3rd generation systems can exploit the technological improvement provided by DSP programmable architectures [4], that are making it possible to embed image processing functions into the cameras themselves The executable software algorithms, written in conventional programming language (e.g. C or C++) and compiled by means of special DSP compilers, can be remotely downloaded by embedded machines working at the sensor level. Such kind of hardware devices, called *smart cameras* [5], are currently able to perform in real-time MPEG and JPEG compression coding and low-level image processing tasks, such as: noise filtering, change illumination compensation and frame-background differences. In the next future it is foreseen that embedded image processing systems will be able to implement higher-level image processing functions, such as: motion detection and tracking, thus allowing to improve the peripheral intelligence of the AVS systems. An actual example of intelligence distribution in advanced AVS applications has been presented in [6], where a non-intrusive system for identifying stolen vehicles on the national highways is dealt. The system approach followed in [6] considered the generation of a target stolen vehicle profile that can be downloaded to each of the remote video-surveillance sites on the highway. The information contained in the vehicle profile is then processed by each remote intelligent camera that employs advanced neural network classification techniques to search the visual scene for a potential candidate of the stolen vehicle.

In this paper, the distributed decomposition of an existing remote video-based surveillance system over a multimedia communication network, regarded as a case of study, has been intensively studied through laboratory simulations. In particular, we considered transferring of a different amount of intelligence from the operator level to the sensor level.
The case of study is related to a remote video-based people counting system developed in the context of tourist and protected sites monitoring applications [7], originally exploiting an existing wired CATV network for information transmission from the sensor level to the operator level. In the present dealing, the transmission of the information from the camera site to the remote control centre is performed by using a wireless network. A wireless urban channel affected by multipath fading has been simulated. The analysis shown in the paper will consider the distribution of the system intelligence in the following way:

- results of a 2nd generation architecture with digital camera and digital wireless modem transmission will be firstly shown, as touchstone for system performance evaluation;
- a 3rd generation architecture with intelligent camera performing low-level image processing tasks (i.e., up to change detection) will be regarded as the further step of peripheral intelligence distribution;

A reconfigurable digital wireless transmission module will be also simulated. Such module is a Direct-Sequence Spread-Spectrum (DS/SS) transceiver [8] with programmable modulation, spreading factor, and channel coding. The single-user DS/SS transmission is actually performed by the IEEE 802.11 digital wireless standard, which allows one to transmit at a bit-rate up to 11 Mb/s using a CSMA/CD multiple access protocol [15]. Many commercial products for wireless surveillance are based on 802.11 standard, whose main drawback is related to the fixed configuration of the radio link. In such sense, an improved flexibility will be provided by the *software radio technology*, which will enable the total reconfigurability of the transmission link in terms of programmable channel coding, spreading factor, modulation, detection algorithms etc. [17].
At the end of the proposed analysis some guidelines for the design of the peripheral sensor level and the remote operator level, together with the digital communication link will be issued. The paper is structured as follows: Section 2 will deal with a brief description of the image processing system for people counting, Section 3 will consider the results provided by a 2nd generation architecture of the proposed system, using different kind of upstream transmission techniques, compared with the ones provided by a 3rd generation architecture with low-level intelligence transferred in the camera, Section 4 will draw the conclusions of the paper.

2. System description

2.1 The Considered Video-based Surveillance Application

The actual AVS system chosen as a case of study for the present dealing is related to a video-based people counting system for tourist flows estimation. The actual system implementation performed during the EU ESPRIT 28494 AVS-RIO project (1998-1999) [7] [10] led to a people counting demonstrator using fully analog information transmission of the visual information from the camera site to a remote control centre through an existing coax-cable network for TV broadcasting service. In this work, we will consider a digital wireless network as communication link between the monitored site and the remote control centre, instead of the wired network originally considered. The aim of the proposed analysis is to test an eventual adaptation of the proposed AVS system also for tourist zones not reached by cable networks (e.g., in Italy only small areas are locally served by CATV) or for other kind of people counting applications not restricted to the tourist flows estimation. In every case, the camera site should be chosen in order to overlook a central people passage point. At the present moment, a single-camera system is proposed. Evolution towards a multi-camera system is currently studied at research level. The modular structure of the image processing system for people counting is depicted in Figure 1.

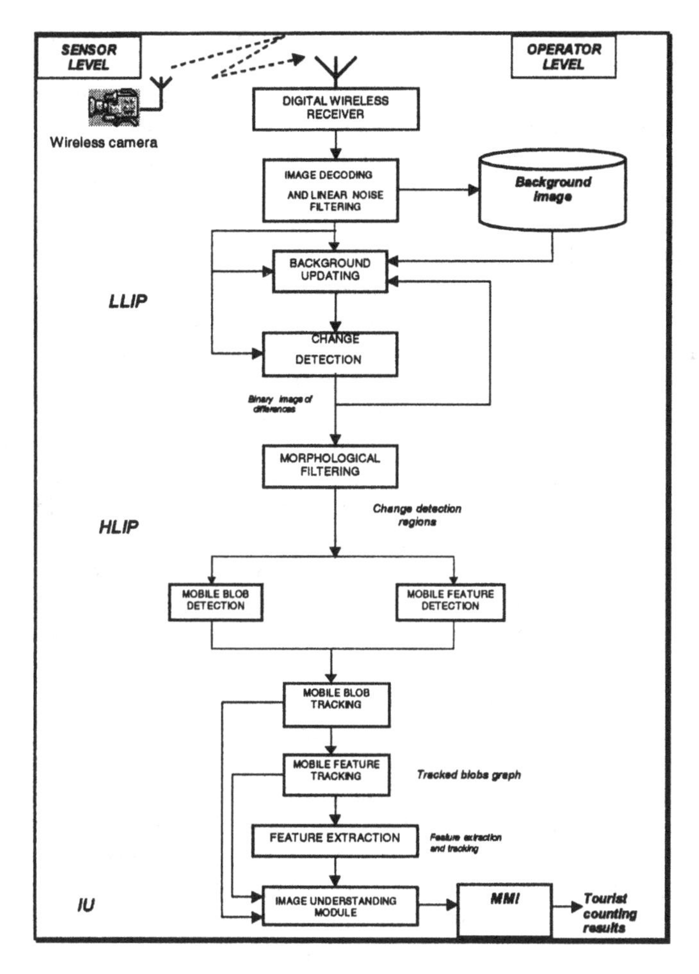

Figure 1. Modular diagram of the image processing system for people counting

The image processing system shown in the above figure can be subdivided into three separated modules:

- The *Low-level Image Processing (LLIP) module*, implementing the image decoding, noise filtering, background updating, and change detection functions [11]. The processed image sequences are monochromatic with resolution equal to 512x512 pixels and 8 bit for each pixel. The minimum frame–rate required by the application equals to 1 frame/sec;
- *the Higher-level Image Processing (HLIP) module*, devoted at labelling and tracking the mobile areas detected in the current image (i.e., the *blobs*) [12], and at extracting some useful features concerning the detected blobs;

- The *Image Understanding (IU)* module, addressed at *classifying* the mobiles areas in detected the current frame by the lower-level modules. The class attributed to each blob is substantially the number of persons grouped in the blob itself. This module is based on a neural-network classifier [13], aided by some other information achieved by the blob tracking sub-modules, which are needed for regularising in time the people counting results [10].

The scheme of Figure 1 is related to a typical 2nd generation AVS system implementation, working in remote modality. The wireless camera, installed close to the guarded site, acquires images and transmits them in digital compressed format to the remote control centre (JPEG and MJPEG formats are generally employed for video-surveillance applications [2]). The image processing tasks for people counting are entirely performed at the remote operator level. In a distributed 3rd generation implementation, some of the image processing tasks will be transferred by means of downloading operations to the sensor level and executed by intelligent smart cameras, which will transmit to the remote control centre the processing results. The advantages of using distributed video-based surveillance architectures, instead of centralised ones can be listed as follows:

- Reduction of the amount of information to be transmitted in real-time through existing wireless networks. Therefore, the lack of bandwidth problems affecting 2nd generation AVS systems should be overcome;
- Complying with privacy legal issues, as the processing results transmitted from the sensor level to the operator level do not generally contain any useful information for malicious users or fraudulent detectives;
- Improvement of the processing capabilities of the central architecture located at the operator level, which could be devoted only at implementing higher level image understanding functions, thus avoiding the waste of computational resources for low-level tasks, and making a step ahead in the direction of the implementation of more and more intelligent systems;
- Improved evolution towards multi-sensorial and remotely configurable AVS architectures (more sensors can be used within the same bandwidth);
- Improved evolution towards fully-autonomous AVS systems, without needing any intervention and/or interaction by a human operator, explicitly devoted at surveillance tasks.

The disadvantages are mainly related to the increased costs of embedded DSP-based image processing architectures (i.e., smart cameras), with respect to the more conventional (and commercial) PC-based ones, which are very cheap now.

In the specific context of the people counting application considered here, the actual DSP devices available on the market can allow one to embed at the sensor level at least the entire low-level image processing (LLIP) module.

In the following, we will compare the results provided by a state-of-the art 2nd generation system with digital transmission of compressed image sequences from the sensor level to the operator level, with the ones provided by a 3rd generation system with low-level image processing tasks embedded at the sensor level.

2.2 The Wireless Transmission Simulation

The wireless transmission of the information from the sensor level to the operator level is performed by using a Direct-Sequence Spread-Spectrum (DS/SS) programmable transceiver, with settable parameters such as: processing gain, trellis coding, and modulation. We considered for simulation a programmable Spread Spectrum transceiver with QPSK modulation, e.g. like the ones supplied by SIRIUS corporation (Belgium), who recently launched on the market some wireless modem products based on DSP and software radio technologies (see e.g. DATASAT™ system [18]). The transmission is performed over the ISM band at 2.4 GHz (2.4-2.4385 GHz), which is the bandwidth portion licensed by ETSI for Spread Spectrum experiments. The receiver is implemented by using a *rake filter* [9], able to extract the faded signal components due to multipath fading effects usually affecting wireless channels [14]. The channel model chosen for the considered test is a *frequency-selective multipath fading* channel, with *Rayleigh-distributed* fading [14] (we suppose that no direct path is available from the wireless camera to the remote control centre. This would be true in the actual site of the camera installation). The entire transmission system (from source to sink) has been simulated by exploiting the capabilities of the MATLAB® SIMULINK® 5.3 software libraries, following the idea of using the SIMULINK® blocks for realistic simulation of digital communication systems already proposed in [16]. In particular, the SIMULINK® diagram of the multipath wireless channel simulator is shown in Figure 2. Each branch of the structure corresponds to a different channel path.

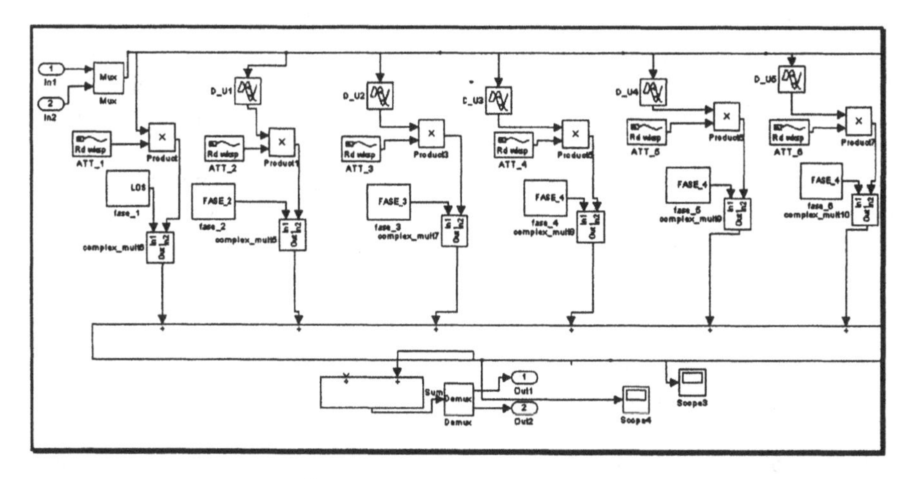

Figure 2. SIMULINK® diagram of the wireless multipath channel simulator

3. Performance evaluation of different video-based surveillance architectures

3.1 Transmission Link Settings

First of all, we fixed the parameterisation of the digital wireless radio system simulations. The bit-rate before spreading has been fixed at 1 Mb/s. Such a value can meet the constraints about the frame-rate mentioned in Section 2. Accordingly with the measurements on urban multipath fading channels in pedestrian environment presented in [19], the Doppler spread has been settled to 1.9 Hz. and the coherence bandwidth equal to 3.5 MHz. In order to focus in a better way the effects of multipath fading on the quality of service provided by the transmission link, we neglect the Gaussian noise term, as often done in the literature. In the simulations, we shall consider three different settings of the spreading factor N and trellis coding rate:

1) Very low spreading factor ($N = 7$) and FEC rate = 1/3 (signal bandwidth occupation about 4 MHz);
2) Spreading factor $N = 15$ and FEC rate $R_c = 1/2$ (employed binary AO-LSE m-sequences, signal bandwidth occupation about 8 MHz.);
3) Spreading factor $N = 63$, and FEC rate $R_c = 1/2$ (employed binary Gold sequences, signal bandwidth occupation about 32 MHz.).

In the case 1) the transmitted signal experimented the heaviest degradation due to the frequency selective fading, as the rake receiver can extract and combine only two faded replicas, corresponding to one separated path. For this reason, we try to protect the information by means of a more robust FEC coding. In the case 2), the rake receiver can extract and combine three faded replicas of the signal, corresponding to two separated paths [9]. In the case 3), the rake receiver can extract and combine ten faded replicas of the signal, corresponding to nine separated paths. In each case, we applied a low-complexity channel equalisation algorithm, that provide a very good – however not ideal – estimation of the channel coefficients. This is a realistic situation occurring in most of actual digital transmission applications.

3.2 Performance Evaluation of a 2nd generation Video-based Surveillance Architecture for People Counting

The first task of performance evaluation is related to a 2nd generation video-based surveillance architecture, with continuous transmission of digital image sequences from the wireless camera to the remote control centre. In this case, the only kind of processing performed at the sensor level is related to the source coding of the acquired visual information. To this aim, JPEG coding is preferred to MPEG one in remote video-surveillance applications due to the better quality provided after image decoding. In our simulation, the transmission of a monochromatic image sequence of 15 frames with JPEG compression rate equal to 20 has been considered. The numerical results shown in this section are related to the number of persons entering the monitoring scene during a given time interval (or, equivalently, for a given number of frames). As expected, the low-spreading wireless radio link configuration seen at point 1) is severely affected by multipath fading and provides very poor results in terms of BER (BER $\approx 2 \cdot 10^{-1}$ before FEC decoding and BER $\approx 10^{-1}$ after FEC coding). The introduction of a very robust FEC coding is not useful to recover an acceptable quality of the transmission. Only four frames of the JPEG-coded received bit-stream could have been decoded, however with heavy degradations, as shown in Figure 3a. JPEG decoding process failed for the other 11 frames. Therefore, the people number returned by the system versus the processed frame number is absolutely incorrect. The radio link configuration considered in the case 2) is more robust with respect to the detrimental effects of multipath fading than narrowband configuration (BER $\approx 10^{-2}$ before FEC decoding, and BER $\approx 10^{-3}$ after FEC decoding). In this case, only few decoded images are severely degraded by huge artifatcs (see Figure 3b), whereas most of the transmitted frames are successfully decoded. However, the people counting precision will noticeably suffer by these degradations, as some persons can be missed in the destructed parts of the corrupted frames. Indeed, the people counting software will return incorrect results when process such kind of frames. The error-free transmission is obtained only by the wideband radio link configuration considered in the case 3), which is able to provide a perfect quality of the reconstructed frames, only degraded by the negligible loss of quality due to JPEG coding (see Figure 3c).

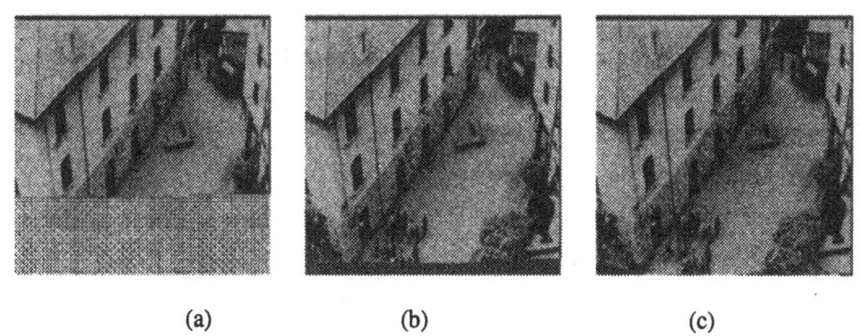

(a) (b) (c)

Figure 3. (a) Received and decoded JPEG frame obtained by the wireless transmission with spreading factor N = 7, (b) received and decoded JPEG frame obtained by the wireless transmission with spreading factor N = 15, (c) received and decoded JPEG frame obtained by the wireless transmission with spreading factor N = 63

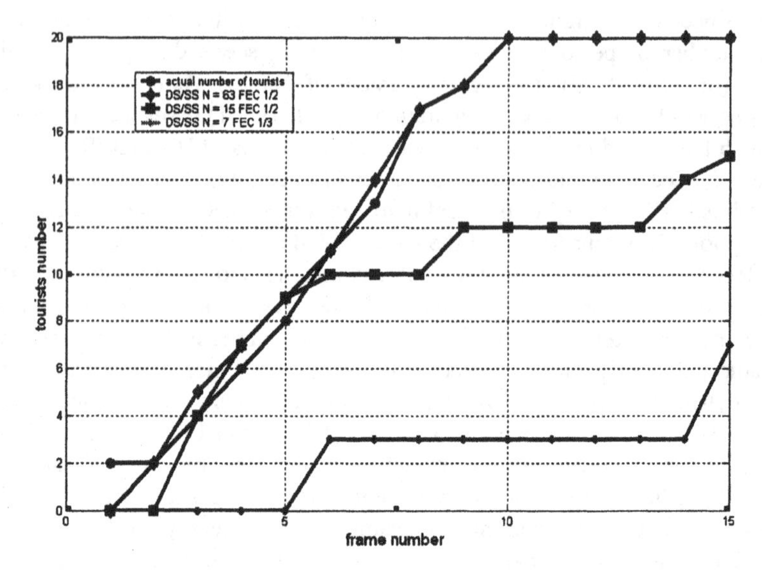

Figure 4. People counting results provided by a remote AVS architecture with centralised intelligence (2nd generation AVS system)

The graphs reporting the results in terms of counting precision are shown in Figure 4. The expected counting precision in terms of absolute mean error (less than 10% required) is reached only by the wideband transmission link configuration with spreading factor $N = 63$, needed for protecting the transmitted signal against fading effects. The results provided by the low-spreading transmission link ($N = 7$) are disastrous, and also those ones provided by the radio link configuration with processing gain $N = 15$ are almost below the required precision threshold.

3.3 Performance Evaluation of a 3rd generation Video-based Surveillance Architecture for People Counting with Low-level Processing Transferred at the Sensor Level

Now, we consider the transferring of intelligence from the operator level to the sensor level. In particular, we have simulated the embedding of the low-level image processing (LLIP) module (i.e., image acquisition, noise filtering, background updating and change detection) in an intelligent camera. The other image processing modules of the people counting system (i.e., HLIP and IU ones) works at the operator level. The output of the image processing performed at the sensor level is the binary image of differences, as shown in Figure 5a. Such image, source encoded by means of a simple binary coding (i.e., 0--> 0, 255 --

>1), is transmitted to the remote operator level by using the same radio link configurations shown in section 3.1.

Also in this case, the multipath fading channel has destructive consequences on low-spreading transmission. Received and decoded binary images are not useful for processing, as clearly shown in Figure 5a. The people counting tasks performed at the operator level return identically zero values, as blobs of unacceptable dimensions, such the ones present in the change detection image of Figure 5a, are not processed by the software.

The wideband digital radio link, with processing gain $N = 63$, allows one to obtain an error free transmission (see Figure 5c). Therefore the results provided by the people counting software are the same achieved in the 2nd generation system implementation shown in section 3.2.

Probably, the most interesting result concern with the Spread Spectrum radio link with processing gain $N = 15$. The effects of the channel errors over the received and decoded binary images are not as heavy as the one experimented by the JPEG transmission. Indeed, such errors only involves the presence of some isolated noisy points in the binary image of differences (see Figure 5b). In most situations, the morphological operators working in the HLIP module can filter off the isolated noisy points, without compromising the people counting precision, as shown in Figure 6.

In such perspective, one can consider that an improvement of the intelligence at the peripheral level of the system can involve a reduced amount of bandwidth required for a noise-robust information transmission.

4. Conclusions

The availability of intelligent sensors and configurable wireless radio links by means of software radio technologies will allow in the next future the effective distribution of the system intelligence across video-surveillance networks. By exploiting the current state-of-the-art smart camera devices, it is already possible to transfer some relevant image processing tasks at the peripheral sensor level, thus allowing a reduction, both of the amount of bandwidth for the transmission of visual information over the wireless network, and of the computational weight of the centralised PC-based architecture working at the operator level. These improvements with respect to the 2nd generation AVS systems will allow the evolution of the AVS systems toward multi-camera and multi-function systems and the discharge of low-level image processing tasks from the central HW/SW processing architectures working in remote. In this way, it will be possible to improve the higher level understanding capabilities of the central processing architecture and therefore of the entire system. Moreover, the transmission of visual information already processed in remote is fully complying with the restrictive normative on privacy. Even though a malicious intruder succeeded in intercepting information such the one shown in Figures 5, he could never exploit it for illegal use. Future developments of the present work, will consider further steps of intelligence transferring from the operator level to the sensor level, in order to provide a further reduction of the amount of information to be transmitted in remote, enhancing both the security and the understanding potentialities of the proposed AVS architecture.

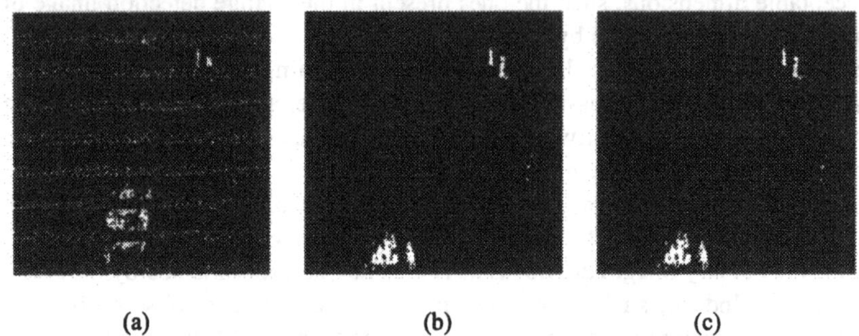

(a) (b) (c)

Figure 5. (a) Received and decoded binary image of difference obtained by the wireless transmission with spreading factor N = 7, (b) received and decoded binary image of difference obtained by the wireless transmission with spreading factor N = 15, (c) received and decoded binary image transmission obtained by the wireless transmission with spreading factor N = 63

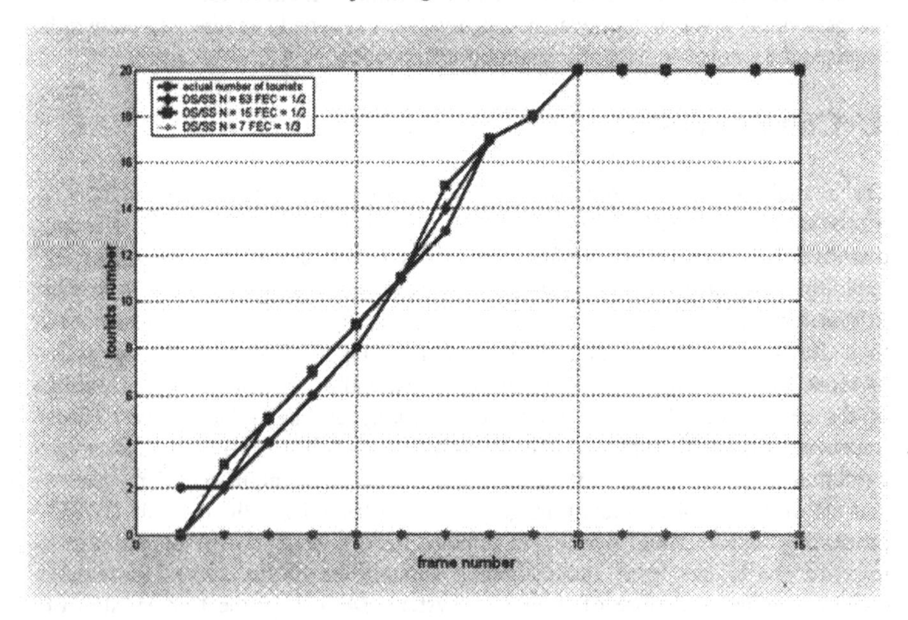

Figure 6. People counting results provided by a remote AVS architecture with low-level distributed intelligence (3rd generation AVS system)

5. References

[1] C.S. Regazzoni, "Synergic image transmission, processing and understanding in AVS systems", *IEE BMVA Workshop*, invited lecture, London (UK), 8th March 2000, available on the web at: http://www.dibe.unige.it/department/imm/ISIP/events.html.

[2] P. Mahonen, "Integration of Wireless Networks and AVS", in: *Advanced Video Based Surveillance Systems*, C.S. Regazzoni, G. Vernazza and G. Fabri, eds., Kluwer Academic Publishers, Norwell (MA): 1999, Chapter 4, pp. 144-153.

[3] C. Sacchi, C. S. Regazzoni, "A Distributed Surveillance System for Detection of Abandoned Objects in Unmanned Railway Environments", *IEEE Trans. on Vehicular Technology*, paper accepted, in press.

[4] E. A. Lee, "Programmable DSP Architectures: Part I", IEEE ASSP Magazine, October 1988, pp. 4-19.

[5] A. MacLeod et. al., "Applications of intelligent cameras", *Proceedings of the SPIE*, Vol. 1989, 1993, pp. 88-102.

[6] S. S. Shyne, "Distributed surveillance network utilises neural networks for stolen vehicle detection", *Proceedings of the SPIE*, Vol. 2938, 1997, pp. 186-190.

[7] C. Sacchi, C.S. Regazzoni, and C. Dambra, "Remote Cable-based Video-surveillance applications: the AVS-RIO project", *Proc.10th Internat. Conf. on Image Analysis and Processing (ICIAP99)*, Venice (I), 27-29 September 1999, pp. 1214-1215.

[8] R. L. Pickholtz, D. L. Schilling, L. B. Milstein, "Theory of Spread-Spectrum Communications – A Tutorial", *IEEE Trans. on Comm.*, Vol. Com – 30, No. 5, May 1982, pp.856-884.

[9] J. G. Proakis, *"Digital Communications"*, 3rd Edition, McGraw-Hill International Editions, New York: 1995.

[10] C. Regazzoni, C. Sacchi, et. al., *"AVS-RIO project: Public Final Report"*, Tech. Rep. University of Genoa, Department of Biophysical and Electronic Engineering (DIBE), January 2000, http available at: http://spt.dibe.unige.it/ISIP/Projects/avsrio.html.

[11] L. Marcenaro, G. Gera, C. S. Regazzoni, "Adaptive change detection approach for object detection in outdoor scenes under variable speed illumination changes", *European Signal Processing Conference 2000 (EUSIPCO 2000)*, Tampere (SF), paper accepted, in press.

[12] A. Tesei, A. Teschioni, C.S. Regazzoni, and G. Vernazza, Long memory matching of interacting complex objects from real image sequences, in: *Time Varying Image Processing and Moving Object Recognition, 4*, V. Cappellini (ed.), Elsevier, 1997, pp. 283-288.

[13] F. L. Luo and R. Ubenhauen, *"Applied Neural Networks for Signal Processing"*, Cambridge University Press, Cambridge (UK), 1997.

[14] E. Biglieri, J. Proakis, S. Shamai. "Fading channels: Information-Theoretic and Communications aspects". *IEEE Trans. on Information Theory*, Vol. 44, No. 6, , October 1998, pp. 2619-2692.

[15] *"IEEE 802.11 Std., Part 11: Wireless LAN Medium Access Control (MAC) and Physical layer (PHY) specifications"*, edited by IEEE Standard Board, approved 26 June 1997.

[16] P. Piccardo, C. Regazzoni, C. Sacchi, G. Sciani, e A. Teschioni, "Software Design and Simulation of a DS/CDMA Multimedia Transmission System for Remote Video-Surveillance Applications", in: *Multimedia Communications*, pp. 557-568 F. De Natale and S. Pupolin eds., Springer, 1999.

[17] S. Srikanteswara, J. H. Reed, P. Athanas, and R. Boyle, "A Soft Radio Architecture for Reconfigurable Platforms", *IEEE Communication Magazine*, February 2000, pp. 140-147.

[18] *"DataSat™ technical information"*, edited by SIRIUS corporation (B), http available at: www.siriuscomm.com.

[19] M. Tarkiainen, I. Niva, P. Kemppainen, "Performance of WCDMA system with space diversity and power control in slowly fading channel models", *Proc. of 1999 IEEE Wireless Communications and Networking Conference (WCNC)*, 21-24 September, 1999 New Orleans (USA), Vol. 2, pp. 674-678.

Digital Receiver Architecture for Multi-Standard Software Defined Radios

N. Benvenuto, G.A. Mian and F. Momola

Università di Padova, Dipartimento di Elettronica e Informatica, Via Gradenigo 6/A
I-35131 Padova, Italy

Abstract: Multi-standard software-defined radios, which are capable of operation according to a variety of different mobile radio standards, represent an extremely powerful tool for evolution towards the next generation of cellular mobile radio systems. This is particularly true with the variety of emerging UMTS air-interfaces which need to be accompanied with some degree of backward compatibility with the well established GSM system. This paper examines a number of architectural issues involved in the design of receivers for wideband multi-standard digital radios. In particular, an examination of the filtering and analog-to-digital converter technology requirements for their implementation is presented[1].

1. Introduction

The mobile communication industry is testifying the parallel emergence of a wide variety of radio standards throughout the world. This has resulted in a growing interest on multi-standard software-defined radios (SDR) [1], whose goal is to process signals corresponding to a wide range of frequency bands, channel bandwidths and modulation schemes.

A primary criterion in the design of a multi-standard radio receiver is the digital implementation of as much signal processing functions as possible, in the attempt to extend the digital boundary further towards the antenna. These functionalities include operations such as channel-selection filtering, frequency conversion and power control, which have been traditionally performed in the analog domain [2].

A second design criterion is the exclusively software-based implementation of the data demodulation, i.e., digital filtering, equalization and data detection.

A very short list of standards supported by a SDR implementation is presented in Table 1, where f_p is the passband frequency and f_s the stopband frequency of a radio channel (baseband equivalent).

The structure of a "general" SDR is sketched in Fig. 1 and includes an analog section, where a wideband signal $x(t)$ (about 2 MHz) is selected [10]. The signal $x(t)$ then enters an analog-to digital converter (ADC) whose output $y(nT)$ is further filtered to select the narrow-band channel of interest.

Two approaches of the ADC structure are possible [3]:

[1] Work carried out with the contribution of Telit S.p.A., Sgonico (TS), Italy.

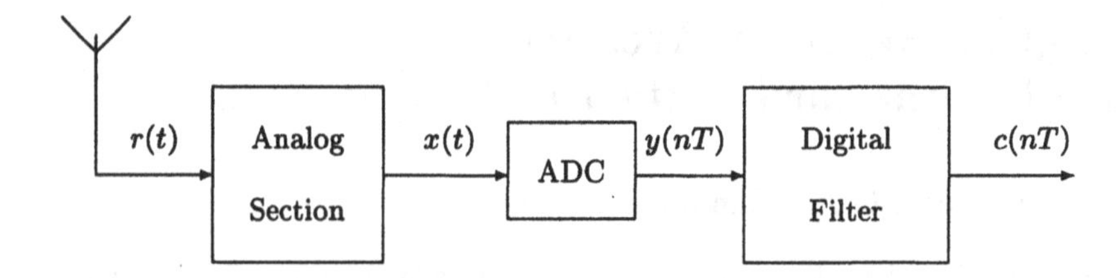

Fig. 1. General receiver structure.

1. *Full Band Digitization:* the whole bandwidth of the SDR service is digitized, i.e., all channels of all standards to be supported. This bandwidth can easily extend to some 100 MHz. Taking into account the radio interferers characteristics, the dynamic range of the ADC should be higher than 100 dB. Hence, although it is the most general one, due to power consumption and cost, this solution cannot be realized with today technology.
2. *Partial Band Digitization:* this approach uses a digitization bandwith equal to the widest channel bandwidth of the supported standards; in the case of Table 1, the required bandwidth is approximatly 2 MHz.

Since our intention is to present a SDR architecture realizable with today technology, we will focus on the second approach and demand to the analog section the task of band limiting the signal to approximately 2 MHz.

Table 1. Standards supported.

PARAMETERS	DECT	GSM	IS-54
f_p (kHz)	576	100	12
f_s (kHz)	1728	200	30

2. Received Signal Characteristics

The wireless communications environment, especially in urban areas, imposes severe constraints upon transceiver design, the most important being the limited spectrum allocated to each user (e.g. 30 kHz in IS-54 and 200 kHz in GSM) and the disturbance due to interference.

Here we are interested mainly on the interference, because it impacts the design of both the RF section and the digital front-end of a SDR.

As an illustrative example of a possible received signal composition $r(t)$, we will refer to the worst case scenario of the GSM system; in Fig. 2 the minimum GSM received signal power spectral density level (solid) is sketched together with the possible interference level (dashed) [3]. We note that within a bandwidth of 2 MHz, the power spectral density of the narrow-band channel of interest may be up to 70 dB under the level of interference. The whole signal of 2 MHz bandwidth to be digitized will be called ***wideband signal.***

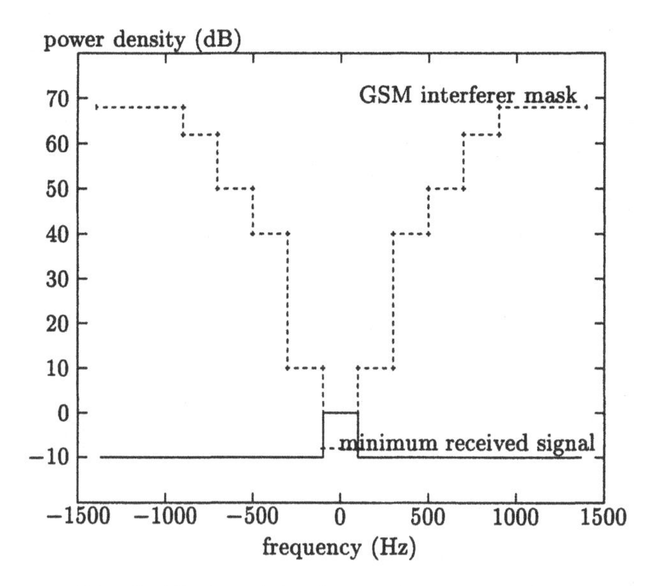

Fig. 2. GSM-mask in the worst case scenario.

3. A/D Converter Architecture

3.1 Converter Requirements

Classical solutions proposed for the A/D conversion in SDR consist of 2nd order Σ-Δ converters, i.e., an oversampled Σ-Δ modulator followed by a decimation filter [3,6]. The high SNR given by this configuration is due to the fact that the modulator shapes the quantization noise out of the narrow-band channel of interest and the noise is partially removed by a digital filter. The weakness of this modulator is that it does not introduce any kind of shaping on the wideband input signal: in fact for a 2nd order Σ-Δ modulator the signal transfer function is simply a unit delay. Hence, taking into account that the useful radio signal over the bandwidth of interest (about 2 MHz) is affected by interference (see Fig. 2 for the GSM case), the requirements on the decimation filter may be quite high. In other words, on top of the quantization noise, also interference has to be removed by the decimation filter.

Based on the desired narrow-band signal of interest and interference mask, it would be useful to reduce interference levels already within the Σ-Δ modulator (SDM), by introducing for example a partial filtering in the ADC.

The general solution for a SDR consists in the synthesis of a converter for each standard supported, nevertheless using always the same architecture and making it "programmable".

3.2 Converter Linear Model

As in the case of the Σ-Δ, the converter is composed of a modulator followed by a low pass filter. The modulator, whose model is sketched in Fig. 3 [7,8], consists of a single-bit quantizer and a 4th-order linear loop filter, working at a sampling frequency $F = 20$ MHz.

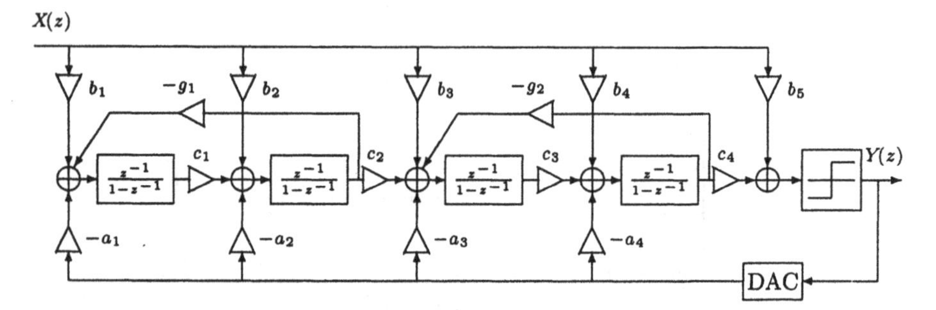

Fig. 3. Discrete time model of a $4-th$ order Σ-Δ modulator.

For analysis purposes, we assume that the quantizer of Fig. 3 can be modelled as an additive noise source. Hence the model becomes linear and the modulator output can be described as the sum of the input signal, $X(z)$, multiplied by signal transfer function $H_X(z)$, and the quantizer noise, $N_q(z)$, multiplied by the noise transfer function $H_N(z)$. The structure is a two input, one output, linear system described by the following four matrices, $\{A, B, C, D\}$:

$$\begin{aligned} \mathbf{s}(n+1) &= A\mathbf{s}(n) + B \begin{bmatrix} x(n) \\ y(n) \end{bmatrix} \\ y_{NQ}(n+1) &= C\mathbf{s}(n) + D \begin{bmatrix} x(n) \\ y(n) \end{bmatrix} . \end{aligned} \tag{1}$$

In (1) $\mathbf{s}$ is the state vector whose components are the four accumulator outputs, while y_{NQ} represents the output of the system before quantization. The choice of $H_X(z)$ and $H_N(z)$ should be tailored to the particular standard under consideration and it determines the coefficients a_i, b_i, c_i and g_i of the system.

From (1) it turns out:

$$Y_{NQ}(z) = [C(zI - A)^{-1}B + D] \begin{bmatrix} X(z) \\ Y_{NQ}(z) + N_q(z) \end{bmatrix} . \tag{2}$$

and

$$H_X(z) = Y_{NQ}(z)/X(z) \qquad \text{for } N_q(z) = 0, \tag{3}$$

$$H_N(z) = Y_{NQ}(z)/N_q(z) \qquad \text{for } X(z) = 0. \tag{4}$$

At first we assume $c_i = 1 \ \forall i$ and $g_i = 0 \ \forall i$. Hence, it is:

$$H_X(z) = \frac{\sum_{i=1}^{5} b_i(z-1)^{i-1}}{(z-1)^4 + \sum_{i=1}^{4} a_i(z-1)^{i-1}} \tag{5}$$

$$H_N(z) = \frac{(z-1)^4}{(z-1)^4 + \sum_{i=1}^{4} a_i(z-1)^{i-1}} \tag{6}$$

The linearization of the modulator scheme has reduced the modulator design to a linear filter design problem, and we should find the best trade-off in the design of the two transfer functions $H_X(z)$ and $H_N(z)$. We note that both transfer functions have the same poles, while the zeros of $H_N(z)$ are all at $z = 1$ and the zeros of $H_X(z)$ are free parameters.

The noise transfer function is the most important modulator parameter since it is responsible for minimizing the in-band noise. Let $\tilde{H}(f) = H(e^{j2\pi fT})$, and let *in-band* denotes the frequency interval $[0, f_p]$ and *out-band* the frequency interval $[f_s, F/2]$.

As for the noise transfer function, it should have a large in-band attenuation, i.e.,

$$|\tilde{H}_N(f)| \simeq 0, \qquad in - band. \tag{7}$$

$H_N(z)$ must also produce a working modulator, and so it must meet other design constraints. For example, for modulator stability it is advisable to set an upper value of 2 to the out of band gain of $H_N(z)$[4]. In our case we maintained a safety margin and set the constraint

$$|\tilde{H}_N(f)| \leq 1.5, \qquad out - band. \tag{8}$$

Another constraint is realizability, which is equivalent to the condition

$$|H_N(\infty)| = 1. \tag{9}$$

The signal transfer function spectrally shapes the input signal and it should have a unity in–band gain, i.e.,

$$|\tilde{H}_X(f)| \simeq 1, \qquad in - band. \tag{10}$$

Moreover, it is desirable to introduce some filtering already in the ADC process; hence it should be

$$|\tilde{H}_X(f)| < 1, \qquad out - band. \tag{11}$$

3.3 Design Procedure

The design is split into two steps:

1. firstly, the poles of $H_N(z)$ (hence of $H_X(z)$) are determined by using, e.g., the design optimization tool [11], while the zeros of $H_N(z)$ are all set at $z = 1$ ($g_i = 0\ \forall i$);
2. the zeros of $H_X(z)$ are selected to obtain a low-pass behaviour by using the following cost-function:

$$J = \gamma \left[\int_0^{f_p} \left(|\tilde{H}_X(f)| - 1 \right)^2 df \right]^{\frac{1}{2}} + (1-\gamma) \left[\int_{f_s}^{F/2} |\tilde{H}_X(f)|^2 df \right]^{\frac{1}{2}}, \quad (12)$$

where γ is a weighting parameter.

Such design procedure was used to synthesize the Σ-Δ modulator of Fig. 3 for GSM. The corresponding magnitude responses of the noise and signal transfer functions are shown in Fig. 4. The in-band attenuation of the noise level is more than 62 dB. At the same time, as apparent from Fig. 4, the modulator introduces significant interferers attenuation (more than 15 dB) in $\tilde{H}_X(f)$; moreover the $\tilde{H}_X(f)$ response exibits a transmission zero, which helps rejecting adjacent interferers.

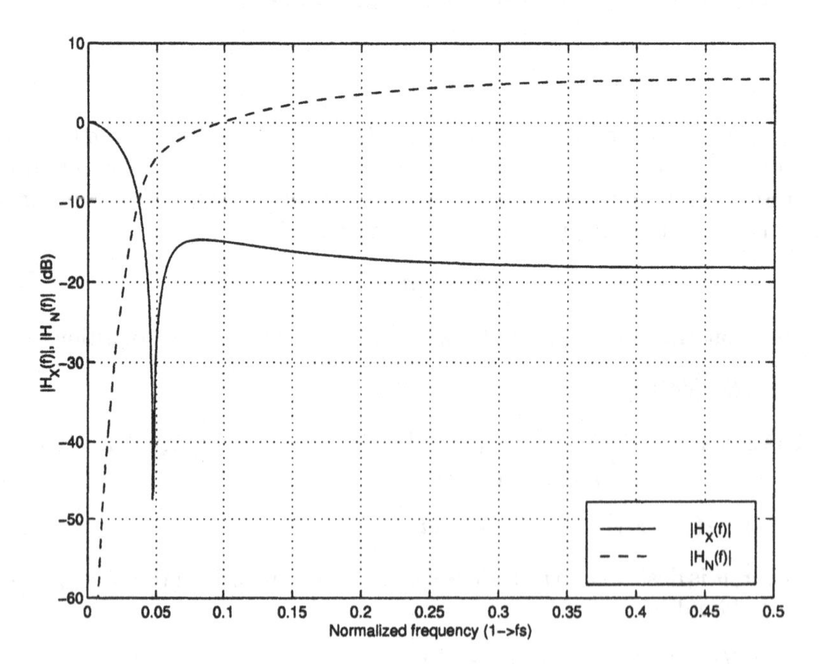

Fig. 4.: GSM fourth order SDM magnitude responses of the noise and signal transfer functions.

At this point it is important to see how the structure works with the other standards of Tab. 1. Firstly, the choice $g_i = 0$ $\forall i$ is acceptable also for the IS-54 signal. In fact both in GSM and IS-54 the normalized band of interest is very small and it is useless to optimize the zeros position of $H_N(z)$. However, in the DECT case, whose bandwidth is quite high (see Tab. 1), the situation is different and there is the need of optimizing the zeros position of $H_N(z)$; otherwise we are unable to shape the quantization noise out of the useful channel. Hence we should re-introduce the $\{g_i\}$ coefficients. Fortunately, in this case the best choice for $H_X(z)$ is to set the zeros in the same positions of poles and obtain an allpass function for the input signal. In fact there are no interferers, since they are already been totally rejected by the analog section. Overall the modulator design becomes very simple and only step 1 should be used.

4. Digital Front-end Architecture

We have to solve, now, the problem of low-pass filtering the output of the SDM. Considering that an SDR should be able to cope with signals of different radio standards, there is a need for sample rate adaptation between the digitization rate and the various standards symbol rates. In this section we will present an architecture that allows a flexible design of the decimation filters [9].

4.1 Decimation Filters Structure

For the three standards of interest of Tab. 1, the digital filtering architecture appears as in Fig. 5, where the input signal is the output of the SDM, at a sampling rate $F = 20$ MHz (with the use of a 4th order modulator). The structure provides the correct decimation necessary for the SDM to work correctly and also for narrowband channel selection. In effect it is a multistage realization of filters needed to choose the narrowband channel of interest; the global filter for a given channel (for instance GSM) is obtained as the cascade of a chain of decimation filters, that progressively remove out of band signal, followed by a post filter that precisely delimits the channel of interest and attenuates interferers.

In the structure, the task of the main chain is to provide a series of "exit points", where the desired signal is sufficently decimated with respect to the input wideband signal and almost frequency shaped. The channel of interest is extracted by means of a final post filter.

The main advantage of this structure is its extreme modularity: one can choose the standards of interest and determine the right "exit point" in the chain; another advantage is the possibility of optimizing the various filters with the goal of a computationally efficient realization.

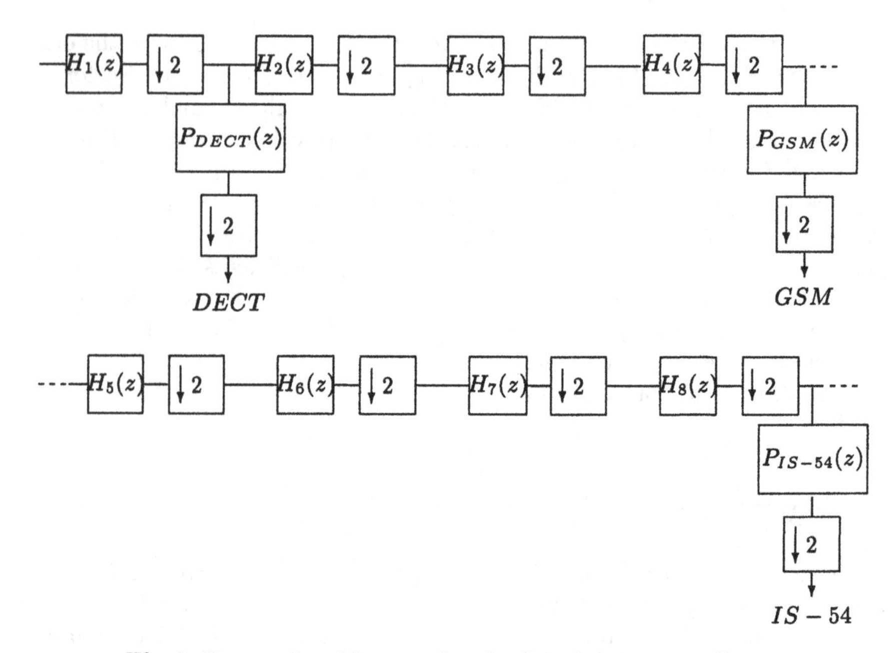

Fig. 5. Proposed architecture for the digital decimation filters.

4.2 Filter Design

The design of the general structure of Fig. 5 consists in determining the parameters of linear phase FIR filters $H_i(z)$ and $P_i(z)$, i.e., their passband ripple ($\delta_{p,i}$), stopband attenuation ($\delta_{s,i}$) and transition band ($B_{t,i}$). The order N_i of a low-pass linear phase FIR filter is approximately given by:

$$N_i \simeq -\frac{10\log_{10}(\delta_{p,i}\delta_{s,i}) + 13}{14.6\Delta f_i}, \tag{13}$$

where Δf_i is the transition band normalized to the sampling frequency F_i. The filters parameters must be chosen with the goal of mimimizing the global computational cost given by $\sum_i N_i F_i$.
The key optimization principle is that the first filters in Fig. 5 are computationally more expensive than the last stages filters because they work at an higher sampling rate. Therefore filters should have lower order in the first (high rate) stages, allowing if necessary an higher order for (low rate) next stages and for post-filters.

Passband Ripple Distribution. Since in a cascade of digital filters the passband ripples sum, there are a lot of possibilities for the passband ripple distribution; nevertheless, the best should be in accordance with the optimization principle stated in the previous section that allows higher order filters only in last stages of the main chain. In this case, since the filter order raises with the precision required in passband, the passband error distribution should permit great ripples in first stages and small ripples in last stages, while assuring for the output signal an overall ripple of $-20\log_{10}(\delta_p) = 0.2$ dB.

The choice for the passband ripple $\delta_{p,i}$ of filter $H_i(z)$ that minimizes the overall computational cost, is

$$\delta_{p,i} = \frac{\delta_p}{2^i}. \tag{14}$$

where 2^i is the overall decimation factor of the signal at stage i with respect to the input.

It is worthnoticing that, at stage K where we extract the signal for the final post filtering, the total ripple is given by

$$\sum_{i=1}^{K} \delta_{p,i} = \sum_{i=1}^{K} \frac{\delta_p}{2^i} = \delta_p - \frac{\delta_p}{2^K}. \tag{15}$$

Hence the ripple allowed for the K-th post filter is $\delta_{p,K} = \delta_p/2^K$.

Stopband Attenuation Choice. The choice of stopband attenuation depends on the accuracy required for the output signal and on the adopted ADC. We observe only that for an out-of-band attenuation of $-20\log_{10}\delta_s$, globally the filters that the signal encounters from input to output should yield an attenuation not less than δ_s. In the case of the fourth order SDM we impose $-20\log_{10}(\delta_s) = -80$ dB.

Transition Bands Optimization. The knowledge of the structure of output channels allows for further optimization of filters orders by extending transition bands.

For post filters $P_i(z)$ there is no much we can do: their transition bands are imposed by radio standards as in Tab. 1. However we can optimize the transitions bands of the main chain filters $H_i(z)$. Unfortunatly there is a limit due to aliasing. The algorithm that maximizes Δf_i for $H_i(z)$ is sketched in Fig. 6. Let J be the number of the output channels of interest and let $f_{p,j}$ and $f_{s,j}$, $1 \leq j \leq J$, be the passband and stopband frequencies of output channel j (see Tab. 1). In the design of the i-th filter, H_i, its cutoff frequencies F_{p,H_i} and F_{s,H_i} are chosen as solutions to the problem:

$$F_{p,H_i} = max\left(f_{p,j} \mid i \leq j \leq J\right) \tag{16}$$

$$F_{s,H_i} = max\left(f \mid \left(\frac{F}{2} - f \geq f_{s,j}\right) \ , i \leq j \leq J\right) \tag{17}$$

In this way the aliasing introduced by successive decimations will not fall into the useful band of the narrowband channels of interest.

For the three standards of Tab. 1 the resulting structure needs about $20\,F$ op/s, while three structures, each taylored to one standard, overall would require about $18\,F$ op/s.

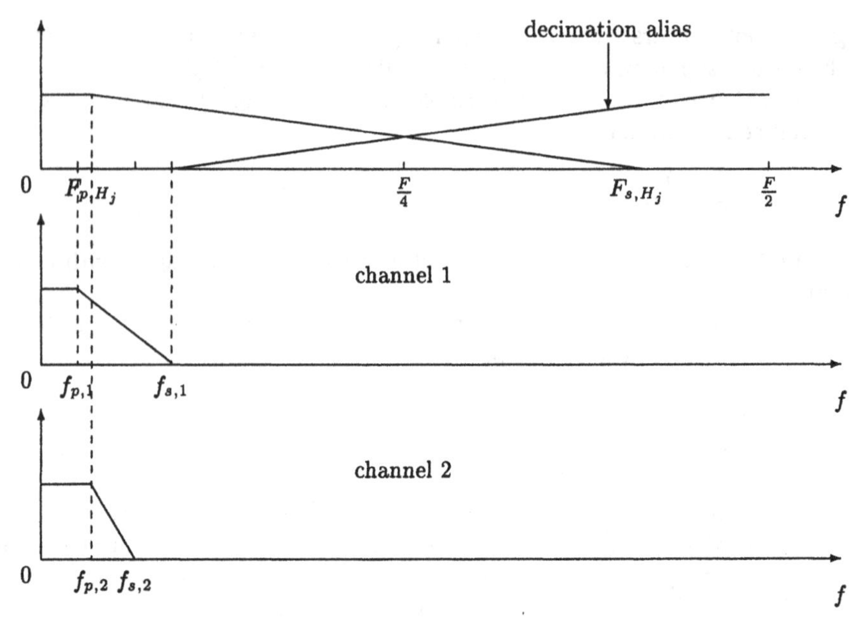

Fig. 6. Transition band maximization.

5. Results

To test the proposed receiver, and to choose its final structure, we resorted to a simulation of ADC and filters bank behavior with a tipical GSM signal as the input signal.

In Fig. 7 the simulation structure is sketched.

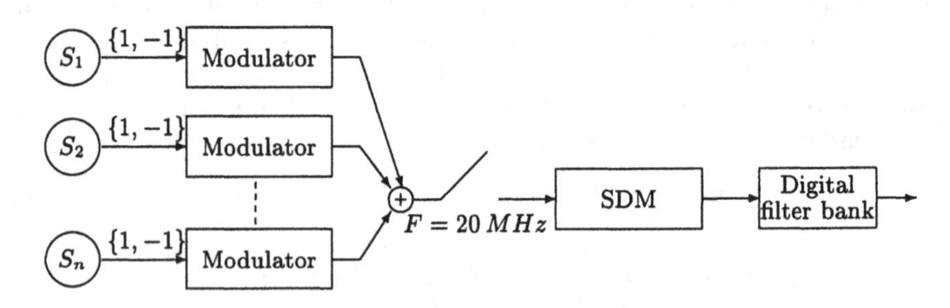

Fig. 7. Simulation configuration.

Each source generates random symbols in the alphabet $A = \{-1, 1\}$; these symbols form the input a GMSK modulator that produces a GSM signal [5]. S_1 is the source of the useful channel, the other sources being interferers that are amplified before the summing node to reflect the worst case scenario of Fig. 2 over a band of 2 MHz.

The global signal enters the SDM and the digital filter structure described in the previous section. The output narrowband signal is analyzed.

Fig. 8 shows the power spectrum of the input signal $v_r(t)$ and the spectrum of the signals at the output of different SDM structures with 1-bit quantizer. In particular

- $y_2(nT)$ is the output signal of a classical 2nd order SDM;
- $y_3(nT)$ is the output signal of a 3rd order SDM;
- $y_4(nT)$ is the output signal of a 4th order SDM (Fig. 3);

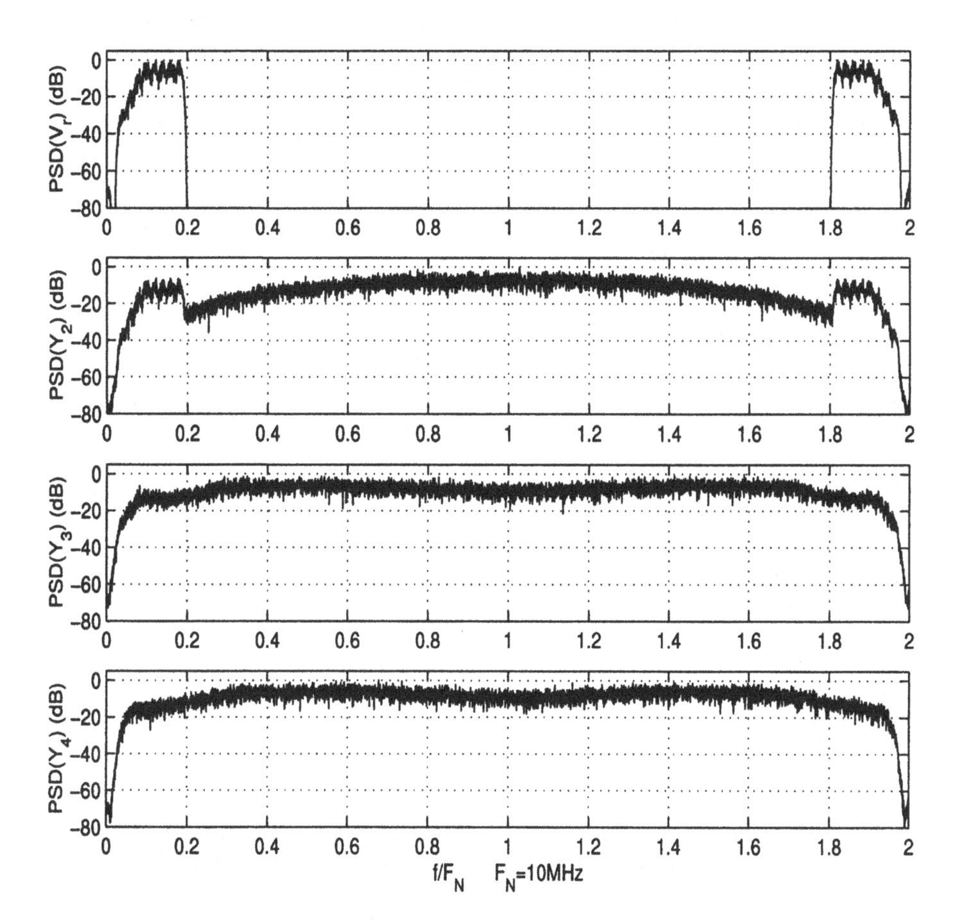

Fig. 8.: Power spectral density of the input signal and of its quantized versions. $F_N = F/2$ is the Nyquist frequency.

Let us note that the useful signal is around DC and has a very small amplitude. From this figures it is apparent the usefulness of the 4th order scheme on attenuating interferers. Moreover, experimental results (not reported for space limitations) show that the proposed digital structure works correctly, and represents a good architecture with any SDM.

6. Conclusions

The final structure of the proposed receiver is composed of three parts:

1. the analog front end consists of a wideband IF with double conversion, which selects a signal with a bandwidth of about 2 MHz;
2. the A/D interface is constituted by a 4th order 1-bit SDM, operating at a sampling rate of 20 MHz, with programmable coefficients. This structure shapes the quantization noise out of the useful narrow band and may prefilter interferers in the 2 MHz band;
3. the digital decimation chain of Fig. 5, tries to solve two problems: decimation of the oversampled signal of the $\Sigma - \Delta$ modulator, and output channel selection.

We observe that the structure is very modular, and allows the introduction of many radio standards. Moreover it is computationally efficient, because optimization can be introduced in all phases of the project.

Simulation has been performed for a GSM system, with good results in terms of signal to noise ratio at the output.

7. References

[1] N.J. Drew, P. Tottle, "IC Technologies and Architectures to Support the Implementation Of Software Defined Radio Terminals", 3rd ACTS Mobile Communications Summit, Rodos, Greece, pp. 38-44, June 1998.

[2] N.W. Anderson, H.R. Karimi, P. Mangold, M. Wezelenburg, "Software Definable Implementation of a Dual Mode TD-CDMA/DCS1800 Transceiver", 3rd ACTS Mobile Communications Summit, Rodos, Greece, pp. 52-59, June 1998.

[3] T. Hentschel, G. Fettweis, M. Bronzel, "Channelization and Sample Rate Adaptation in Software Radio Terminals", 3rd ACTS Mobile Communications Summit, Rodos, Greece, pp. 121-126, June 1998.

[4] S. Janzi, C. Ouslis, A. Sedra, "Transfer function Design for Delta-Sigma Converters", ISCS , pp. 433-436, 1994.

[5] M. Rahnema "Overview of the GSM System and Protocol Architecture", *IEEE Communications Magazine*, vol. 31, pp. 92-100, April 1993.

[6] P. M. Azizi, H.V. Sorensen, J. Van der Spiegel, "An Overview of Sigma-Delta Converters", *IEEE Signal Processing Magazine*, pp. 61-84, Jan. 1996.

[7] K.C.H. Chao, S. Nadeem, W.L. Lee, C.G. Sodini, "A Higher Order Topology for Interpolative Modulators for Oversampling A/D Converters", *IEEE Transactions on Circuits and Systems*, vol. 37, N. 3, pp. 309-318, March 1990.

[8] P.F. Ferguson, A. Ganesan, R.W. Adams, "One Bit Higher Order Sigma-Delta A/D Converters", Proc. IEEE International Symposium on Circuits and Systems, New Orleans, pp. 890-893, May 1990.

[9] R.E. Crochiere and L.R. Rabiner, *"Multirate Digital Signal Processing"*, Prentice Hall, 1983.

[10] B. Razavi, *"RF Microelectronics"*, Prentice Hall, 1997.

[11] R. Schreier, The Delta-Sigma MATLAB Toolbox, July 1997 (ftp://next242.ece.orst.edu/pub/delsig.tar.z)

Part 3

Software Radio Architecture

SDR Architecture for US Tactical Radios

J. Mitola III

The MITRE Corporation, McLean, VA (jmitola@mitre.org)

Abstract: Software-defined radio (SDR) has emerged as a focus of academic research, commercial development, and military investment for future wireless systems. This paper describes architecture migration for military SDR. It characterizes the tradeoffs among core software-radio technologies including wideband RF hardware, a flexible digital signal processing (DSP) platform, and software architecture. Object-oriented analysis by industry participants defined architectures for the Joint Tactical Radio System (JTRS) and the SDR Forum (the "reference architectures"). This paper analyzes those architectures in terms of layered virtual machines. The goal is to clarify technical aspects that cost-effectively support simultaneous hardware and software evolution. Research issues include layering, software objects and virtual machines. Potential benefits are pointed out along with research, development, and deployment issues that must be addressed in order to meet warfighter needs.
Keywords: Software radio, SDR, architecture.

1. Background

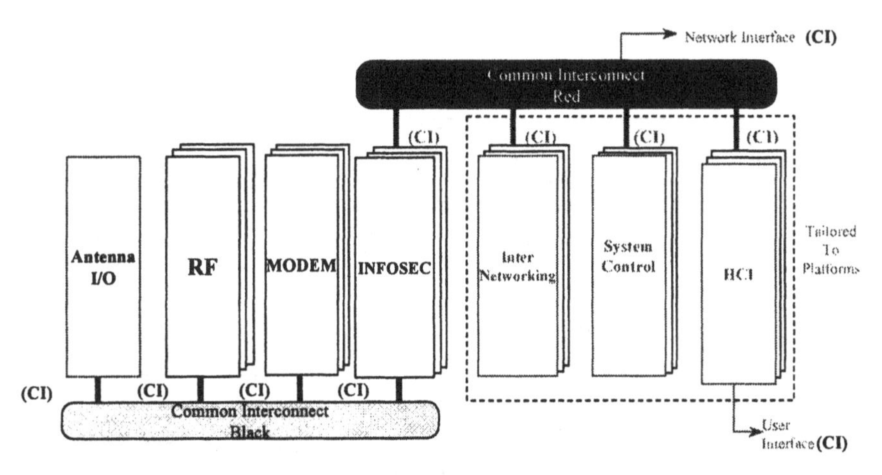

Figure 1. PMCS Entity Reference Model

The Programmable Modular Communications System (PMCS) integrated process team recommended that the US DoD replace its 200 families of aging discrete radio systems with a single family of SDRs. The PMCS guidance document outlined an architecture that included a list of radio functions, categories of hardware and software components ("entities"), and design rules [1]. The PMCS entity reference model is illustrated in Figure 1.

This architecture represents the air interface personality of a radio (a "waveform" in military jargon) as a specific set of six functional entities with interfaces defined by this architecture reference mode. This model was adopted with slight modification by the global Software-Defined Radio (SDR) Forum in its [MMITS] Functional Interface Diagram [2]. Industry has called this a fine-granularity architecture because it requires the definition of ten or more interfaces among the eight functional entities of the reference model.

At present, the SDR Forum has not agreed to these interfaces to the level of detail of an interface control document (ICD), nor has PMCS' successor, the JTRS Joint Program Office (JPO). Instead, attention has focused on defining common infrastructure software ("middleware") that would allow any set of such entities to be hosted on a variety of hardware-software platforms. Since about 1/3 of the software of SPEAKeasy I and II consisted of such infrastructure code, the focus on middleware has high leverage. This focus occurred in part because of industry consensus that the desired portability of plug-and-play components requires interfaces between the radio entities and the radio platform.

2. An Integrated View

To appreciate the reasons for the focus on middleware better, consider the distributed layered virtual machine reference model of Figure 2. This provides an integrated view of the technologies needed to achieve open-architecture plug-and-play for SDR. It consists of six layers.

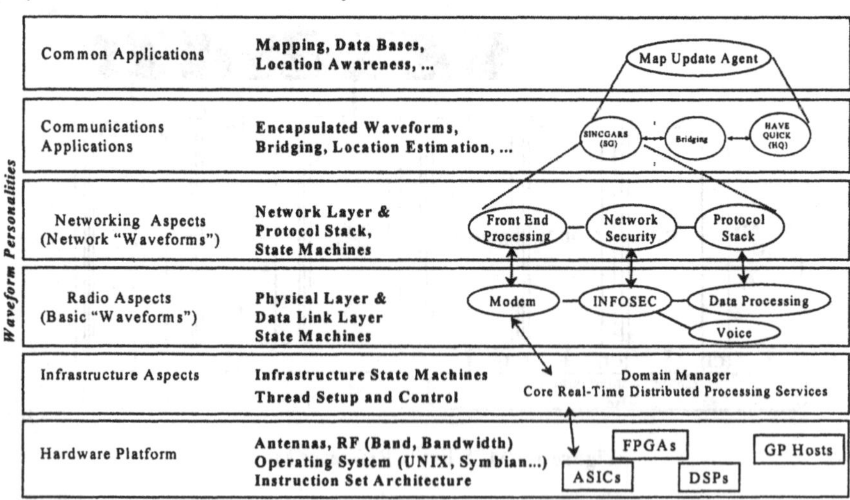

Figure 2. The Distributed Layered Virtual Machine Reference Model Relates Common Applications to Open-Architecture SDR Communications

The bottom layer, hardware platform, includes antennas, RF, IF, and software-programmable components such as ASICs, FPGAs, DSP chips and general purpose host processors. The platform may also include analog RF or IF processing, e.g., interference suppression. Each successively higher level of abstraction is differentiated by computational and topological properties [3].

The infrastructure-aspects layer provides common real-time distributed processing services. These include setting up and tearing down high performance pathways in the heterogeneous signal processing hardware for the flow of isochronous signals, such as digitized speech. System resources include software components for the four higher levels. These resources may be loaded, allocated to processors, initialized, and administered by a domain manager. The domain manager assures that computational and other constraints needed for successful communications applications are met, possibly using a constraint language [4]. The domain manager may also arbitrate over-the-air download of software modules.

The radio-aspects layer consists of software objects that employ infrastructure and platform capabilities to realize the programmable capabilities of an SDR on the physical and link layers of the protocol stack. Wireless media access (e.g., frequency, time, code and/or space-division multiple access), channel coding, power control, equalization, forward error control, and the like may be encapsulated in radio-aspect objects such as the modem.

The networking-aspect layer includes link management, host adaptation, and conventional network-layer to presentation-layer protocol stacks. Mobility support may include network probing, registration, mobile IP, multicast, and possibly dynamic or resource-discovery protocols [5]. Software objects that deal with networking aspects realize these higher levels of the protocol stack(s) on behalf of communications application. For example, a front-end data processing object may rely on modem support to route encrypted packets reliably in a dynamic network. In military jargon, a waveform is a consistent set of software objects that delivers radio- and network-aspect capabilities. Historically, waveforms have been implemented in monolithic hardware-software configurations. Industry is moving towards layering, however, because this is necessary for plug-and-play wireless.

The communications-applications layer, then, includes encapsulated radio applications built on the four lower layers, e.g., waveforms. Bridging allows the host radio node to act as a gateway between diverse networks (e.g., SINCGARS and HAVE QUICK). The software entity realizing this capability exists on the communications application layer. Such agents may synthesize Opportunity Driven Multiple Access (ODMA). Other communications-related services at this layer may include location finding and mode adaptation. In the commercial sector, location finding may be a byproduct of the radio network architecture for Emergency 911 reporting. Mode adaptation extracts a low data rate subset from a higher data rate stream and structures it in ways suitable for disadvantaged users, e.g., supporting a collaborative workspace conference call.

Finally, common applications (e.g., collaboration tools, maps, logistics, *etc.*) are assigned to the highest layer. The DoD's Common Operating Environment (COE) and commercial applications exist on this layer. At present, voice communications is not part of COE. Given the potential of voice over IP (VoIP), this seems likely to change soon. It may be advantageous, then, to define a common user interface to voice services that could be used for VoIP and SDR applications enabled by JTRS.

3. JTRS Middleware

The purpose of the JTRS program is to acquire a family of affordable, high capacity radios for "interoperable C4ISR capabilities" [6]. C4ISR includes communications, command, control, computers, intelligence, surveillance, and reconnaissance. The scope of DoD interest includes all layers from applications to hardware platforms. During FY00, the emphasis of the JTRS JPO has been on defining an industry-based open architecture that will shape implementations as illustrated in Figure 3. The current emphasis of the JPO program is on infrastructure middleware so that future military radio needs may be met more affordably using an industry-standard plug-and-play architecture for waveform portability among fewer types of hardware platforms. This may be more affordable than fielding a unique radio hardware platform for each new waveform. The strategy of Figure 3 is based on defining classes of hardware and software to be developed and integrated according to design rules.

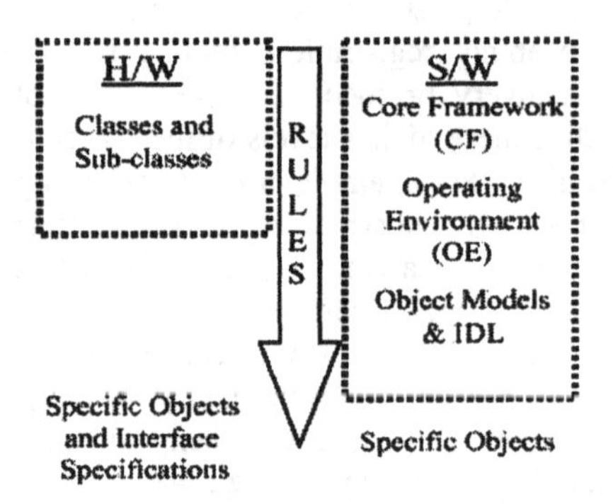

Figure 3. Architecture & Implementation [7]

The structure of the software is illustrated in Figure 4. At present, the architecture includes a core framework (CF), the infrastructure middleware intended to achieve the platform-independence necessary for waveform portability. Hardware and software entities below the CF are part of the host platform. The JTRS program, Step 2A is acquiring an architecture specification, open-standard interfaces to the CF, and four prototypes that implement the CF (among other things).

Since the CF is based on the Common Object Request Broker Architecture (CORBA), it employs the industry standard Interface Definition Language (IDL). The Step 2A contractor, the Modular Software-programmable Radio Consortium (MSRC), defined the CF using the Unified Modeling Language (UML) [8].

Consequently, the CORBA IDL for the CF was generated automatically by the Rational Rose software tool [9]. At present, the CF includes the capabilities listed in Table 1.

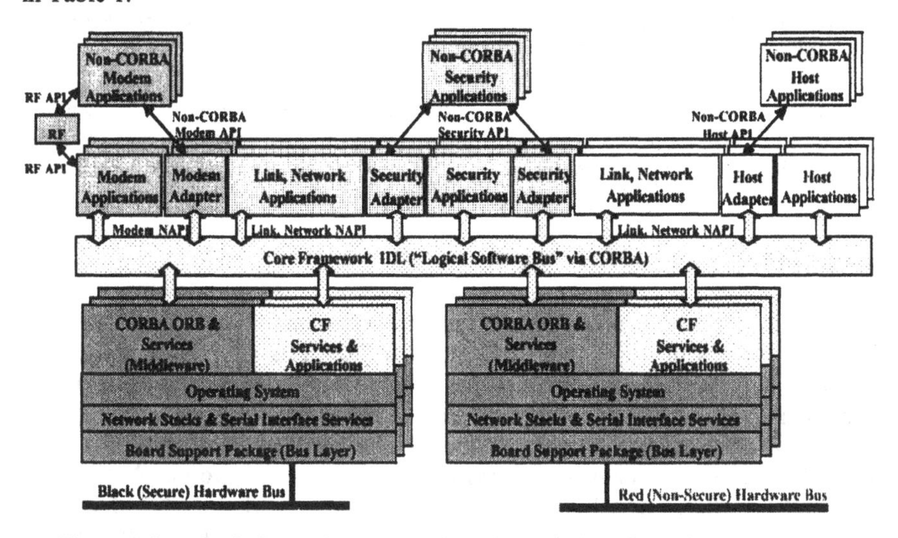

Figure 4. Step 2A Software Structure Isolates Radio Entities from Host Hardware [7]

Table 1. Major CF Interfaces (May 00)

Interface	Remarks/Highlights	Interface	Remarks/Highlights
Device	Struct holding any basic type	Testable Object	Interface to run self-test
Device Manager	Installs devices	Property Set	Configure, query component
Domain Manger	HCI, registration, *etc.*	Life Cycle	Initialize, release component
Application Factory	Create an application from a set of devices	Resource	Minimum interface for SW: start(), stop(), getPort(), *etc.*
Application	Control an application	Resource Factory	Create, release, or destroy a resource
File Manager	Mount, unmount file system	Port	Message passing interface
File System	Remote copy, remove, mkdir	File	Read, write like POSIX/C

Devices contain state and status information. The interface is based on the ITU X.731 state management function. Capacities from device profiles are checked and exceptions thrown if exceeded. The device manager/ resource manager manipulate defined resources that are registered with the Domain Manager.

The CF also defines Boolean, octet, character, double, float and long sequences. The string consumer and related facilities support publish-subscribe mechanisms. For example, the logger interface keeps track of system events and pushes messages to registered consumers.

Together, these CORBA interfaces provide the services needed to configure and run SDR software in a heterogeneous mix of processors that provide POSIX-compliant operating system facilities.

4. Hardware Platform Tradeoffs

DoD stakeholders have divided the acquisition programs into items of common interest and items that are unique to a "domain," such as the hardware platform suitable for airborne applications [10]. The JTRS JPO is addressing items of common interest by acquiring the architecture. The military services must acquire domain-specific hardware platforms. Classes of hardware platform may be derived by partitioning the required waveforms into narrowband waveforms (NBW) and wideband waveforms based on the instantaneous RF bandwidth (=< 3 MHz) and aggregate data rate (=< 1 Mbps). This yields about forty narrowband waveforms (e.g., SINCGARS, HAVE QUICK, EPLRS, Link 11, *etc.*) and a half dozen wideband waveforms (e.g., JTIDS, HCLOS, new high data rate (HDR) waveform, *etc.*). The narrowband waveforms, which comprise the majority of the requirements, are compatible with industry-standard radio platforms like the single superheterodyne radio with IF ADC illustrated in Figure 5. A given domain, such as ground vehicles, for example, could employ such a hardware platform for its narrowband needs. In such situations, the software capabilities of the CF may be fully employed in support of waveform portability among radio hardware platforms.

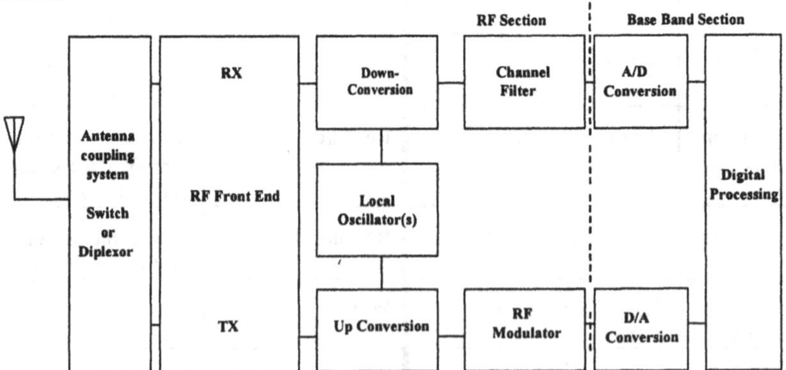

Figure 5. Illustrative Single Channel Narrowband Vehicular Radio Platform

The ground domain also needs HDR radios, e.g., Tactical Operations Center radio relays with aggregate data rates of 2 Mbps or more. With today's technology, such HDR implementations, even in vehicles where power is not as constrained as handsets, may require ASICs and/or FPGAs. Those acquiring the hardware platforms therefore have to trade off the higher data rates afforded by less programmable ASICs (or FPGAs) versus the greater flexibility of DSP or microprocessor-based digital signal processor pools. The alternatives to resolve this tradeoff include those listed in Table 2.

Table 2. Hardware Platform Alternatives

Alternative	**Remarks**
Separate HDR, NBW hardware platforms	Potential hardware platform proliferation
HDR hardware platform with NBW modes	High cost for NBW-only applications; excess NBW capacity; volume HDR production

The fielding of distinct HDR hardware platforms that support only the HDR waveform adds another radio platform type to the DoD inventory that now numbers near 200. If, however, the introduction of this platform facilitated the consolidation of a large number of NBR radios into a single standard NBR platform, then the economies of scale and reduced cost of ownership needed by the DoD might in part be realized. In addition, if the HDR platform included NBW compatibility, then the HDR platform could also serve as a gateway among NBW and HDR users.

5. Architecture Evolution

In the SDR Forum and MSRC architectures, the networking- and radio-aspect layers are not differentiated (see Figure 4). The entities perform a mix of physical layer and networking functions. The future partitioning of these entities into layers should facilitate the independent evolution of networking and radio access technology. The CF facilitates this process. The construction of these two layers using re-hostable radio objects based on the CF will also be beneficial. NBW applications of the CF achieve cross-platform portability, permitting DoD to consolidate current NBW platforms into a few types per domain. By focusing on the CF, industry and the JTRS JPO are focusing on the aspects of radio software with maximum leverage for reduced cost of ownership.

6. References

[1] Programmable Modular Communications System (PMCS) Guidance Document (Washington, DC: US Department of Defense) 31 July 97

[2] Modular Multifunction Information Transfer Systems (MMITS) Revision 0.9 MMITS Technical Report in Support of Standards Recommendations (Rome, NY: SDR Forum) June 1997

[3] J. Mitola, "Software Radio Architecture: A Mathematical Perspective" IEEE JSAC (NY: IEEE Press) April 99

[4] M. Adams "SDR Software Architecture: Managing Generic Resources" (Melbourne, FL: Exigent International) 25 February 1999
[5] C. E. Perkins, "Mobile-IP, Ad-hoc Networking, and Nomadicity" IEEE 0730-3157/96 (NY: IEEE Press) 1996
[6] JTRS Mission (www.jtrs.sarda.army.mil) April 00.
[7] *Software Communications Architecture Specification, MSRC-5000SCA V0.4* (Ft Wayne: MSRC Consortium) April 00.
[8] UML reference material and software tools (www.rational.com/uml) April 00.
[9] CORBA Tutorials and Specifications (www.omg.org) April 00.
[10] Operational Requirements Document (ORD) For Joint Tactical Radio (JTR) (Washington, DC: US DoD), 23 March 1998

A Multiband, Multirole and Multimode Suited Radio Architecture

Karlheinz Pensel

Radiocommunications Systems Division
ROHDE & SCHWARZ GmbH & Co. KG
Mühldorfstraße 15
D-81671 Munich
GERMANY
e-mail: Karlheinz.Pensel@rsd.rsd.de

Abstract: Software Radios strive for maximum flexibility in terms of multiband, multimode and multirole capabilities. In turn, the radio architecture has to show flexibility by a software driven modular structure that is able to cope with high computational power demands. The programmable Software Radio M3TR covering the complete frequency range from HF up to UHF and providing various fixed network or air interfaces sets a good example how to fulfill these requirements.

1. Introduction

High performance digital signal processing devices like DSPs, FPGAs and ASICs ensure increasing radio flexibility. On going technological advances are certain to reduce formerly required analog processing steps, like mixing, filtering to a minimum level. In the end, a radio platform will behave like a freely programmable computer, where any arbitrary waveform described by typical air interface parameters like modulation, carrier frequency, and any application based on can be mapped onto the specific radio platform through appropriate programming interfaces (cf. Figure 1).

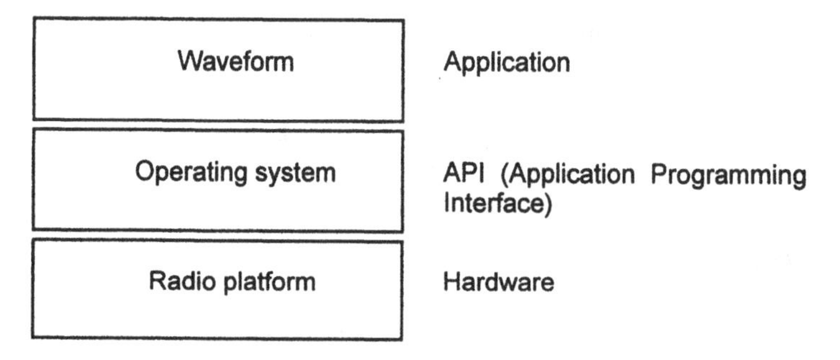

Figure 1. Horizontal architecture layering separates (software) applications strictly from the hardware platform

2. Architecture

A typical Software Radio architecture falls into various logical or physical components interconnected by Radio Control Buses (RCBs) as shown in Figure 2 [2],[6].

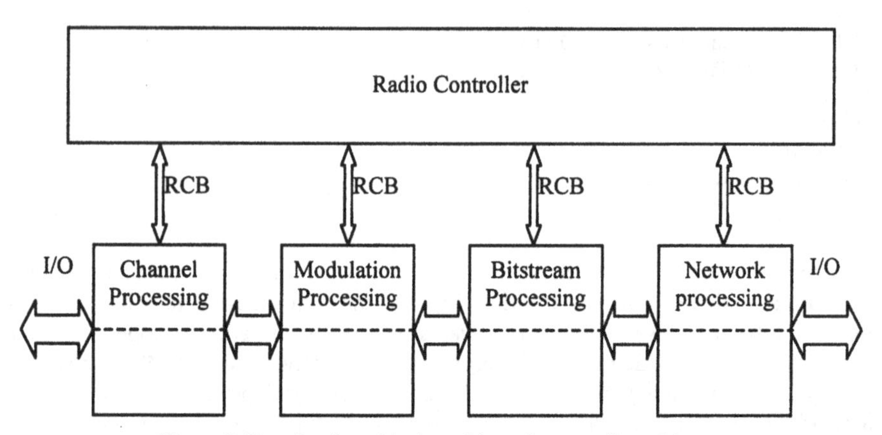

Figure 2. Functional partitioning of the software radio architecture

A channel processing module performs RF filtering, IF filtering, mixing, AGC, while the modulation processing module carries out e.g., modulation, demodulation, equalization, digitization, symbol tracking. In order to improve the BER encoding, decoding is added in the bitstream processing module. Network processing includes the Medium Access Control (MAC) functionalities, like acquisition and multiple access. Network processing also provides an interface to adjacent fixed networks like LAN or PSTN.

In each module, hardware and software are separated from each other by the horizontal layering in Figure 1. The software stored in PROMs and incarnated by DSPs and programmable FPGAs controls the main operating parameters and offers an extreme flexibility. The modular structure reduces requirements [7] in terms of computational speed and memory the modules have to provide. Well-defined and standardized interfaces between the modules prepare some kind of "open architecture", and make the radio platform scalable to e.g., handheld or base station applications. The modular structure therefore eases the trade-off between desired flexibility and hardware efforts (and costs).

3. Flexibility

The software radio flexibility falls into three individual domains, these are multiband, multimode and multirole operation.

As for multiband operation, a Software Radio should cover a maximal frequency range starting from HF (about 1.5 MHz) up to several GHz because of numberless reasons: The increasing demand for information exchange and its involving broadband waveforms are facing sharply limited frequency resources and are shifting applications to higher currently not used frequency ranges. Secondly, each country has its individual frequency assignment scheme. Finally, ITU refarming procedures place civil, commercial communication in formerly military occupied frequencies. GSM or the security service system TETRA for example cover a broad range of frequencies. Therefore, multiband operation is of extremely importance to cope with mobility requirements across international borders.

A software radio, however, must not place unrealistic demands on linearity, image rejection, dynamic range and interference reduction [3]. From the current technology point of view some analog preprocessing by mixing and filtering helps a lot to meet requirements in terms of power consumption and size [8].

Multimode operation requests the Software Radio to be compliant with various air interfaces. Throughout the whole world there exists a large number of different standards, e.g., GSM in Europe, IS-95 and AMPS in the US and there will be no convergence of wireless standards in the future due to political/commercial (see for UMTS) and technical reasons.

Multirole operation addresses the question, which roles a software radio has to play. Besides the capability to handle voice and data transmission as a so called multimedia terminal (required by UMTS [5]) does, a Software Radio has to answer to different operational scenarios particularly placed by security services. It has to be scalable from handheld with stringent power consumption requirements up to a base station with several communication lines in parallel. Besides typical hierarchical networks (between mobile and base) mobile radios should be able to establish links among themselves, like TETRA's direct mode. A Software Radio has to cope with lots of different access and network control schemes, introduced by advanced supplementary services (e.g., Access Priority, Dynamic Group Assignment, Late Entry, Remote Disable/Enable) or redundant, multi-hierarchy networks. That applies to fixed networks such as ISDN/PSTN, LAN, WAN and to moving networks on the air as well. A Software Radio should be able to establish point-to-point and point-to-multipoint links and to provide broadcast services. Consequently, there is a need to build radio families based on scalable platforms to meet the requirements in terms of size, weight power and functionality.

4. M3TR Features

The M3TR (Multimode Multirole Multiband Tactical Radio) of Rohde & Schwarz represents a completely new generation of high-performance Software radios. Contrary to conventional radios with fixed architecture the M3TR features maximum flexibility in terms of frequency bands, waveforms and functions satisfying the requirements of various user domains. M3TR is not restricted to military networks, but serves via loading the appropriate software also as a terminal in civilian PMR (Professional Mobile Radio) networks.

By forecasting technology trends the platform is designed in advance to cope with future applications, frequency ranges additional functions and future COTS products. Evolutionary updating of modules fully exploits the technological advance of semiconductors and keeps the radio up-to-date, an implementation of the ETSI standard TETRA for example is planned. In fact, software configurability and upgradability by Pre-Planned Product Improvement (P^3I) is a key asset of a modular hardware and software architecture in order to reduce technology refresh insertion time and to lower costs.

4.1 Multiband

The two manpack transceivers MR3000H and MR3000U providing seamless coverage of the transmission range from 1.5 MHz up to 108 MHz (model H) and from 25 MHz up to 512 MHz (model U) form the core of the M3TR transceiver family. In total, both units are designed for transmission and reception from 1.5 MHz to 512 MHz. So, with just two transceivers (MR3000H and MR3000U), the M3TR transceiver family covers the whole spectrum from short wave through to the UHF band.

4.2 Multimode

Thanks to optimized protocols and waveforms M3TR attains high data rates for digital voice, real-time video and visual display data. Beyond Line of Sight (BLOS), e.g., HF offers up to 5.4 kbps user rate per 3 kHz channel, while in the Line of Sight (LOS) case VHF/UHF provides up to 64 kbps per 25 kHz channel suited for real-time data, video and Internet / Intranet access via the radios integrated Ethernet interface. In command systems this ensures among other things automated data exchange, for example for online position display and distribution. P^3I (Pre-Planned Product Improvement) enables to subsequently integrate planned and future methods in the equipment through simple software upgrades.

Different communications standards exist even within NATO and new ones are still being prepared. Examples are HAVE QUICK I and II, the future SATURN for UHF or STANAG 4444 for the shortwave band. As a software-defined radio, M3TR can be made compatible with almost all existing EPM (Electronic Protection Measure) radios. It is interoperable with legacy communication systems and supports growth for new requirements.

4.3 Multirole

The use of open system standards, like TCP, Ethernet, and well defined interfaces within the radio makes M3TR scaleable to match the communication requirements of different users and furthermore extendible to support further growth and changes. M3TR radios are suited for all applications like manpack, for vehicular or stationary use. An optional switching unit provides interfaces for in principle all land-based communication networks. Connections to ISDN/PSTN, TCP/IP, UDP as well as to serial and optical interfaces of data terminals are provided.

Comprehensive multirole features allow its easy integration into communication networks, e.g., as a functional terminal in a subnet, e.g., CNR (Combat Net Radio: voice and data half-duplex transmission in combat networks) or PRN (Packet Radio Net: multi-hop functionality for packet data transmission, adaptive routing of messages in case of jamming or relocation). But M3TR can also act as an interface between the subnets, REN (Range Extension Node: for user voice and data services established among radios out of range). Playing the role of a RAP (Radio Access Point) it establishes the interface to fixed networks, e.g., ISDN/PSTN, LAN, WAN, and standardized bus systems, e.g., RS485 (see Figure 3), and to data interfaces, e.g., RS232, RS422 and MIL-STD-188-114A. It also offers intelligent gateway and relay functions.

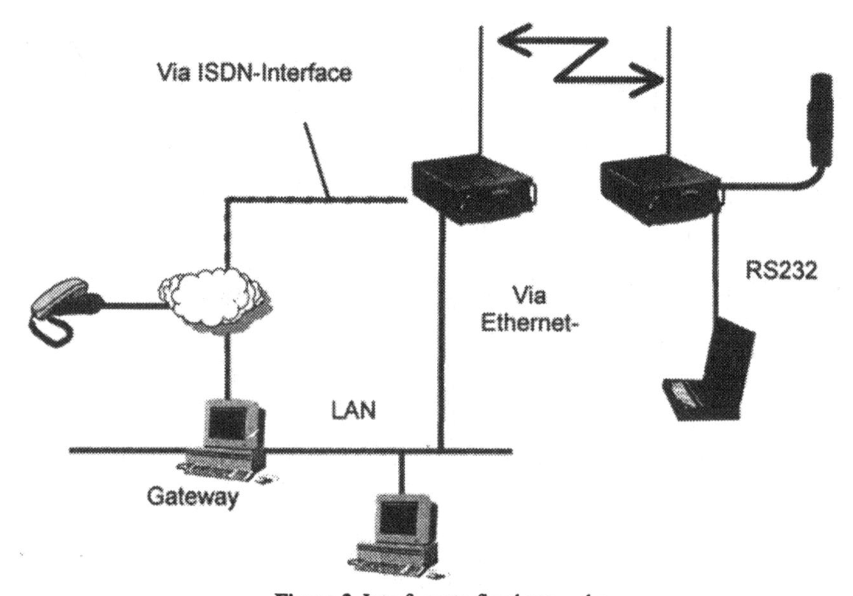

Figure 3. Interfaces to fixed networks

5. M3TR Architecture

The waveforms, protocols, encryption, and communications processes are implemented in a joint modular software architecture as depicted Figure 4.
The Radio Controller (RC) steers the radio and handles all user voice/data services supported by several links to the radio components. The most important link is the Radio Control Bus (RCB), which addresses all the modules inside the radio including an interface for external test controlling and monitoring at the connectors of the modules.

The RCB-Master within the Radio Controllers FPGA is configured by the central control microprocessor after Power On. The RCB delivers for example information control data like bandwidth, carrier frequency and so on. Further additional external components (e.g., 400 W HF Power Amplifier, 50 W UHF Power Amplifier, Antenna Tuning Unit (ATU), Switching Unit (RS232/485, Ethernet, ISDN interfaces), co-site-Filters for interference free communication lines) can be connected to the rear of the radio. The data port supplies three serial control interfaces (2 x RS-232, 1 x RS-232/ RS-485) for interoperation with PCs, data terminals *etc.*

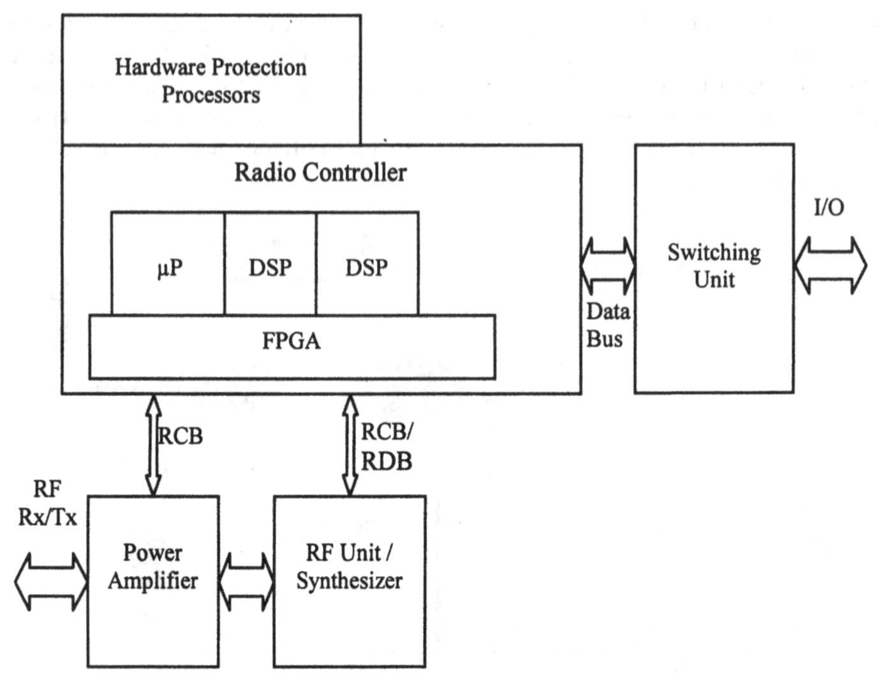

Figure 4. M3TR radio architecture

In case of EPM (Electronic Protection Measure) protected radio links the Hardware Protection Processor (HPP) guarantees tap-proof links through its embedded ciphering procedures and synchronization signaling data.

The overall Signal Processing Power consists of one central microprocessor (µP), two DSPs and up to two ciphering processors, interlinked and initialized by FPGAs. A total storage capacity of 25 Mbytes ensures up to 99 default setting including several EPM methods to be saved simultaneously. The presets store all variable parameters of the selected method, such as hopsets, modulation modes and addressing.

The Power Amplifier (PA) amplifies the transmit signal and filters out harmonics of the transmit signal. The internal Antenna Tuning Unit (ATU) as an optional part of the PA is responsible for antenna matching. In receive mode the harmonic filters increase the image rejection and improve the preselection performance. In addition to that, the receiver input protection is a part of the PA.

The Radio Frequency Unit (RFU) is divided into 3 main sections, the Front-End (FE), the Synthesizer (ST) and the RFU controller.

The front-end is mainly composed of input image and IF rejection filters, mixers, channel filters, amplifiers and the AGC. In the receive mode the incoming signal is low pass filtered for HF (0...30 MHz) or tuned band pass filtered for VHF/UHF and mixed in 3 steps down to an IF, where it is sampled and digitized by means of an A/D converter. While the estimated received signal strength adjusts the variable gain within the AGC loop, the received data is handed over to the first DSP in the Radio Controller. In transmit mode a direct digital synthesis chip generates modulated IF signal according to the data from the Radio Controller. After filtering, amplification and mixing the final transmit signal is filtered by the tuned HF/VHF/UHF-pass bands to reduce spurious emission and noise.

The synthesizer controlled by means of a reference clock oscillator supplies all local oscillator and clock signals of the radio, e.g., all mixer frequencies, data bus clock. The synthesizer monitors the PLL and provides test frequencies according to built in test procedures.

The RFU controller processes the signaling and user data from the Radio Controller and delivers it to the Front-end (e.g., for switching, tuning, pre-selection control, power management, AGC) or synthesizer (e.g., for gain loop control) and vice versa, e.g., signal strength measurement and test report.

6. Conclusions

A software-defined platform offers a broad flexibility and results in the great Software Radio potential of multiband, multimode and multirole operation. In spite of analog components, like A/D converters, that still represents a bottleneck in Software Radio development efforts, a programmable radio architecture is not only feasible, but as shown by M3TR an already achieved solution.
As for future general programming languages [4], a parametric approach seems to be favorable in order to adjust the waveform to varying QoS requirements or propagation conditions [1].

7. References

[1] Software Defined Radio Forum, Mobile Working Group, "Distributed-Object Computing Software Radio Architecture v1.1", July 2, 1999

[2] SDR Forum Mobile Working Group, "Software Framework Architecture", 30 March 1999

[3] B.A. Sharp, "The Design of an Analogue RF Front End for a Multi-Role Radio", MILCOM`98

[4] A. Wiesler, and F. Jondral, "Software Radio Structure for Second-Generation Mobile Communication Systems", VTC`98, Ottawa

[5] A. Wiesler, H. Schober, R. Machauer, and F. Jondral, "Software Radio Structure for UMTS and Second-Generation Mobile Communication Systems", VTC`99 (Fall), Amsterdam

[6] G. Hering, "Multi-Mode Radios - The Way Forward to Flexible Mobile Communication on the Battlefield", RTO Information Systems Technology Symposium, Lillehammer, 1999

[7] K. Pensel, and R. Bott, "Software Radio Requirements from Dual Use Point of View", 1. Karlsruhe Workshop on Software Radios, Karlsruhe, March 29/30 2000

[8] J.R. MacLeod, P.A. Warr, and M.A. Beach, "Dimensioning of a Software Defined Radio Receiver", 1. Karlsruhe Workshop on Software Radios, Karlsruhe, March 29/30 2000

Software Radio Implementability of Wireless LANs

Antony Jamin, Petri Mähönen, Zach Shelby
VTT, Technical Research Center of Finland

Abstract: While a significant amount of work on software-defined portable terminals has been carried out, little interest has been shown on the advantages of using this technology for emerging WLAN standards. We present in this article the technical challenges raised by the design of a highly flexible, software-defined radio interface for WLAN modem. Two different design approaches are investigated, for which technological requirements, implementation complexity and power consumption have been evaluated. Performance of the required components are compared with state-of-art technology capacity and, whenever needed, evolution trends have been used to estimate their probable time of availability.

1. Introduction

The fast-changing mobile communication industry has already seen the advantages a flexible radio interface could bring to their devices. Especially, the continuously increasing number of competing standards is reducing the life-cycle of products so dramatically that it is difficult to stay at the technology cutting edge. The solution that has been used till recently was to design *multi-modes* ASICs, where device functionality could be switched accordingly to the selected mode of operation. This leads to complex device whose long design cycles are incompatible with the fast changing telecommunication world. This demand for flexibility increased interest on digital world, much more flexible by nature. The recent availability of fast digital signal processors (DSP) and reconfigurable logic has triggered an amazing number of publications where digital design is to replace and improve most of today's analog and mixed-signal technology. If no one can deny the improvement for systems such as mobile telephony base station, little has been shown on some of the drawbacks for portable devices.

We focus in this article on the implementability of flexible, software-defined wireless LAN modems. Several reasons make those devices an adequate domain of study. First, the recent interest in wireless communication led to introduction of many (complementary or competing) standards. Thus, the possibility of having a single interface capable of accessing any of them is attractive. Secondly, wireless connections are by nature very often operated on portable devices. This adds to our design supplementary constraints: low power consumption to ensure long system autonomy and low complexity for maximum integration and reduced cost.

We will first analyze the requirements that a WLAN modem must fulfill in order to be multi-standards capable. We will review current WLAN standards and underline their differences in term of air interface, baseband processing and channel coding. From those requirements, we then derived a WLAN generic model. Its radio-frequency section, analog digital interface and baseband functional blocks are discussed. Two alternative implementation approaches are proposed and their respective performance is analyzed.

2. Wireless LAN Standards Overview

In term of frequency band, the 2.4 GHz Industrial, Scientific and Medical (ISM) band is currently the most widely used by WLANs. Leaving out the proprietary protocols supported by independent manufacturers, we will limit our analysis on currently available standards: IEEE802.11, IEEE802.15, Bluetooth, and SWAP-CA (from the HomeRF committee). We review in this section the basic parameters for air interfaces and baseband sections of the WLAN standards. They are summarized in Table 1. The interested reader can refer to the standard texts themselves for complementary information on each WLAN [1], [2], [3] and [4].

IEEE 802.11 standard [1], completed in 1997, is currently the most widely used WLAN in the world. The base standard includes both Frequency Hopping (FHSS) and Direct Sequence (DSSS) spread spectrum methods. The FHSS mode hops over up to 79 channels[1] in the 2.4 – 2.5 GHz band with channel spacing of 1 MHz. Hop rates are actually not defined within the standard, but held to a minimum of 2.5 hops/sec. The DSSS mode is based on a 11 - chip Barker spreading code. There are in total 14 overlapping channels of 22 MHz. Differential Binary and Quaternary Phase Shift Keying modulations (D-BPSK, D-QPSK) can be selected in order to achieve 1 or 2 Mbps.

The Bluetooth standard [3] was released in 1999 and is on the rise as a wireless personal area network standard for all types of low power devices. Moreover, the IEEE standardization group working on the 802.15 Wireless Personal Area Network (WPAN) standard seems to head toward a subset of Bluetooth specification. Bluetooth operates with FHSS with parameters quite like in IEEE802.11. The channelization and bandwidth are the same, while only the 1 Mbps, 2-GFSK mode is defined. The frequency hop period, equal to one TDMA slot, is rather high: 1600 hops/sec.

Wireless networking for the home has pushed the HomeRF consortium to release their SWAP-CA [2] standard for WLAN. This resembles IEEE802.11 very closely, although being slightly simpler and cheaper. Offering only a FHSS mode, it is identical to 802.11 with 1 Mbps and 2 Mbps modes. The frequency hopping rate is fixed at 50 hops/sec. This standard adds a DECT mode using TDMA to give voice capability over the interface.

[1]In some country, it is limited to a lower number of channels due to local regulation.

Table 1: Comparison of WLAN standards requirements

	IEEE 802.11	*IEEE 802.15*	*Bluetooth*	*HomeRF*
RF Freq.	2.4 GHz ISM band (2402-2497)			
Spreading	FHSS, DSSS	FHSS		
Bandwidth	1 MHz, 22 MHz	1 MHz		
Modulation	GFSK, PSK	GFSK		
Power	1 – 100 mW	.1 – 50 mW	.25 – 100 mW	1 – 100 mW
Throughput	1 – 2 Mbps	50 – 721 kbps	50 – 721 kbps	1 – 2 Mbps
MAC	CSMA/CA	Link Based	Link Based	CSMA/CA

3. Generic Software Radio Architecture

This section introduces the requirements of a generic modem architecture that would permit to comply with any of the WLAN standards we have discussed. Ideally, this architecture would be similarly capable of interfacing totally different standards providing their complexity is of the same order of magnitude.

An ideal software-defined modem would be fully implemented digitally, so that it could be completely reconfigurable via software. Thus, it would require using direct conversion method, converting digitally the whole band of interest without preliminary conversion to intermediate frequency. Since it is not yet a realistic option for mass market product, we will only focus here on digital conversion at intermediate frequency.

3.1 Radio-frequency Front-end

Our modem front-end is very classical except that it downconverts the full ISM band to intermediate frequency. Figure 1 shows the overall radio-frequency section. Let us analyze now the standards requirements for the transmitting section. Transmission power range is especially critical for power-limited devices. From Table 1, we can see that the highest transmit power is 100 mW (20 dBm). A power control capacity ranging from 0 to −20 dB, permits suitable transmitted power adjustment and thus minimal consumption and minimum disturbance to other devices.

To evaluate the maximum noise factor of the front-end section, we must determine once again what are the standard requirements. A closer look at the WLAN standards gives us the information summarized in Table 2. Using well known BER versus SNR relation [5], [6] for those modulations and adding an arbitrary implementation margin of 5 dB, we obtain the minimum signal-over-noise ratio at the ADC input, as indicated in the Table 3. The results for each standard are given in the last column of Table 3, where we see that the 15.8 dB maximum front-end noise factor is set by IEEE802.11 standard in its 2-GFSK version. This is perfectly realistic with what current technology

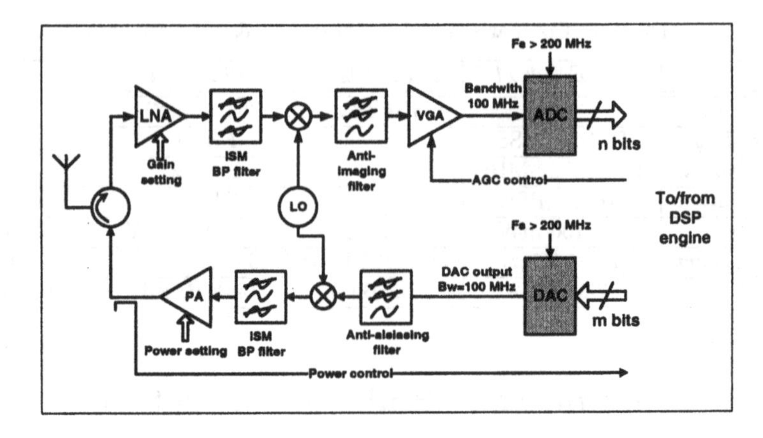

Figure 1: Generalized WLAN air interface

can currently provide. Intersil PRISM II chipset for instance, claims a value of 8.6 dB in low-level signal reception condition [7].

3.2 Analog/Digital Converters

Being the interface between the analog and digital word, AD/DA converters are critical blocks. They are subject to many constraints and largely responsible of modem performance. Signal SNR is linked to converter resolution by the well-known relation [8]:

$$SNR_{A/D} = 1.76 + 6.02b + 10log\left(\frac{2B_w}{F_{sampling}}\right) \quad (1)$$

where b is the DAC resolution in bits, $F_{sampling}$ the sampling frequency and B_w the bandwidth of the signal of interest. Nyquist theorem gives us a lower bound for $F_{sampling}$ of twice the bandwidth of the signal to reconstruct. Adding a 20 % margin to compensate for frequency drift, the band of interest of 97 MHz

Table 2: WLAN standards demodulation requirements

Standard	*Modulation*	*BER/FER*	*Signal power*
Bluetooth	*GFSK* 1 *Mbps*	$BER = 10^{-3}$	-70 *dBm*
IEEE802.11 DSSS	*BPSK/QPSK* 1/2 *Mbps*	$FER = 8.10^{-2}$ (*MPDU*=1024 *bytes*)	$-80/-80$ *dBm*
IEEE802.11 FH	2/4 *GFSK* 1/2 *Mbps*	$FER = 3.10^{-2}$ (*MPDU*=1024 *bytes*)	$-80/-75$ *dBm*
SWAP-CA	2/4 *GFSK* 1/2 *Mbps*	$FER = 3.10^{-2}$ (*MPDU*=111 *bytes*)	$-76/-62$ *dBm*

Table 3: Front-end minimum noise factor

Standard	*SNRmin at ADC input*	*In-band noise*	*Minimum noise factor*
Bluetooth	14.0 *dB*	−84.0 *dB*	29.8 *dB*
IEEE802.11 DSSS-BPSK	5.1 *dB*	−85.1 *dB*	16.3 *dB*
IEEE802.11 DSSS-QPSK	8.1 *dB*	−83.1 *dB*	18.3 *dB*
IEEE802.11 FH-2GFSK	16.0 *dB*	−96.0 *dB*	15.8 *dB*
IEEE802.11 FH-4GFSK	19.0 *dB*	−94.0 *dB*	17.8 *dB*
SWAP-CA FH-2GFSK	16.0 *dB*	−92.0 *dB*	19.8 *dB*
SWAP-CA FH-4GFSK	19.0 *dB*	−81.0 *dB*	30.8 *dB*

leads to a sampling frequency of roughly 240 MHz. The signal quality goals remains to be defined. This is done by collecting the requirements set by standards.

In the transmit path, the DAC is responsible of modulated signal reconstruction with as less in-band spurious noise as possible. The most limiting standard requirement comes from IEEE802.11-DSSS that requires a maximum in-band spurious emission level of −60 dBc [1]. Using Equation (1), it gives a minimum resolution of 10 bits. Assuming the dynamic range of the transmitter is handled by analog devices, the system can make full use of the DAC dynamic range and thus keep the SNR at its highest value.

The receiver front-end parameters being defined, we need to calculate the noise added by the quantization process. In the case of a digital radio receiver, it is actually more interesting to evaluate the SNR degradation introduced by the sampling process. Assuming that the automatic gain control (AGC) is ensuring a full scale signal level at the ADC input and no other perturbation than thermal noise is received, we can effectively calculate the loss in term of SNR due to the ADC. With those assumptions and the worst-case values of Table 3, we can calculate that with 8 bits of resolution, the signal quality is more often *thermal* noise than *quantization* noise limited.

However, in the more realistic case where interfering signals are present in the ISM band, the AGC adjusts itself so that the highest signal power in the band is 0 dBm. For instance, let us assume the desired incoming signal is received together with an interfering signal of power 40 dB higher than the signal of interest. To obtain the same level as before, we thus need 7 extra bits (i.e. 1 bit per 6 dBc), leading to a total of 15 bits. This value is really high, although the case assumed is realistic. Typically, this might be for instance a desktop computer communicating simultaneously with a Personal Digital Assistant (PDA) using Bluetooth and a remote WLAN using IEEE802.11. If the WLAN link requires full transmit power, the receiver of the bluetooth modem will receive a large part of it due to the short distance between modems, and will have to reduce consequently its front-end gain.

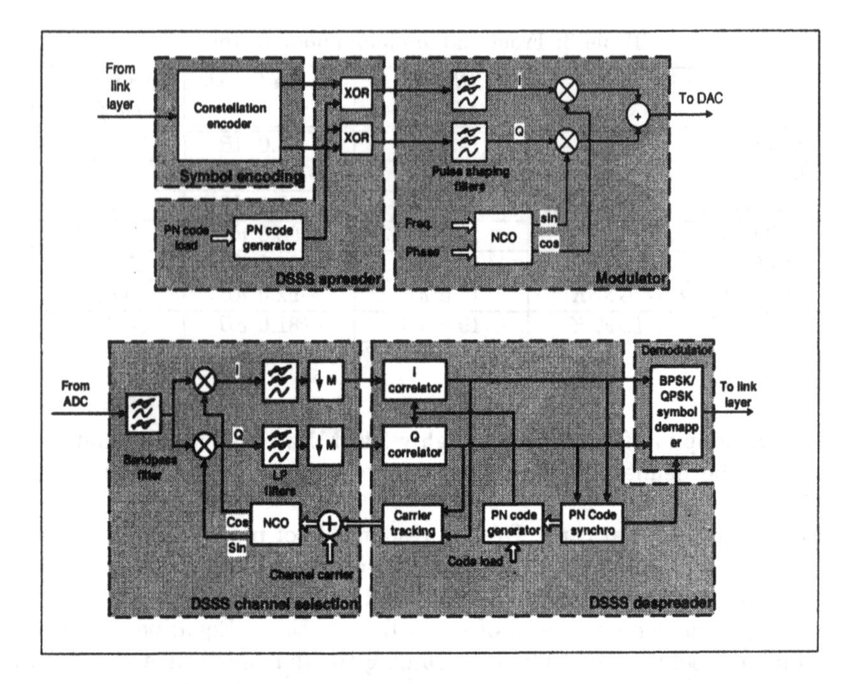

Figure 2: DSSS modem block diagram

3.3 Baseband Signal Processing

The modulations we are interested in are simple but yet robust: FSK with FH and PSK with DS spread spectrum. Hence, optimum architecture are well described in literature [5], [6]. Those structures, shown in Figures 2 and 3 are nevertheless given here for completeness first, but especially because they will be the base of our complexity evaluation.

4. Implementation and Complexity Analysis

Now that we have drawn up the sketch of a generic WLAN modem, we will analyze in this section its complexity and power consumption. For software-defined radio, we have already seen that the added value is flexibility. But this is valuable only if we are not loosing the interesting features of existing and available technology. In the specific case of WLAN modem, low complexity is required in order to achieve high integration as well as low development cost.

Power consumption is another valuable point of comparison. Wireless LAN connection loose most of its interest when not operated by a portable device. Such equipment is usually battery-powered hence power-limited. Evaluating

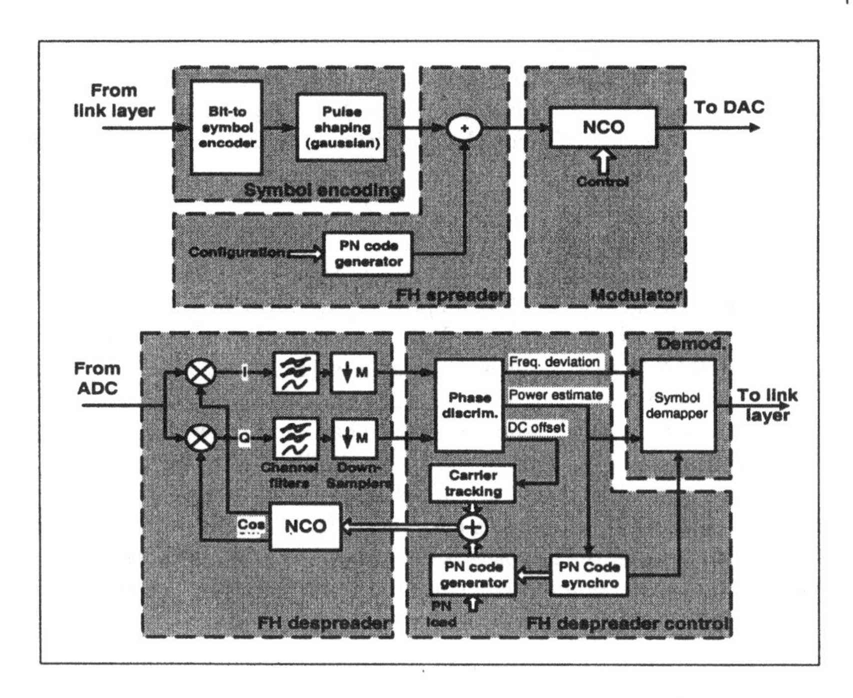

Figure 3: FH modem block diagram

power consumption of a system is not an easy task if the system implementation is not totally done. Consequently, we will only give here an order of magnitude that is meaningful enough to allow comparison with currently available devices.

Since we have seen the front-end requirements are almost identical to the current analog implementations, we will not discuss it further. We can however have a good approximation of the radio-frequency section power consumption using available components data sheet [7].

4.1 Analog/Digital Converters

Previous calculation indicated our modem would require a 10-bit resolution DAC. Several of the highest quality commercial devices could match our demand. Among those devices, Analog Devices is for instance offering the *AD*9751 and *AD*9753, both with a 300 MHz maximum update frequency and respectively 10 and 12 bits of resolution, for a consumption of approximately 200 mW. In this case, even the first one meet our requirements since it presents a typical Single Frequency Dynamic Range (SFDR) of 86 dB.

We have calculated our modem requires an ideal ADC resolution of up to 15 bits in order to comply the most difficult case. Thus, according to [9], a device with an actual resolution of 16-bit should compensate for the distortion due to non-linearity. However, we can see from the ADC technology state-of-art review done in [9], that such a device is far ahead of the current technology performance. The author of this highly valuable survey also took into account power consumption of ADC. For this, he defined an ADC *figure of merit* F defined by:

$$F = \frac{2^{SNRbits} F_s}{P_{dis}} \tag{2}$$

where $SNRbits$ is the SNR equivalent number of bits of the ADC, F_s its maximum sampling frequency and P_{dis} its power consumption. Using the average value the author has determined for F, we can extrapolate the consumption of our 16 bits, 240 MHz ADC built with today's technology: 200 W! Reiterating the calculation with best value of F, we still get 10 W! These values emphasize very well the extremely high requirements we have and show clearly that even with the actual state-of-art technology, no ADC fulfilling our requirements will be available so soon. The same author actually reported improvement in resolution of 0.25 bit per year. At this pace, our ADC might be available around year 2032. Once again, the hypothesis we selected for our calculation (i.e. the 40 dBc disturber) represents an extreme case and a more detailed study might show that if was too pessimistic.

4.2 Modem Digital Section Implementation

For the modems functional blocks we discussed till now, there was no technological choice on how their could be implemented. For the digital signal processing section however, implementation has to be considered since several options are available. We can actually see two possible alternatives on how to digitally process the modem signals: 1) Programmable logic or 2) Digital signal processor (DSP) based. For both, we give an estimate of the signal processing core complexity and map it into current state-of-the art technology.

Using programmable logic, though a far less flexible option than the other, allows performing most computational-demanding operations in dedicated hardware sub-modules running in parallel. As it can be seen from Figure 4, the lack of flexibility of the FPGA is compensated by using a microprocessor for less time demanding tasks such as baseband interfacing and FPGA setup and control. For the DSP based approach, the signal processing chain is fully software-defined. A fully generic, microprocessor-based architecture using ROM, RAM and I/O block optimized for speed is sufficient to implement a complete modem. Despite this very attractive view, one can already foresee that the computational load is extremely high and thus far beyond the capacity of current processors.

4.2.1 Simplified Reference Model

We use block diagrams of Figures 2 and 3 as the bases of our estimations. However, some approximation is done in order to keep the analysis simple. For instance, synchronization and decoding blocks have been voluntarily omitted. Such choice might look arbitrary and simplistic but the results obtained are meaningful enough to justify them *a posteriori*. Moreover, no optimization of any kind has been done: word length is equal to 16 bits in all implementation, NCO are simple look-up table in ROM, and filter orders are typical. Obviously, our complexity evaluation is using an extremely simplified model. Those assumptions done, we can now make a conservative estimation of complexity for both implementations.

4.2.2 DSP-based Implementation

For digital signal processing computation load evaluation, it is usual to use millions of multiply-accumulate (MMAC) operations to evaluate processing power requirements. Table 4 summarizes the number MMAC per seconds required by each elementary block. As expected, demodulators are the most demanding. Despreading the direct sequence modulated signal is extremely computationally intensive, due mainly to the correlation operation. On the other hand, the FH modulator requires only few operations per sample, the modulation at full DAC clock rate being the most demanding function.

For reference, the fixed point DSP processor 'C6202 from Texas Instruments provides 600 MMAC per seconds at 300 MHz. Hence, a single one should be able to perform full FH modulation. However, full software implementation of the DSSS demodulator could be achieved using at least 23 of those, or equivalently, using a single one having a clock frequency of 6.9 GHz. With a typical consumption of about 1.5 W per device, this leads to a total of 34.5 W for the DSSS modulator. Moreover, this very high consumption does not take into account peripheral components consumption such as RAM and ROM for instance. These results shows clearly how irrealistic fully software-defined radio design is with the current technology. For comparison, HARRIS PRISM II chipset has a total power consumption of about 0.5 W (150 mA at 3.3 V). Such a difference can not in any case be justified only by the functionality improvement.

4.2.3 FPGA-based Implementation

For FPGA, the number of gates commonly used by hardware engineers is replaced by the number of elementary cells. This is somehow a subjective measure since manufacturers have their own proprietary cell structure. For this analysis, we chose to refer to Xilinx products in reason of their cutting edge performance and large availability. For equivalence with the software option, adder and multiplier chosen are 16-bits signed, optimized for speed. The number of elementary cell used by a multiplier is 250 using fast, parallel architecture on Xilinx devices, as defined in [10]. Using Xilinx software development suite, a 16-bit adder can be build with 10 cells[2]. Using this value

[2] Xilinx is using the term CLB to design its FPGA elementary cell.

together with the elementary number of add and multiply operations we obtained for the software implementation, we reach the values mentioned in the third column of Table 4. We can have a better idea of what those values are by converting in equivalent number of FPGA devices. Once again, this has to be considered as a lower bound since resource allocation limits the use of

Table 4: Digital modems hardware complexity

Block	*RAM (bits)*	*Operations (Mult/Add)*	*Clock freq. (MHz)*
DSSS modulator			
Symbol coder	32	1	F_{symbol} (1)
PN code generator	11	2/2	F_{chip} (11)
Spreader	2	1/1	F_{chip} (11)
Pulse-shaping filters (order $N = 20$)	1280	40/38	$8F_{chip}$ (88)
NCO	$\simeq 1$ *Mbit*	1	F_{DAC} (240)
Total	$\simeq 1$ *Mbit*	4569 *MMAC/sec*, 10970 *CLB*	
DSSS demodulator			
Input BP filter (order $2 * N = 16$)	512	16/15	F_{ADC} (240)
NCO	$\simeq 1$ *Mbit*	1	F_{ADC} (240)
Down-convertor		2/0	F_{ADC} (240)
Pre-decim. filters (order $N = 20$)	640	40/38	F_{ADC} (240)
Correlator	768	24/24	$2F_{chip}$ (22)
Total	$\simeq 1$ *Mbit*	14688 *MMAC/sec*, 22210 *CLB*	
FH modulator			
Symbol coder	32	1	F_{symbol} (1)
Pulse-shaping filter (order $N = 20$)	640	20/19	$8F_{symbol}$ (8)
PN code generator	632	1/1	F_{TDMA} (1.6)
Spreader	1	1/0	F_{DAC} (240)
NCO	$\simeq 512$ *kbit*	0/1	F_{DAC} (240)
Total	$\simeq .5$ *Mbit*	404 *MMAC/sec*, 5020 *CLB*	
FH demodulator			
Down-convertor	2/0		F_{ADC} (240)
Pre-decim. filters (order $N = 20$)	640	40/38	F_{ADC} (240)
Discriminator	64536	0/4	$8F_{symbol}$ (8)
NCO	$\simeq 512$ *kbit*	0/1	F_{ADC} (240)
Total	$\simeq .5$ *Mbit*	10355 *MMAC/sec*, 10460 *CLB*	

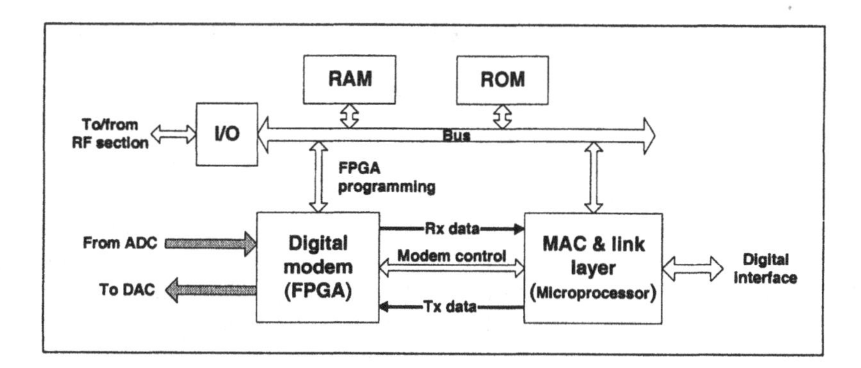

Figure 4: FPGA-based modem block diagram

one device to less than its full capacity. The device referenced XCV3200E from Xilinx Virtex-E family [11], one of the top range products, has a total of 16224 CLBs. Hence, one single device would be sufficient to realize any of half-duplex modem, except for the DSSS demodulator that would require two. Note that those values are likely to be one-half or one-third of what a complete, workable implementation would use.

From the design complexity evaluation, we would like to extrapolate the consumption of the programmable logic block. However, the design-assisting and prediction tools are not able to provide any significant results and thus, we can not give any valuable information on this. Nevertheless, total consumption is likely to be of the order of few watts, thus several folds under the DSP-based one. This should be confirmed once a workable prototype will be available.

Those values, even with the large approximations done, show clearly that programmable logic is today a more realistic choice than a DSP for software define radio. Unfortunately, re-configuring large FPGA can take few seconds, time during the device is not able to perform its normal function. Thus, dynamic changes of protocol might not be feasible, limiting the network access choice to initialization time. One is in position to wonder if such a little improvement would be worth what it costs. As a last technical issue, design engineer would have noticed that the data blocks required to re-program the FPGA are rather large (few hundreds of kilobytes) and need to be store on-board in order to be accessed at any time.

Consequently, many issues remain on the way of software-defined radio terminals deployment. Even the most feasible alternative based on FPGA is not yet able to outperform optimized, protocol specific ASIC-based implementation.

5. Conclusion and Related Research Topics

We have reviewed the main issues on implementing a software-defined WLAN modem. The flexibility provided by the software-defined radio technology has a high cost and none of the architecture options we proposed is well suited. Thus, the best compromise would be to combine a programmable logic block together with a DSP-core. Simple, repetitive and thus computationally demanding tasks could be performed within the logic block under DSP control. This type of architecture, if well-dimensioned, would probably present the best flexibility over consumption ratio. Semiconductor manufacturers have envisioned this long ago and several of them propose devices combining DSP, RISC and peripheral cores within the same chip. This last type of product is perfectly suited for application requiring both signal processing and low protocol layers implementation. A generic software-defined platform design and realization based on those is currently under consideration by the authors.

Out of the implementation issues, we have essentially discussed here the improvement that software-defined modems could bring and the issues their realization present. Such drawbacks can only be justified if real advantage of the added flexibility is taken. This is under the responsibility of higher layers that have to be smart enough to use it. Those issues have started being studied and discussed [12]. However, the processing power required by such "smart agent" might very well be out of the range of what technology is capable to provide before at least the next decade. This will nevertheless motivates lots of research and development from areas of intelligent mobile networking management to very low power integrated circuits design without forgetting flash re-configurable devices.

References

[1] IEEE Standardization body, *"IEEE802.11-1997: Local and metropolitan area networks"* , August 1997.

[2] The HomeRF Committee, *"Shared wireless access protocol (cordless access) specification (SWAP-CA)"*, December 1998.

[3] Bluetooth Committee, *"Specification of the bluetooth system"*, July 1999.

[4] IEEE802.15 Standardization body, *"IEEE802.15 personal wireless area network draft specification"*.

[5] John G. Proakis, *"Digital communications"*, Mac Graw Hill, New York, third edition, 1995.

[6] R. E. Ziemer and W. H. Tranter, *"Principles of communications"*, John Wiley and Sons, New York, fourth edition, 1988.

[7] *"PRISM II chip set overview, 11 Mbps SiGe"* , Application note AN9837, Intersil, February 1999.

[8] J. A. Wepman, *"Analog-to-digital converters and their applications in radio receivers"* , IEEE Communications Magazine, May 1995.

[9] R. H. Walden, *"Analog-to-digital converters survey and analysis"*, IEEE Journal on Selected Area of Communications, April 1999.

[10] Russell J. Petersen, *"An assement of the suitability of reconfigurable systems for digital signal processing"*, M.S. thesis, Dept. of electrical and computer engineering, September 1995.

[11] *"Virtex-E field programmable gate arrays"*, Advance products specification, December 1999.

[12] G. Q. Maguire and J. Mitola, *"Cognitive radio: making software radios more personal"*, IEEE Personal Communications, August 1999.

[13] G. Fettweis, T. Hentschel and M. Henker, *"The digital front-end of software radio terminals"* IEEE Personal Communications, August 1999.

JCIT, A Production-Ready Field Tested Non-Proprietary Software Definable Radio

Robert F. Higgins and Charles C. Herndon

Naval Research Laboratory Naval Center for Space Technology Space Systems Development Department Washington, D.C.

Abstract: The Naval Research Laboratory (NRL) has developed a multi-channel, multi-mode, multi-band, modular communications terminal called the Joint Combat Information Terminal (JCIT). The JCIT is the heart of the communications system for the Army Airborne Command and Control System (A2C2S). The production-ready JCIT has been field tested by the US Marine Corps (USMC) in command and control variants of the Light Armored Vehicle (LAV) and Amphibious Assault Vehicle (AAV). The functionality, expandability, and modularity of JCIT provide state-of-the-art communications for military and civil agencies. Interoperability with SINCGARS, HAVE QUICK, HF, National Satellite Intelligence Broadcasts, military precision GPS, IEEE-802.11, Direct Broadcast Satellite (DBS/GBS), Air Traffic Control, Police, Fire, and Civil Maritime has been demonstrated in the field and in the laboratory.

The JCIT architecture was designed to meet current and future military command, control, communications, computers, and intelligence (C4I) requirements. It provides a total communications system solution from the antenna to the user. The modules developed for the JCIT are not proprietary. The architecture is built from a common set of hardware and software modules. Software allows the hardware to interoperate with other legacy or future radio systems on operator command. Hardware and software insertion to meet future requirements is supported by the architecture. The architecture is readily adaptable to new platforms, utilizing unchanged hardware and software modules. Although the architecture is not considered "open" by the industry standards, full Government owned Interface Control Documents exist that allow any third party to build hardware or software components for the system. Eight-channel vehicular/airborne units and two-channel vehicular/manpack units have been built. These units share hardware modules and software applications without modification.

1. Background

For the past 15 years, the NRL has been developing communications terminals that significantly advance the state of the art. NRL initially developed this expertise in combining tactical data processing with INFOSEC and radios in the dissemination of tactical reconnaissance data via satellite to terrestrial users. This heritage includes the development of:

- multi-mission Advanced Tactical Terminal (MATT);

- improved Data Modem (IDM);
- commanders Situation Awareness Workstation (CSAW);
- briefcase MATT (BMATT);
- joint Combat Information Terminal (JCIT);
- army Airborne Command & Control System (A2C2S).

The NRL technology path has emphasized total-system solutions for tactical users, rather than just replacing radios. Development of multi-mode, programmable communications terminals has provided the user with automatic message data fusion and correlation, routing between channels, data/voice routing, multi-message protocol handling, automatic protocol conversions, and situation awareness processing. The variety of input/output protocols allows these new systems to be "drop-in" replacements for the legacy communication systems without requiring modifications to the user data processor.

The USMC and Civil Agencies will capitalize on the Army's A2C2S program, leveraging the multi-million dollar development that was initiated in FY 94. The A2C2S provides a mobile airborne or ground based command, control, communications, computers, and intelligence capability for the 21st century maneuver commander. The voice and data connectivity provided by the communications segment is equivalent to and in some cases exceeds an Army's Brigade Tactical Operations Center.

The A2C2S communications suite consists of three subsystems: JCIT, the Radio Frequency Manager (RFM), and the Antenna Interface Modules (AIM).

2. JCIT Architecture

The JCIT architecture was designed to meet current and future military C4I requirements. It was designed to provide a total system solution from the antenna to the warfighter/decision maker. The modules developed for the JCIT are not Proprietary. The modular, architecture is built from a common set of hardware and software modules. Software allows the hardware to interoperate with other legacy or future radio systems on operator command. Hardware and software insertion to meet expanding needs is supported by the architecture. The architecture is readily adaptable to new platforms, utilizing unchanged hardware and software modules. The JCIT has implemented thirteen of the Joint Tactical Radio System (JTRS) waveforms. IEEE-802.11 and Direct Broadcast System (DBS) and Government Broadcast System (GBS) have also been implemented. The JCIT has received endorsement and mandate from Under Secretary Gansler to become JTRS compliant. When this is accomplished, JCIT will be the first and only Government owned implementation of the JTRS architecture. Selected waveforms are being provided to the JTRS from the JCIT development. Eleven of the JTRS waveforms can be implemented in software with no hardware modifications. The remainder of the JTRS requirements can be met with the architecture by the addition of hardware and software. The architecture supports these future software and/or hardware additions while preserving and interoperating with the initial hardware and software that has already been completed.

The overall JCIT hardware architecture is shown in **Figure 1**. The antenna interface, transmit power amplification, receive filtering, and cosite mitigation functions are housed in the Radio Frequency Manager (RFM). An Antenna Interface Module (AIM) provides RF connectivity between the band specific RFM assets and the fully programmable radio terminal functions. The JCIT houses the radio, modem, INFOSEC, and processing functions. JCIT, RFM and AIM functions are interconnected via the IEEE-1394. Using IEEE-1394 busses, multiple chassis are chained together, and act as a single unit. In addition, digital intercoms, video equipment, and other IEEE-1394 compliant devices can be connected. The RFM, AIM, and JCIT are all 19" rack mount units suitable for airborne, ground, vehicular, or shipboard use. Other mounting options and module configurations can be accommodated.

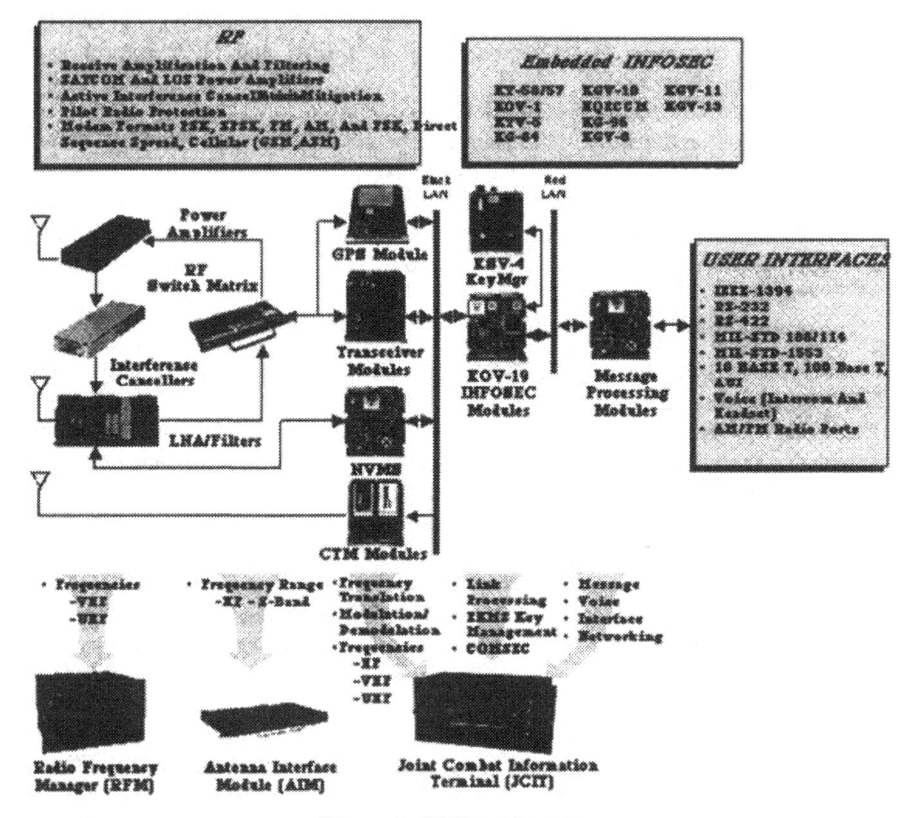

Figure 1. JCIT architecture

3. RF Manager (RFM)

The RFM is modular and tailored to the host platform. It is responsible for providing band preselection, transmit amplification, and managing the cosite interference on the platform. The cosite problem, simply stated, is that the user must have the ability to simultaneously transmit and receive on multiple links in a hostile electronic environment without loss of performance. This is accomplished utilizing several techniques, some of which are in the RFM and some of which are in the JCIT Multi-Band Transceiver (MBT). The tuned low noise power amplifiers (RFM)) makes the cosite environment less harsh than traditional amplifiers. Broadband and narrow band filtering on the receive path (RFM) prevents undesired signals from impacting the receiver. Active transmit signal cancellation (RFM) removes the effect of the local transmitters on the receivers by 50 dB. High performance receivers (MBT) have large spurious-free dynamic range. Intelligent receive signal search/acquisition algorithms (MBT) can distinguish the desired signal intelligently from the noise.

There are two different modular RFM units, a VHF RFM and an UHF RFM. The VHF RFM chassis contains the following re-useable modules:

- LNA/Filter assembly for VHF receive channels (6 FM, 2 AM), 1 in chassis;
- VHF low noise High Power Amplifier (HPA) and HPA control logic assembly, 4 in chassis;
- 50 Watts output power, adjustable;
- low noise architecture, with internal hopping and harmonic filtering;
- controller, to execute commands received from JCIT via the IEEE-1394 bus, 2 in chassis for redundancy;
- cosite control assembly, for active transmit signal cancellation, 6 Channels;
- protects receiver section from the four VHF FM transmitters and other radios;
- Signal Distribution assembly, to combine cancellation signals.

Also required by the VHF RFM are four VHF transmit antennas, and a single VHF receive antenna. The USMC command variants and the A2C2S use Electronically Tuned Helix (ETH) antennas for transmit and a broadband dipole antenna for receive. The RFM provides tuning words for external antennas. The RFM design can operate with any antenna technology.

The UHF RFM chassis contains re-useable modules. The SATCOM tuned low noise HPA and HPA control logic assembly is in a 1 in chassis. It has 100 Watts of output power, adjustable. The low noise architecture has internal hopping and band filtering. The SATCOM Filter assembly has 4 channels heavily filtered to prevent interference (2 in chassis). There is one UHF Power Amplifier and one UHF Filterper chassis. The controller implements commands received from JCIT via the IEEE-1394 bus. The Wideband T/R module has 10 Watts of output power, adjustable. The HF/VHF/UHF band selection chooses transmit or receive. It includes LNA/Diplexer assemblies with a high compression point, low noise figure LNA. The Diplexer allows simultaneous SATCOM transmission and reception on the same antenna. The A2C2S unit was modified to support UHF LOS (bypass diplexer with two relays). The UHF RFM operates with standard SATCOM and LOS UHF antennas.

4. The Antenna Interface Module (AIM)

The AIM provides electronic switching between each RFM's band specific transmit and receive paths and the full band functional MBTs in the JCIT. This allows a single MBT in the JCIT to connect to the HF, VHF or the UHF HPAs and receive filters. The proper switch configuration is determined on the fly automatically by the JCIT. The AIM is designed to support 2-2500 MHz transmit or receive signals.

5. The JCIT

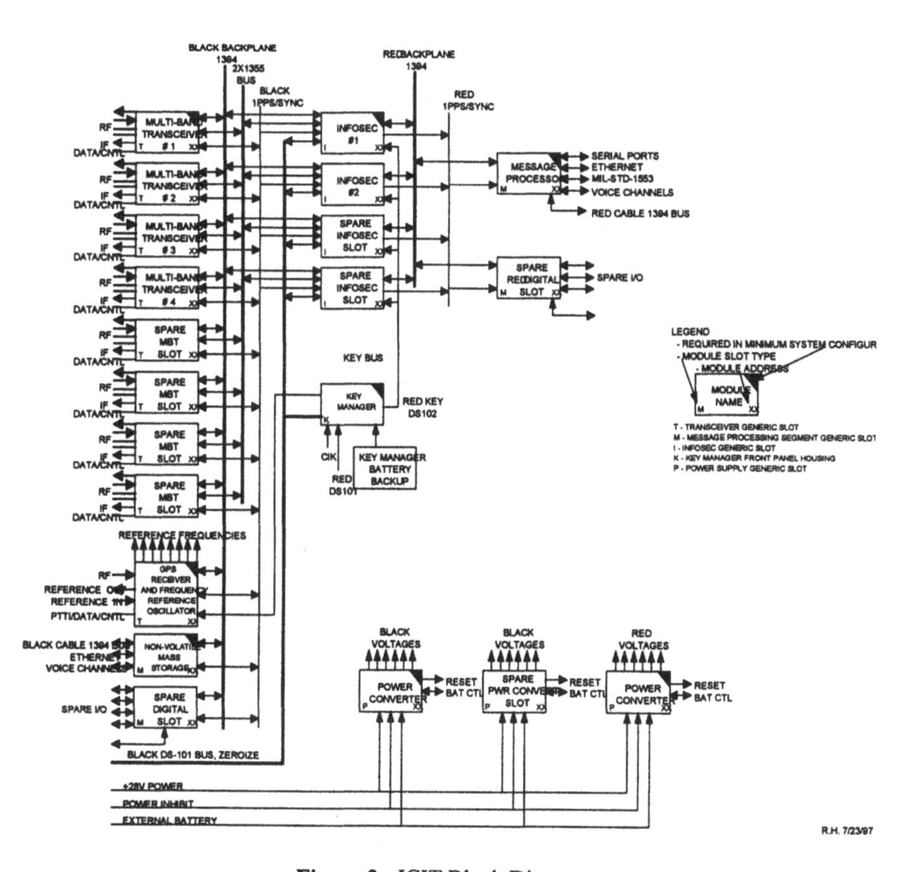

Figure 2. JCIT Block Diagram

The JCIT is the specific hardware-software unit within the A2C2S communications segment that provides the RF conversion, demodulation, modulation, decoding/encoding, TRANSEC, COMSEC, message processing, networking, inter-networking, user interface, and control functions. The JCIT is a modular design, where functionality can be expanded or contracted to meet user requirements. Overall structure of the JCIT architecture is based on a "red" bus for classified data and "black" bus that contains only unclassified or encrypted data. The modules on each bus transfer commands, status, and data over the IEEE-1394 high speed serial bus. The radios reside on the black bus, the classified processing on the red bus, and the INFOSEC functions utilize both the red and the black busses.

The building blocks for the JCIT consist of the Multi-Band Transceiver (MBT), the Commercial Technology Module (CTM), the GPS/Reference Module (GRM), a Non-Volatile Mass Storage Module (NVMS), the INFOSEC module, the Key Manager Module (KMM), and the Message Processor Module (MPM)

A JCIT block diagram reflecting the number and arrangement of modules is shown in Figure 2.

The MBT is a multi-band transceiver that can replace legacy radios. Its main function is to convert RF to useful baseband data bits or voice samples and to take baseband data bits or voice samples and convert them to RF. The salient features of this module are as follows:

- single width SEM-E module (IEEE-1101.4 compliant form factor, 5.3" x 6.1" x 0.6");
- 2 – 512 MHz transmit or receive;
- 100 dB Spur Free Dynamic Range;
- 13 KHz maximum hop rate;
- half duplex operation;
- cosite IF filtering;
- intelligent signal acquisition and tracking;
- modem functions: PSK, SPSK, FM, AM, FSK;
- Data Rates: 1.2 – 76.8 Kbps;
- 68K general purpose CPU, 56002 DSP;
- IEEE-1394-1995 100 Mbps backplane interface;

The CTM is a general purpose module that can host commercial PCMCIA modules to incorporate functionality quickly and inexpensively into the JCIT architecture. The CTM can operate in any MBT or GPS slot. This module is currently used for the IEEE-802.11 wideband TCP/IP wireless networking function and simultaneously for the DBS/GBS receive and processing function. The salient features of this module are as follows:

- single width SEM-E module (IEEE-1101.4);
- Motorola PowerQuick @ 200 MHz;
- 128 Mbytes SDRAM;
- 2 universal serial ports (RS232, RS422, MIL-STD-188/114);
- Ethernet (10bT and 100bT);
- IEEE-1394-1995 100 Mbps backplane interface;
- 1 Type II PCMCIA slot;
- 1 Type III PCMCIA slot;

The GRM is a module that contains an off-the-shelf Trimble military precision GPS receiver and an ovenized stable reference oscillator. This module provides time, position and frequency references to the system. The salient features of this module are as follows:

- single width SEM-E module (IEEE-1101.4).

GPS:

- PPS/SM military precision receiver;
- L1 and L2 frequencies;
- 6 simultaneous channels;
- GPS JPO approved TRIMBLE product.

Reference:

- 8 High accuracy, low phase noise reference outputs for MBTs;
- 0.1 ppm over temperature;
- designed for high vibration environment;
- system synchronization signals;
- system timing source.

Other:

- 68K general purpose CPU, 56002 DSP;
- IEEE-1394-1995 100 Mbps backplane interface.

The NVMS provides storage for the operating system and application code for all of the modules. Black modules have their code stored on the NVMS in non-encrypted form. Red modules have their code stored on the NVMS in encrypted form. This and other features allow the JCIT architecture to be CCI when un-keyed. Black user interfaces are also provided on this module. The salient features for this module are as follows:

- single width SEM-E module (IEEE-1101.4);
- Motorola PPC603 @ 200 MHz;
- 32 Mbytes SDRAM;
- 96 Mbytes non-volatile flash memory storage, upgradeable via mezzanine module;
- 3 TMS320C549 DSPs @ 100 MHz each;
- 1 RS232 serial port;
- 4 universal serial ports (RS232, RS422, MIL-STD-188/114);
- Ethernet (10bT and 100bT);
- 4 analog headset microphone interfaces;
- 4 analog intercom interfaces;
- 4 analog external radio interface (IDM like);
- IEEE-1394-1995, 200 Mbps cable interface;
- IEEE-1394-1995 100 Mbps backplane interface.

The INFOSEC module performs data processing, TRANSEC, and COMSEC functions. Data processing functions consist of the following representative items: voice coding/decoding encoding/decoding, interleaving/de-interleaving, multi-pattern correlation, data parsing, and network timing. TRANSEC functions consist of the following representative items: HAVE QUICK and SINCGARS frequency hopping calculations, pattern mixing, and COMSEC initialization sequencing. Four simultaneous TRANSEC function are supported on this module. COMSEC functions include 6 simultaneous channels of crypto functions. The salient features of this module are as follows:

- single width SEM-E module (IEEE-1101.4);
- red side Motorola PPC @ 200 MHz;
- black side Motorola PPC @ 200 MHz;
- 16 Mbytes SDRAM on red and black side;
- 1 Red side TMS320C549 DSPs @100 MHz each;
- 4 Black side TMS320C549 DSPs @ 100 MHz each;
- IEEE-1394-1995 100 Mbps backplane interfaces on red and black side;
- 2 RS232 serial ports;
- 1 external crypt port;
- uses NSA approved crypto chips, providing 6 channels of crypto simultaneously.

The KMM provides storage and protection for JCIT's INFOSEC keys. The KMM is a repackaging (new form factor) of an approved NSA hardware and software design called the ASKMM. This design supports NSA's Electronic Key Management System as well as traditional key delivery systems. Keys are protected using a removable Crypto Ignition Key (CIK). Over one hundred keys can be stored in this device. The use of the CIK and the encryption of classified operating code, allow the JCIT to be CCI when un-keyed or when the CIK is removed.

The MPM provides overall system control, resource allocation, and status. Additionally the MPM provides user interfaces to access classified data and voice. The salient features of this module are as follows:

- single width SEM-E module (IEEE-1101.4);
- Motorola PPC603 @ 200 MHz;
- 32 Mbytes SDRAM;
- 3 TMS320C549 DSPs @ 100 MHz each;
- redundant MIL-STD-1553B data bus;
- 1 RS232 serial port;
- 4 universal serial ports (RS232, RS422, MIL-STD-188/114);
- Ethernet (10bT and 100bT);
- 4 analog headset microphone interfaces;
- 4 analog intercom interfaces;
- 4 analog external radio interface (IDM like);
- IEEE-1394-1995, 200 Mbps cable interface;
- IEEE-1394-1995 100 Mbps backplane interface;

6. Unit Weight and Volume

The following reflect weight and size of the JCIT Architecture components:

JCIT: 19" rack mount, 8.5" tall, 8.4" deep; 34 pounds
AIM: 19" rack mount, 1.75" tall, 8.4" deep; 9 pounds
VHF RFM: 19" rack mount, 12.25" tall, 15.2" deep; 80 pounds with cancellers
UHF RFM: 19" rack mount, 10.5" tall, 15.2" deep; 60 pounds

7. Software

The software in the RFM and JCIT is developed using high level languages. The majority of the software is written in C. Where legacy code existed in ADA, it was ported to the JCIT. This comprised less than 1% of the total code. A trivial amount of well-documented assembly code is used in time critical interrupt routines. The software architecture is formulated in such a way as to provide a robust infrastructure to support a waveform developer. Standard services such as inter-task and inter-processor message routing, generic crypto interfaces, generic radio interfaces, system timing access, scheduling, and resource allocation are handled by the infrastructure software. Software libraries contain commonly-used functions such as encoding/decoding, interleaving/de-interleaving, correlation, voice coding, and message translation. The major challenge in developing the software architecture was to keep the functions generic but still useful. The infrastructure functions are neither waveform nor hardware implementation specific.

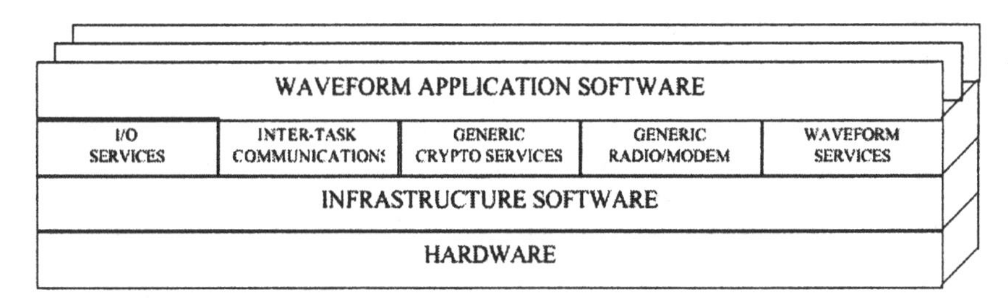

Figure 3. JCIT Software Architecture

All of the JCIT modules use the POSIX-compliant VxWorks commercial operating system. A JCIT applications user manual is also maintained, which includes waveform examples and templates. Detailed information allows new waveform-developers to efficiently use the infrastructure software. Waveforms are expressed as calls to a large standard library of JCIT services. These provide access to RF, modulate the signals, encrypt them, access local I/O, and manage the radio as illustrated in the figure.

Multiple copies of the same waveform can simultaneously execute in the JCIT. Waveforms are started and stopped without affecting the other waveforms running in the system. Waveform application code is not affected by the number of modules present in the chassis.

The life cycle of the software is much longer than the life cycle of the hardware. Hardware will continue to evolve and current technology will become obsolete in just a few years. The application software (HAVE QUICK, SINCGARS, *etc.*) needs to be designed to prevent breakage when the hardware changes. Thus, the infrastructure software was designed to isolate the application software from the physical hardware implementation. The application software uses hardware assets through the following representative isolation mechanisms: device drivers, function calls, inter-task/inter-processor communications via

POSIX Queues, and TCP/IP boot and file access protocols. The applications code is not allowed to communicate directly with hardware assets.

System resource allocation and control is managed by the Resource Manager software. This code dynamically determines the resources available. If at some point a box were made with 50 radios in it, the Resource Manager would recognize and configure those resources as required by the user's configuration commands. No changes are required to the Resource Manager to support such varied module configurations. The Resource Manager responds to high level user commands. The command "SINCGARS clear voice at 45 MHz" is an illustrative command. The Resource Manager picks the best mix of the available resources to bring up that waveform. Resources can be assigned based on user or waveform priority.

The Resource Manager incorporates four levels of configuration mechanisms. The lowest level of configuration (1) gives the operator the same control as the front panel of a discrete radio. The middle level of configuration (2) gives the operator the ability to access networks by their name, rather than by manually entering the radio parameters. Next, the user may access the entire Communications Signals Operating Instructions (CSOI) catalog through the JCIT without cheat-sheets or hardcopy books (3). The highest level of configuration (4) allows the user to set up all radios in the system with a single action. Multiple system pre-sets can be created, each with different radio parameters or radio types. These mechanisms provide a simple method for the user to configure the system.

Application software has been designed, coded, debugged, documented, demonstrated and field tested in the JCIT architecture. Waveforms include SINCGARS, SINCGARS SIP, TRAP, TADIXS-B, TIBS, HAVE QUICK , Civilian Aviation, Civilian Maritime, Civilian Law Enforcement, Generic AM (2 MHz to 512 MHz, voice), Generic FM (2 MHz to 512 MHz, voice), HF upper or lower sideband AM, GPS, Military precision, IEEE-802.11, 11Mbps TCP/IP networking, DBS/GBS (DVB-S) Audio, Video, Data, Voice Over IP (VOIP), bridging and Gateway, any voice waveform to any other, VOIP to any voice waveform.

The SINCGARS and TIBS waveforms were not well documented originally. The JCIT development team therefore invested substantial staff time in the reverse-engineering of the dozen modes of this waveform. Investments made by waveform developers in documentation, simulation tools, *etc.*, that reduce the reverse-engineering effort tend to enhance waveform portability and thus value to future Software Definable Radio (SDR) projects.

8. JCIT Lessons Learned

The development of these waveforms yielded important lessons regarding the evolution of SDR architecture. First, JCIT partitioned and allocated hardware and software functions jointly. Thus, a JCIT module consists of a hardware host module and an associated software object, the combination of which synthesize a specific radio architecture function (e.g., modem). Next, JCIT uses a "heavy" infrastructure layer. This resulted from up-front analysis of the waveforms to identify common building blocks. In addition to distributed processing building blocks typical of the JTRS infrastructure [ ], the JCIT infrastructure includes radio building blocks such as PSK modulators and Viterbi decoders. Since these blocks occur in many waveforms, they are well documented, coded efficiently, and subject to high reuse.

9. Conclusions

Consistent with over thirty five years of technology innovation and transition to industry, JCIT represents the cutting-edge of Software Radio technology for the DoD and Civilian agencies as well as NRL's continuing ability to invent, productize, and field advanced technology to the warfighters.

Parameter Representations for Baseband Processing of 2G and 3G Mobile Communications Systems

A. Wiesler, O. Muller, R. Machauer, F. Jondral

Institut für Nachrichtentechnik, Universität Karlsruhe
D-76128 Karlsruhe, Germany

Abstract: The international standard for third generation (3G) mobile communications IMT-2000 includes five different air interfaces. It is desirable that all interfaces can operate everywhere in order to reach a global roaming, however this may be not feasible. Furthermore a backward compatibility to the second generation (2G) mobile systems such as GSM, IS-136 and IS-95 is required. So at least the hand-helds should be able to perform the main 2G and 3G air interfaces to operate in the available systems. A parametrized structure for the base band processing is proposed in this paper, which guarantees small handsets, the possibility of updating and of seamless system handover, e.g. between UTRA and GSM. Since the baseband functions will be implemented on reconfigurable hardware like DSPs and FPGAs, it can bee viewed as a software radio or software defined radio. The way of parameterization is explained in detail by the example of a common modulator in this paper. Also some implementation aspects of the MAP-algorithm used for channel decoding and equalization are discussed.

1. Introduction

Recently the International Telecommunication Union (ITU) has specified five different air interfaces, developed by different countries, to be included into the global 3G mobile communication system IMT-2000. These five modes are: direct spread (UTRA-FDD), multi carrier (cdma2000), time code (UTRA-TDD), single carrier (UWC-136) and frequency time (DECT). It is desirable that all interfaces can operate everywhere to reach a global roaming. Also, the backward compatibility to 2G systems must be integrated into a 3G hand held. In the case of UTRA-FDD/UTRA-TDD a completely new air interface has been developed with very few similarities to GSM. However, the concept of UTRA-FDD enables seamless handover to GSM with only one receiver branch, which may help to keep the cost of receivers modest by simultaneously guaranteeing global coverage [1]. That means a parametrized implementation of the baseband functions includes the possibility seamless system handover, however, no parallel communication with different systems is possible. For commercial mobile communication it is sufficient to restrict the software defined radio to systems which will be used for the next decade. For instance it is not necessary to include all known modulation schemes, since only four different are used in all major systems.

In Section 2 the principle of parameterization will be described by the example of a common modulator. In Section 3 some solutions for a common channel decoding and equalization structure are proposed. In Section 4 implementation aspects of the MAP-algorithm are discussed. Finally, Section 5 draws the conclusions of the paper.

2. Parameterization of a Generalized Modulator

For UTRA-FDD, GSM, DECT and IS-136 the parameterization of a common modulator is taken as an example in order to keep the discussion simple. In the case of UTRA-FDD different modulator structures for up- and downlink are dedicated. As only the hand-helds are under investigation in this paper, only the uplink modulator is considered.
The standards UTRA-FDD and IS-136 use special kinds of QPSK modulators, e.g. $\pi/4$-DQPSK in IS-136, which can be implemented by a linear I/Q modulator structure. In contrast, GMSK, used for GSM and DECT, is a nonlinear modulation technique that cannot easily be realized by a general linear modulator. Since we aim for a common baseband signal processing structure for all standards under consideration, the linear approximation of GMSK has been proposed as modulation for GSM and DECT within a software radio [2], [3]. It has been shown, that a GMSK-signal can be linearly approximated without performance loss as follows:

$$s_{\text{GMSK}}(t) \approx s^{\text{lin}}(t) = \sum_{k=0}^{\infty} z_k C_0(t - kT) \tag{1}$$

with the main pulse $C_0(t)$ and the symbols

$$z_k = \exp\left[j\frac{\pi}{2}\sum_{i=0}^{k} d_i\right] \tag{2}$$

where d_i is a NRZ-bitstream. In Fig. 1 several building blocks of the general modulator are shown which have to be adapted to the different standards. The block **precoder** is used for GSM only since here a differential precoding is performed:

$$\tilde{b}_i = b_i \oplus b_{i-1}, \quad b_{-1} = 1 \tag{3}$$

where $\oplus$ denotes modulo 2 summation and $b_i \in \{0, 1\}$ are data bits. For each new burst the precoder starts from $b_{-1} = 1$. The parameters of this model are *BurstLength* and *Precoder_On_Off*, because no precoding is used in all the other systems.
The block **NRZ** transforms the bits $\tilde{b}_i$ into a NRZ signal d_i. The block is controlled by the parameter *NRZ_On_Off*. For *NRZ_On_Off* $= 1$ a bit $b = 0$ is mapped to 1 and $b = 1$ to -1. *NRZ_On_Off* $= -1$ identifies the inverse operation and *NRZ_On_Off* $= 0$ means no conversion. With **MBIT2Symbol** the input bits are

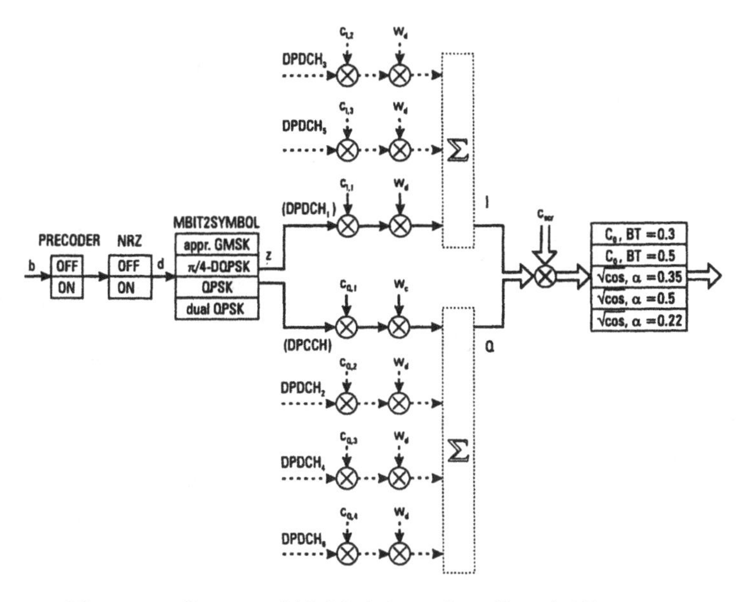

Figure 1. Common I/Q-Modulator for 2G and 3G systems

transformed to complex symbols. The modulation mode is adjusted by the parameter *ModulationNumber*, for which Table 1 defines the values. The parameter *BurstLength* must also be used with **MBIT2Symbol**, since some of the modulation modes have memory and therefore a starting point has to be defined for each burst. In the case of UTRA-FDD (downlink) the control channel DPCCH is routed to the Q-branch and the data bits are routed to the I-branch ($DPDCH_1$). The modulation can be seen as dual QPSK, which performs a serial to parallel transformation with routing *I_Length* bits to the I-branch and *Q_Length* bits to the Q-branch. Clearly, all other systems do not use this parameters. For the spreading factor 4 up to five other data channels can be transmitted in parallel over the I- and Q-branches in UTRA-FDD. This possibility is indicated with dotted lines.

For the CDMA system UTRA-FDD spreading is performed in the next step.

Table 1. MBIT2Symbol Parameters

ModulationNumber	Modulation Mode
1	GMSK
2	$\pi/4$-DQPSK
3	QPSK
4	dual QPSK

This function is indicated by the *Spreadingfactor*. If *Spreadingfactor*= 1, no spreading is performed. The spreading sequences are stored and can be selected by the sequence number. The following weighting of the data and control channels with the weights w_d and w_c and the complex scrambling are only for

UTRA-FDD.
Subsequently complex pulse shaping is done, where various filters can be chosen by the parameter *Filter_Number* (see Table 2.). Table 2. summarizes the para-

Table 2. Pulse Shaping Filters

Filter_Number	Filter
1	main pulse for linearly approximated GMSK with $BT = 0.3$
2	root cosine rolloff with rolloff factor $\alpha = 0.35$
3	main pulse for linearly approximated GMSK with $BT = 0.5$
4	root cosine rolloff with rolloff factor $\alpha = 0.5$
5	root cosine rolloff with rolloff factor $\alpha = 0.22$

meter settings for different systems. For UTRA-FDD the burst lengths differ according to the spreading factor or to the net data rate, respectively. Therefore a set of parameters has to be specified additionally. Table 2. shows the parameter settings for a spreading factor of the DPDCH *Spreadingfactor_I* = 8. The DPCCH channel is always spread with a spreading factor *Spreadingfactor_Q* = 256.

Table 3. Parameter Settings for Different Transmission Standards

	GSM	DECT	IS-136	UTRA-FDD
BurstLength	148	424	312	330
Precoder_On_Off	1	0	0	0
NRZ_On_Off	1	-1	1	-1
ModulationNumber	1	1	2	4
Spreadingfactor_I	1	1	1	8
Spreadingfactor_Q	1	1	1	256
Filter_Number	1	3	2	5
I_Length	-	-	-	320
Q_Length	-	-	-	10

3. Parametrized Receiver Structure for Channel Decoding and Equalization

Obviously there are substantial differences between the CDMA system UTRA-FDD, and the TDMA systems GSM, IS-136 and DECT. However, if we look at the signal processing step by step, it becomes evident that channel coding is quite similar for 2G mobile systems: For speech transmission block coding for the most significant bits is applied, while convolutional coding and interleaving is used for the majority of the bits within the data packet. The cordless phone standards like DECT are designed for indoor and outdoor pedestrian applications. Therefore only block coding for quality control is employed in that standard. The third generation mobile systems permit data rates up to 2 Mbit/s and have to guarantee bit error rates of about 10^{-6} for specific applications. Here turbo

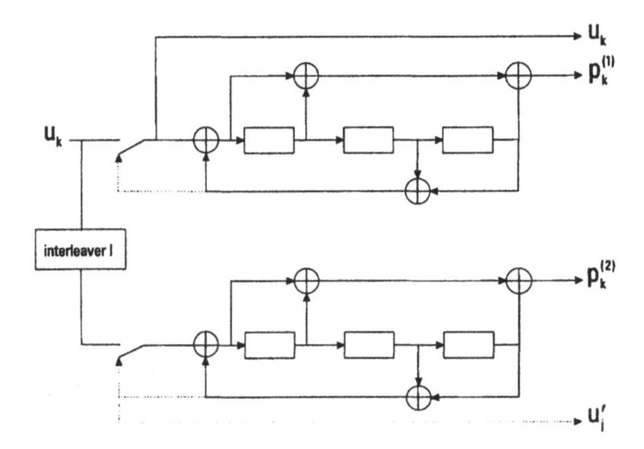

Figure 2. Turbo encoder used in UTRA-FDD

codes are proposed [4].

In UTRA-FDD two parallel recursive systematic convolutional (RSC) codes separated by an intermediate interleaver are used for turbo coding. RSC-codes have the same coding properties as nonrecursive, nonsystematic convolutional codes with the same generator polynomials. The advantage of the systematic coding is that this reduces the coding rate of the turbo code, because the data bits u_k which are included in both code words must only be transmitted once. For the case of two equal convolutional codes of code rate 1/2 as for UTRA-FDD, this results in a coding rate of $R = 1/3$ instead of 1/4. In Figure 2 the used turbo encoder specified from 3gpp [1] is drawn. The transmitted coded bits consist of the data bits u_k and the parity bits $p_k^{(i)}$ produced by the first and second RSC-code. To simplify the implementation the turbo encoder is terminated after a block of N input data bits. The termination is performed by deflecting the recursive bit stream to the entrance of the RSC-encoder. This produces zeros at the entrance of the shift register. After M (equal to the constraint length) zero inputs the shift register reaches the zero state. Therefore N data bits produce $(1/R) \cdot (N + M)$ coded bits including the tail bits for the termination. After channel encoding a second interleaving, consisting of two interleavers, and rate matching is provided, however rate matching is not considered in our simulations.

In Fig. 3 a common receiver for a 2G system like GSM and a 3G system like UTRA-FDD is shown. For GSM sampled symbols y_k are processed according to the dotted lines, UTRA-FDD samples according to the solid lines. The received symbols are corrupted by additive white Gaussian noise (AWGN) n_k and multipath propagation which can be modeled by the channel impulse response h_l:

$$y_k = \sum_{l=0}^{L} z_{k-l} h_l + n_k. \tag{4}$$

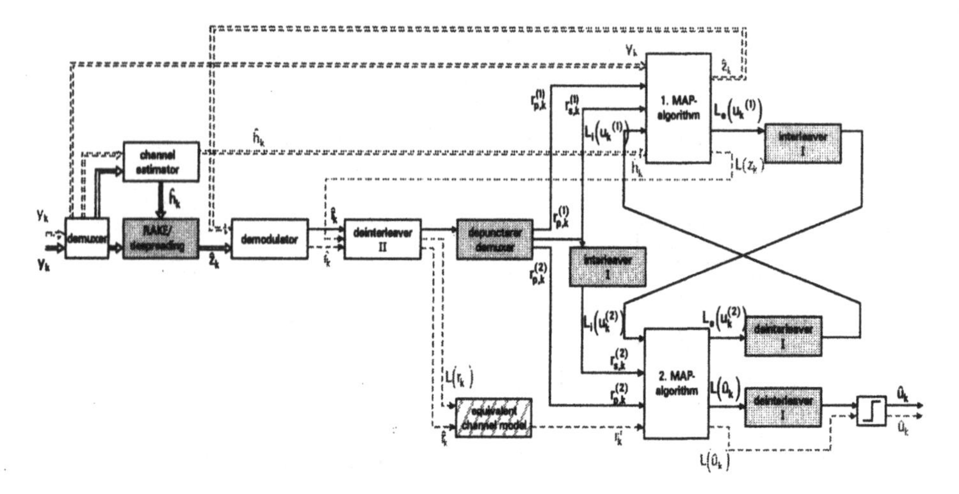

Figure 3. General receiver structure for GSM and UTRA-FDD

The channel impulse response h_l has to be estimated, which can be achieved by correlating the known training or pilot sequence with the received sequence. The sequence y_k may also be interpreted as the output of a convolutional encoder of rate 1. Therefore basically identical algorithms can be used for equalization and for decoding. The best suited algorithm for this task is the MAP symbol by symbol algorithm, which will be discussed in more detail in Section 4. It is based on a trellis similar to the well known MLSE Viterbi algorithm. The trellis shows the possible transitions between succeeding symbols (for equalization) or between convolutionally encoded bits. For equalization, the transitions depend on the modulation used. In a GMSK modulated signal real and imaginary symbols alternate. Fig. 4 shows a trellis for GMSK signals transmitted over a channel with L = 3 taps (see (4)). For channel decoding the possible transitions depend on the generator polynomial of the encoder. The MAP algorithm uses

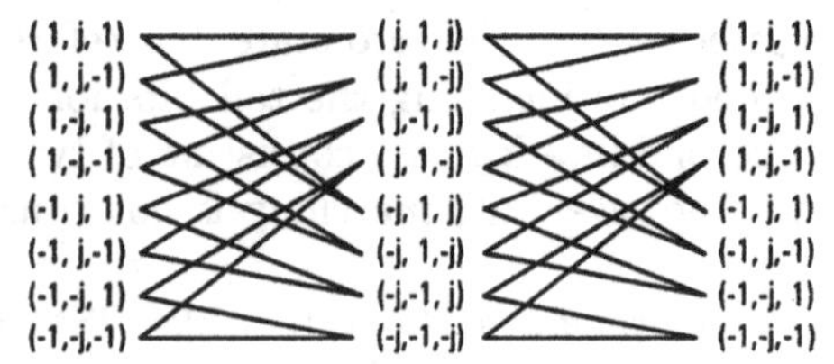

Figure 4. Trellis for GMSK with L= 3

information about the signal to noise ratio and it can additionally take soft input information into account. Furthermore the MAP algorithm can output soft information necessary for turbo decoding.

For GSM signals the estimated channel impulse response is used for equalization with the first MAP algorithm shown in Fig. 3. The MAP-algorithm can be switched from convolutional decoding to equalization by parameter setting. The

outputs of this algorithm are the equalized symbols $\hat{z}_k$ and the corresponding soft output information $L(\hat{z}_k)$. The demodulator transforms the symbols into bits which are deinterleaved under consideration of the soft output information. The data is then sent through the equivalent channel model and the output is fed into the second MAP algorithm for convolutional decoding. The last step is the bit decision with respect to the signs of the outputs $L(\hat{u}_k)$.
For UTRA-FDD signals, equalization and despreading is done by a Rake receiver, which utilizes the (approximately) orthogonality of the different multipath signals generated by using spreading codes with good AKF properties. After demodulation the received data bits r_k are channel decoded. After the outer deinterleaver, two MAP-algorithms are used in the case of turbo decoding as follows. First the received bit stream has to be separated into information bits r_s and parity bits r_p with respect to the first and second RSC-encoder (see Fig. 2). Then the first convolutional decoding is done by the MAP-algorithm, delivering soft outputs $L_e(u_k^{(1)})$. To use this soft output as soft input for the decoding of the second RSC-code, $L_e(u_k^{(1)})$ and r_s have to be interleaved according to interleaver I. The first iteration of the turbo decoder is finished by decoding the second RSC-code and delivering $L_e(u_k^{(2)})$, which is utilized in the next iteration for decoding the first RSC-code again. After a predefined number of iterations a decision is made according to the signs of the outputs $L(\hat{u}_k)$. It is obvious that only approximately statistically independent soft inputs generate an additional coding gain in the following iterations. This independence is provided by interleaver I.

4. The MAP-Algorithm

Basically, detection or decoding means an estimation of the transmitted bit u (or symbol z in the case of equalization) taking into account the corresponding received bit r (or symbols y). In the case of channel decoding ideal equalization is assumed, that means r is only disturbed by additive white Gaussian noise and can be represented by the continuous Gaussian random variable R. In the case of equalization, the received symbols y can be seen as channel coded with coding rate 1 due to the multipath channel and are disturbed by additional white Gaussian noise. To keep the discussion clear only the channel decoding case is discussed in the following, since equalization can be seen as a special case.
The transmitted bits u may either be 1 or -1 and are represented by the discrete random variable U. According to the MAP criterion the following log-likelihood-ratio (LLR) can be derived for one transmitted bit u_k

$$L(U_k|\underline{\mathbf{r}}) = L(\hat{U}_k) = L_R(r_k|U) + L_i(U_k) + L_e(U_k) \tag{5}$$

with the block $\underline{\mathbf{r}} = (r_{s,1}, r_{p,1}, \ldots, r_{s,K}, r_{p,K})$ of received bits. The LLR $L_i(U_k)$ represents the a priori information about U_k. The first term in (5) is

$$L_R(r_k|U_k) = 4E_{cb}/N_0 r_k = L_c \cdot r_k. \tag{6}$$

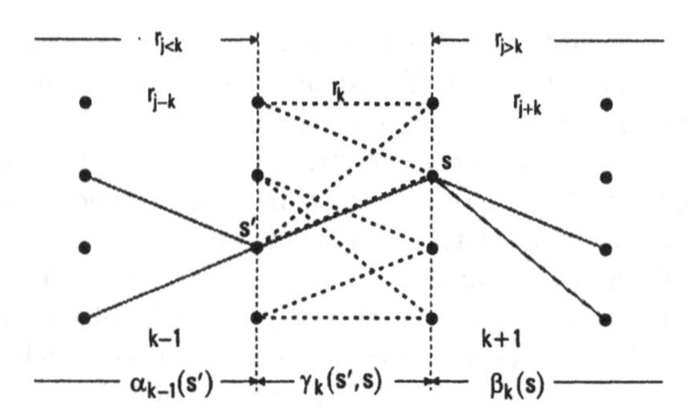

Figure 5. Calculation of $L(\hat{U}_k)$ on the basis of a trellis

Here E_{cb} is the energy per transmitted bit in contrast to E_b used in Section 4.2 which is the energy per data bit. In the case of channel decoding there is also the information about U_k provided by the parity bits which has to be taken into account, which results an additional term $L_e(U)$ in (5). In the case of Turbo decoding $L(\hat{U}_k)$ is computed for every information bit u_k, using the estimated E_{cb}/N_0 and the soft input $L_i(U_k)$, to obtain the soft output

$$L_e(U_k) = L(\hat{U}_k) - L_c \cdot r_k - L_i(U_k) \tag{7}$$

for the next iteration step. For the final decision $L(\hat{U}_k)$ is used after the last iteration (see Fig. 3). The LLR of U_k from equation (5) can be transformed to

$$L(\hat{U}_k) = \ln \frac{\sum\limits_{\substack{\{(s',s)\}\\ U_k=+1}} p(s',s,\underline{r})}{\sum\limits_{\substack{\{(s',s)\}\\ U_k=-1}} p(s',s,\underline{r})} = \ln \frac{\sum\limits_{\substack{\{(s',s)\}\\ U_k=+1}} \alpha_{k-1}(s') \cdot \tilde{\gamma}_k(s',s) \cdot \beta_k(s)}{\sum\limits_{\substack{\{(s',s)\}\\ U_k=-1}} \alpha_{k-1}(s') \cdot \tilde{\gamma}_k(s',s) \cdot \beta_k(s)} \tag{8}$$

Here the sum of all joint densities $p(s',s,\underline{r})$ is generated for every transition from state s' to s with $U_k = +1$ and $U_k = -1$, respectively. $\alpha_{k-1}(s')$ is the probability that the path with the first $k-1$ transitions ends in state s'. The probability $\beta_k(s)$ can be interpreted similarly, but starting from the end of the trellis (see Figure 5). The term $\tilde{\gamma}_k(s',s)$ is the normalized transition probability from s' to s. It can be shown, that

$$\begin{aligned}\tilde{\gamma}_k(s',s) &= \exp\left(\frac{1}{2}\sum_{l=1}^{n} L_c^{(k)} r_{l,k} b_{l,k}\right) \cdot e^{u_k \cdot L_i(U_k)/2}\\ &= \chi_k(s',u_k) \cdot e^{u_k \cdot L_i(U_k)/2}\end{aligned} \tag{9}$$

where $b_{l,k}$ are the channel coded bit values according to the transitions from s' to s in the trellis. The values $\chi_k(s',u_k)$ are unchanged for every turbo iteration and have to be calculated only once. The terms $\alpha_{k-1}(s')$ and $\beta_k(s)$ can then be calculated recursively. For numerical reasons $\alpha_k(s)$ und $\beta_k(s)$ are scaled by a factor which can be canceled in (8). We obtain for $\alpha'_k(s)$ and $\beta'_k(s)$:

$$\alpha'_k(s) = \frac{\alpha_k(s)}{p(r_{j\leq k})} = \frac{\psi_k(s)}{\varphi_k} \tag{10}$$

with

$$\psi_k(s) = \sum_{s'} \tilde{\gamma}_k(s', s) \cdot \alpha'_{k-1}(s') \tag{11}$$

$$\varphi_k = \sum_{\hat{s}} \psi_k(\hat{s}) \tag{12}$$

and

$$\beta'_{k-1}(s') = \frac{\sum_s \tilde{\gamma}_k(s', s) \cdot \beta'_k(s)}{\varphi_k} \tag{13}$$

4.1 The Log-MAP-Algorithm and Max-Log-MAP-Algorithm

Several variations of the MAP-algorithm have been discussed in literature [5], [6]. Transforming (8) by using (10) and (13) yields

$$\begin{aligned} L(\hat{U}_k) = \ln & \sum_{\substack{\{(s',s)\} \\ U_k=+1}} e^{\bar{\alpha}'_{k-1}(s') + \bar{\gamma}_k(s',s) + \bar{\beta}'_k(s)} \\ & - \ln \sum_{\substack{\{(s',s)\} \\ U_k=-1}} e^{\bar{\alpha}'_{k-1}(s') + \bar{\gamma}_k(s',s) + \bar{\beta}'_k(s)} \end{aligned} \tag{14}$$

with

$$\bar{\alpha}'_{k-1}(s') = \ln \alpha'_{k-1}(s') \tag{15}$$

$$\bar{\beta}'_k(s) = \ln \beta'_k(s) \tag{16}$$

$$\bar{\gamma}_k(s', s) = \ln \tilde{\gamma}_k(s', s) \tag{17}$$

To solve this equation mainly operations like $\ln(e^{\delta_1} + e^{\delta_2} + \ldots + e^{\delta_n})$ have to be calculated. In the log-MAP-algorithm the Jacobian logarithm is used:

$$\begin{aligned} \ln(e^{\delta_1} + e^{\delta_2}) &= \max(\delta_1, \delta_2) + \ln(1 + e^{-|\delta_1 - \delta_2|}) \\ &= \max(\delta_1, \delta_2) + f_c(|\delta_1 - \delta_2|) \end{aligned} \tag{18}$$

were $f_c(.)$ is a correction function. If the logarithm of a higher sum has to be calculated this can be done recursively, with $\delta = \ln(e^{\delta_1} + e^{\delta_2} + \ldots + e^{\delta_{n-1}})$ already evaluated:

$$\ln(e^{\delta} + e^{\delta_n}) = \max(\delta, \delta_n) + f_c(|\delta - \delta_n|) \tag{19}$$

If this is done without approximation the log-MAP-algorithm has the same performance as the original MAP-algorithm. The advantage of this form of the MAP-algorithm is that the sums have been turned to maximum functions. However there is still the correction function $f_c(.)$ to be calculated. As this function only depends on $|\delta_2 - \delta_1|$, it has been shown that it is sufficient to store only 8 values of this function for the calculation of $L(\hat{U}_k)$ [7].
It is also possible to use the following approximation of (19):

$$\ln(e^{\delta_1} + e^{\delta_2} + \ldots + e^{\delta_n}) \approx \max_{i \in \{1 \ldots n\}} \delta_i \tag{20}$$

Using this approximation the following equation called max-log-MAP is obtained for $L(\hat{U}_k)$:

$$L(\hat{U}_k) \approx \max_{\substack{\{(s',s)\}\\ U_k=+1}} \left(\bar{\alpha}'_{k-1}(s') + \bar{\gamma}_k(s',s) + \bar{\beta}'_k(s)\right) - \max_{\substack{\{(s',s)\}\\ U_k=-1}} \left(\bar{\alpha}'_{k-1}(s') + \bar{\gamma}_k(s',s) + \bar{\beta}'_k(s)\right) \quad (21)$$

It was shown that this results in two Viterbi algorithms using mainly add, compare and select operations [8], which means less computational effort. However this is traded by poorer bit error rates (see Fig. 6).

4.2 MAP-Algorithm in Quantized Values

The fixed-point representation for the MAP-computation is analyzed by simulating a data transmission over the UTRA-FDD downlink air interface with turbo coding according to Fig. 2. The interleaver depth is very flexible. We chose an interleaver depth of 3037 (one 10 ms UTRA frame using a spreading factor of 8) for interleaver I and 60 ms for the outer interleaver II. The CDMA-signal is transmitted with the chip rate of 3.84 Mchips/s over an time variant multipath channel and AWGN. Different channel profiles have been studied, but mainly the vehicular A channel with a mobile velocity of 120 km/h was employed [9]. For the quantization analysis of the turbo decoding, varying the SNR produced more interesting results than different channel models or varying the mobile velocity. The receiver used for simulations was implemented according to Fig. 3 with a Rake receiver having four fingers performing maximum ratio combining. Ideal channel information including the channel impulse response h and E_{cb}/N_0 generated once per slot is assumed. In Figure 6 simulation results are presented with the original MAP-algorithm without quantization for different iteration numbers. It can be seen that the channel coding gain remains low after four iterations, so mainly four iterations are considered in the following discussions. Also the BER results using the max-log-MAP-algorithm with four iterations are displayed, which show a loss of 0.5 dB at BER=10^{-6} compared with the original MAP.

The necessary word length for the soft output $L_e(U_k)$ soft input $L_i(U_k)$ is analyzed, respectively. The range of these values mainly depend on the iteration number. In Figure 7 the values inside the MAP calculations are still in float accuracy but the soft information is quantized with different resolutions. Here always a quantization performing rounding is used. From the simulations it can be concluded that it is sufficient to use soft input/soft output in the range of $[-8, 8[$ with a word length of 5 bits if only four iterations are performed. Also the quantization of the values inside the MAP-algorithm have been analyzed. It turned out that especially the term $P(U_k = u_k) = e^{u_k \cdot L_i(U_k)/2}$ of $\tilde{\gamma}_k(s', s)$ in (9) is critical in the original MAP-algorithm. Because of the exponential function and the sufficient range of $L_i(U_k)$ discussed above, this value varies between 54.6 and $1.8 \cdot 10^{-2}$, which means a necessary word length of 11 bits for $P(U_k = u_k)$, resulting in a necessary word length of 18 bits for $\tilde{\gamma}_k(s', s)$. It is obvious that

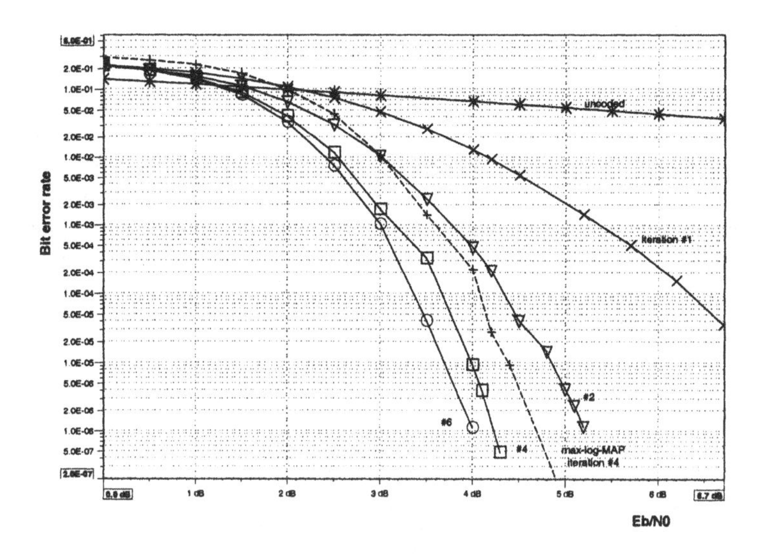

Figure 6. BER of UTRA-FDD with turbo coding

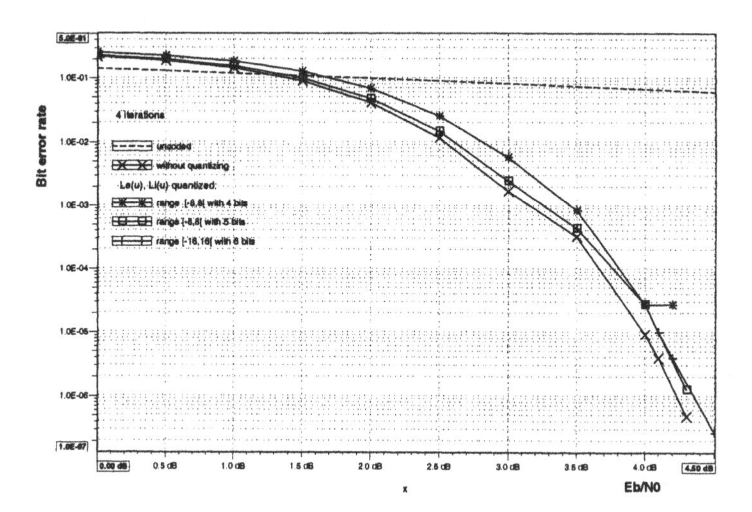

Figure 7. BER of UTRA-FDD with turbo coding with quantized soft information

this in turn results in even higher word length for $\alpha'_{k-1}(s')$ and $\beta'_k(s)$, which makes the original MAP very impractical. On the other hand considering the log-max-MAP or max-log-MAP the ln-function eliminates the exponential function in $P(U_k = u_k)$ and therefor keeps the necessary word length low. Further simulations proved that a word length of 8 bits for every calculated value into the max-log-MAP produce only a loss of about 0.1 dB compared to the exact floating point representation in the case of 5 bit soft information.

5. Conclusion

A parametrized implementation of baseband functions for software radios is discussed in this paper. The principle is described by the example of a common modulator for several 2G and 3G mobile communications systems. Also a common receiver structure for a reduced hardware implementation is proposed. Special aspects of the application of the MAP-algorithm for turbo decoding and consecutive equalization and convolutional decoding are discussed. It has been shown that due to the necessary word length of the fixed-point representation only the variations log-MAP and max-log-MAP are practicable.

References

[1] 3rd Generation Partnership Project (3gpp), *Technical Specification Group Radio Accesss Network*, 3G TS 25.XXX. Mar. 2000.

[2] A. Wiesler, R. Machauer, F. Jondral, "Comparison of GMSK and linear approximated GMSK for use in Software Radio", in *Proceedings of the 5th international Symposium on Spread Spectrum Techniques & Applications ISSSTA '98*, Sept. 1998, pp. 557–560.

[3] A. Wiesler, H. Schober, R. Machauer, F. Jondral, "Software Radio Structure for UMTS and Second Generation Mobile Communication Systems", in *Proceedings of the 50th International Vehicular Technology Conference VTC'99 Fall*, Sept. 1999, pp. 939–942.

[4] C. Berrou, A. Glavieux, P. Thitimajshima, "Near shannon limit error-correcting coding and decoding: Turbo-codes", in *Proceedings of International Conference on Communications*, 1993, pp. 1064–1070.

[5] G. Bauch and V. Franz. "A Comparison of Soft-In/Soft-Out Algorithms for Turbo-Detection", in *Proceedings of International Conference on & Telecommunications (ICT98)*, June 1998, pp. 259–263.

[6] J. Hagenauer, E. Offer, L. Papke. "Iterative Decoding of Binary Block and Convolutional Codes", in *IEEE Transactions on Information Theory*, 42(2):429–445, 1996.

[7] P. Robertson, E. Villebrun, P. Hoeher. "A comparison of optimal and sub-optimal MAP decoding algorithms operating in the log domain", in *IEEE International Conference on Communications*, 1995, volume 2, pp. 1009–1013.

[8] M. Fossorier, F. Burkert, S. Lin, and J. Hagenauer. "On the equivalence between SOVA and max-log-MAP decodings", in *IEEE Communications Letters*, 2(5):137–139, 1998.

[9] ETSI, *Universal Mobile Telecommunications System (UMTS); UMTS Terrestrial Radio Access (UTRA); Concept evaluation (UMTS 30.06; TR 101 146)*, 3.0.0 ed., Dec. 1997.

Digital Compensation in IF Modulated Upconversion Software Radio Architecture

B. Eged, P. Horváth, I. Frigyes

Budapest University of Technology and Economics,
Department of Microwave Communications,
1111 Budapest, Goldmann György tér 3, Hungary,
TEL:(+36)-1-463-3614,
FAX:(+36)-1-463-3289,
e-mail: beged@mht.bme.hu

Abstract: In the transmitter path of software radios a relevant problem is the up-conversion of the baseband or IF signal with high image frequency suppression. For this purpose an up-converter based on a quadrature modulator has been developed utilizing a signal processing method for increasing the image sideband supression. A simplified system has been built to test this new method. For extending the posibble modulation types and bandwidth the development of digital modulation approach using a direct digital synthesizer is in progress.

1. Introduction

The software radio tends to use a simpler radio frequency hardware. That is achieved by a highly sophisticated digital signal processing technique. [1], [2] The optimal border line for the interface between the hardware and the software is to be investigated. [3] Several functions can be realized by combining analog and digital circuitry with a controlling software. In this paper a possible hardware simplification is presented.

The analog circuitry of a software radio has to reflect the demands for an easily reconfigurable transmission system. This flexibility is a very relevant property of the software radio. It means, that the frequency allocation of the transmitted channels, the modulation format and the modulation bandwidth should be variable according to the information content.

2. System Concepts

In the first step the system architectures were studied. The three main methods of transmitter architectures are direct conversion, modulating the IF signal and direct RF modulation. Each method has several advantages and disadvantages.

In our system the digital modulation of the IF signal and its up-conversion by an I&Q modulator was chosen and realized. (Figure 1). Its main advantages are that the digital signal processing unit can be used for generating any type of modulation. In this unit the baseband digital data stream is low-passed and fed to the modulator. The two modulators recive the IF LO in-phase and quadrature-phase, so the modulated IF signal can be up-converted by the I&Q modulator to the RF frequency which is provided by the RF LO synthesizer.

Unfortunately the sideband rejection of the I&Q modulator is limeted by the unbalance of the phase and amplitude of its I&Q signal chain. [4], [5] The typical value of the unbalance is given by the manufacturers of the modulators, generally its depending on the frequency, temperature and the power supply voltage (Figure 2) [6]. This unbalance can be compensated by changing the phase and amplitude difference of the I&Q signals in the digital signal processing unit. The compensation value should be stored in a table and the desired value should be used at a given frequency.

Depending on the required bandwidth and baseband speed the digital IF modulation and quadrature up-conversion would need high-speed digital signal processing and a high performance digital to analog converter which is flexible but expensive. Using a hard-wired DSP block as direct digital synthesizer can be used to extend the IF frequency range and to take over a lot of tasks from DSP unit.

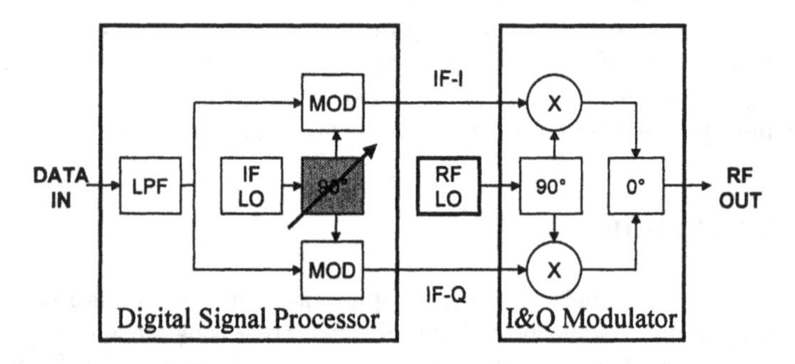

Figure 1. The block diagram of the proposed transmitter arhitecture

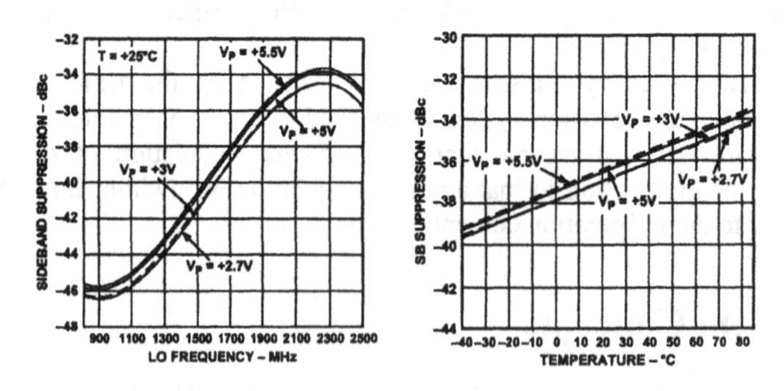

Figure 2. Typical sideband supression of a monolit, wideband I&Q modulator

3. Simulation of Sideband Supression

For validating the concept a small demonstration system was built in a system simulation software environment. (Figure 3). In this simulated system the data source was a 1kbps NRZ pseudo random code seqence which modulates directly the IF LO sources by a multiplier which means a DSB/SC modulation. The IF LO sources were realised by two independent periodic sources with frequency of 5KHz and nominal phase difference of 90 degrees.

The IF-I and IF-Q signals were fed to the I&Q modulator consisting of two multipliers and a quadrature RF LO source with frequency of 12KHz. In this system the output frequency can be 12-5=7KHz with lower sideband and 12+5=17KHz with upper sideband mixing. (Figure 4). The frequencies in the demonstration system are not realistic, they were used by simulation only. Putting the frequencies of IF LO and RF LO close each other leads to a faster simulation.

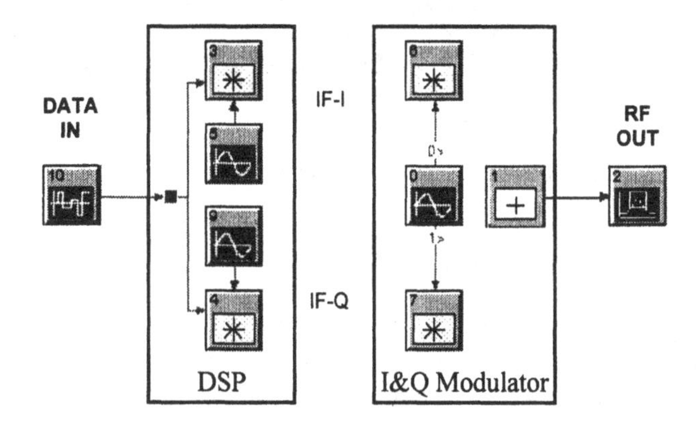

Figure 3. The simulation set-up in the SystemView environment

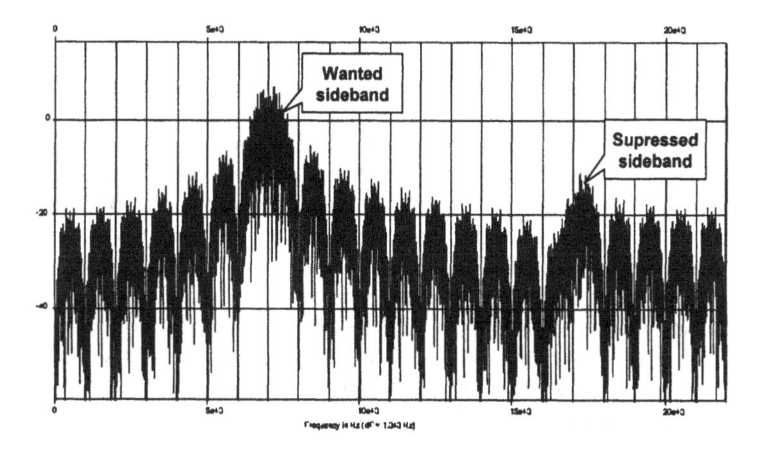

Figure 4. Simulated output spectum of the IF upconverter system

The phase and amplitude difference of the IF LO sources can be varied. In this way the sibeand supression characterisctic of an ideal I&Q modulator was determined. The sideband supression versus phase and amplitude unbalance and the output spectrum of the simulated system can be seen in the pictures. (Figure 5).

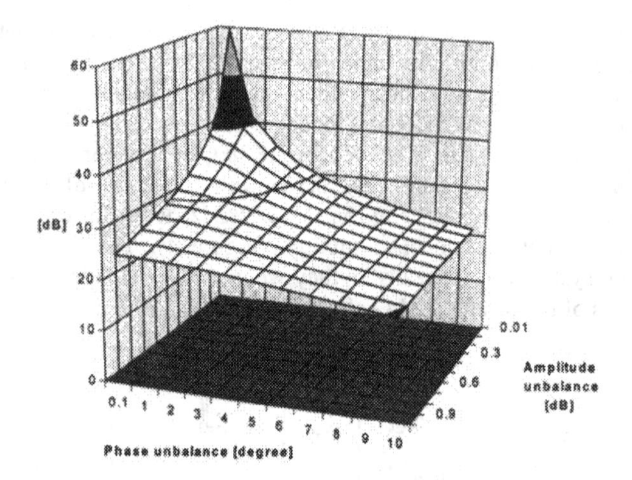

Figure 5. Sideband rejection versus phase and amplitude unbalance

4. Measurements Results

A simple test setup was made to verify the concept and the simulated values. In the test setup an HP ESG series signal generator was used as an I&Q modulator. The quadrature IF signal was generated by DSP card in PC. In the measurements the IF signal was not modulated and its frequency was set to 5KHz. This low IF LO frequency needed by the limitation of the digital to analog converter of the DSP card and it can be seen as a basebandchannel speed of a low-speed radio services. The real frequency of the IF LO is not really interested to demonstrate using digital compensation for increasing the sideband supression. The RF LO at the signal generator was selected with the frequency of 1GHz close to the frequency band of the GSM services.

First measurements were made with equal amplitude and ideal 90 degrees phase difference of the IF LO signals. In this configuration the sideband supression is determined by the I&Q modulator itself, and the phase and amplitude unbalance of the whole IF circuitry. In this case the sideband supression was 18.49dB as it can be seen in figure. (Figure 6).

In the next steps we started to change the phase and amlitude difference of the IF channels digitally and tried to find the optimum values for maximizing the supression of the unwanted sideband. With the optimal settings the supression can be increased to 47.90dB so more than 30dB increasing can be achived by using digital compensation methode. (Figure 7).

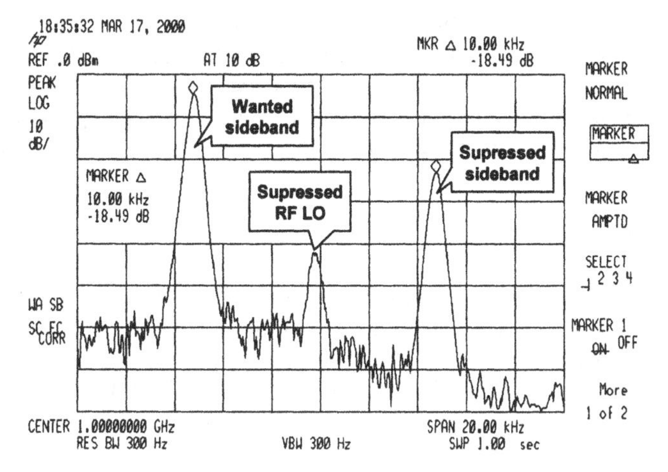

Figure 6. Measured sidebend rejection of the uncompensated system

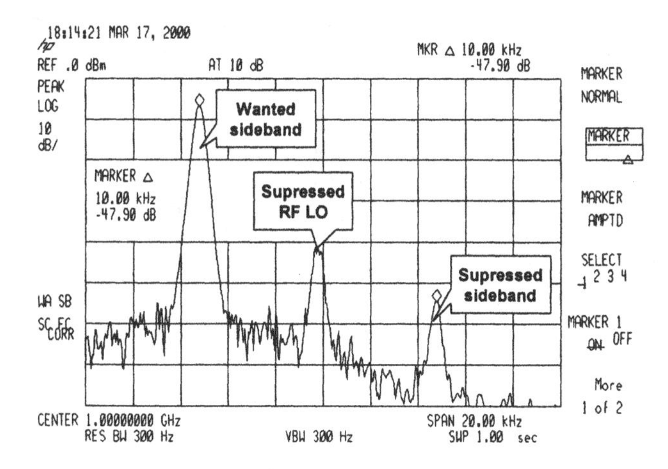

Figure 7. Measured sideband supression of the optimally compensated system

5. Future Plans

We plan to extend the capability of the presented system by using a hard-wired DSP blocks as digital upconverter or direct digital synthesizer. This method makes possibble the usage of a lower speed digital signal processing unit which leads to a cheaper arhitecture.

6. Conclusion

A software radio architecture using single sideband upconversion of the digitally modulated IF signal by an I&Q modulator was developed. The main disadvantage of this method is the limited sideband supression which was avoided by using digital compensation in the DSP module. In this way more than 30dB increasing in the sideband supression was achived. The persented method can be used in he wideband software radios where the supression is depending on the used RF frequency by using a compensation table for digital processing part of the system.

7. References

[1] J.J.Patti, R.M.Husnay, J.Pintar, "A Smart Software Radio: Concept Development and Demonstration", IEEE Selected Areas in Communications, Vol.17, No.4, pp.631-649, April 1999

[2] J.E. Gunn, K.S. Barron, W. Ruczczyk, "A Low-Power DSP Core-Based Software Radio Architecture", IEEE Selected Areas in Communications, Vol. 17, No.4, pp.574-589, April 1999

[3] P. Horváth, T. Marozsák, E. Udvary, A. Zólomy, T. Bánky, M. Csörnyei, B. Klein, F. Kubinszky, Z. Lázár, A. Varga, "Investigation on Hardware-Software Interfacing Problems for Reconfigurables Radios" Proceedings of the 1st Karlsruhe Workshop on Software Radios, March 29-30, 2000, pp.81-84.

[4] Shankar R. Joshi, "Novel I/Q Modulators Mix Celluar Signals", Synergy Microwave Corp., www.sinergie.com

[5] Richard Cushing, "800 to 2500MHz Single-Sideband Up-conversion of Quadrature DDS Signals", Analog Devices, www.analog.com

[6] AD8346 0.8-2.5GHz Quadrature Modulator, Analog Devices, www.analog.com

Part 4

Software Radio Technology Towards Pervasive Appliance

Overview of Japanese Activities in Software-defined Radio

Ryuji Kohno[1], Ryu Miura[2], Hiroshi Harada[2], Shinichiro Haruyama[3], Yukitoshi Sanada[3] and Lachlan Michael[3]

[1] Yokohama National University,
Faculty of Engineering, Division of Electrical and Computer Engineering,
79-5 Tokiwadai, Hodogaya-ku, Yokohama, 240-850, Japan
[2] Communications Research Laboratory,
Ministry of Posts and Telecommunications,
3-4 Hikarino-oka, Yokosuka, 239-0847, Japan
[3] Advanced Telecommunication Laboratory,
Sony Computer Science Laboratories, Inc.,
3-14-13 Higashi Gotanda Shinagawa, Tokyo, 141-0022, Japan

Abstract: In this paper, we will present an overview of the Japanese IEICE Software Radio Study Group's activities since it was established in December 1998. An overview of prototype Software Radio Demodulators developed in Japan will be presented. Details of some papers presented at the most recent IEICE Software Radio are outlined and the paper will be concluded by summarising the work being undertaken in the Software-defined Radio (SDR) field in Japan.

1. IEICE Software Radio Study Group

In the last few years there has been considerable interest world wide in the field of Software Radio [1, 2, 3]. There has been great interest in Japan as well and therefore, the IEICE Software Radio Study Group was established in December 1998 as part of the Communication Society within the IEICE.

The purpose of the IEICE Software Radio Study Group is to promote research and development in the field of Software-defined or Reconfigurable Radio, to allow protocol software and hardware to be integrated easily with future digital radio transceivers, to foster cross-organization collaboration between protocol and hardware developers, and to organize national and international workshops on Software-defined Radio.

The subjects covered by the IEICE Software Radio Study Group include but are not limited to the following:

- theory of software radio;
- software and hardware technology for use in software radio;
- applications to communications, broadcasting, ITS *etc.*;
- research of Application Programming Interfaces (API);
- standardization of software radio;
- information exchange and co-operation with organizations active in other countries.

At present, the main committee members of the IEICE Software Radio Study Group are as shown in Table 1.

Table 1. Main committee members of the IEICE Software Radio Study Group in Japan for 2000

Position	Name	Organization
Chair	Ryuji Kohno	Yokohama National University
Secretaries	Shinichiro Haruyama	Sony Computer Science Laboratory
	Ryu Miura	MPT Communications Research Laboratory
Assistant Secretaries	Hiroshi Harada	MPT Communications Research Laboratory
	Yukitoshi Sanada	Sony Computer Science Laboratory

1.1 Schedule of IEICE Software Radio Study Group in 1999

After the IEICE Software Radio Study Group was formed in December 1998, the first technical committee meeting took place on 27th January 1999. Following that, three technical meetings were held during the year and one panel session was organised during PIMRC '99 in Osaka:

27th January first technical committee meeting (Tokyo);
11th March first technical meeting and workshop (Yokosuka Research Park) together with the SDR Forum (9th – 11th March);
30th June second technical meeting (Nagoya);
September panel session (Osaka): done as a session in the PIMRC '99 conference (10th International Symposium on Personal, Indoor and Mobile Radio Communications);
17th November third technical meeting (Nagoya).

1.2 Schedule of IEICE Software Radio Study Group in 2000

The activities already undertaken in 2000 include two technical meetings and a panel session within the Vehicular Technology Conference (VTC) 2000 (Spring) conference held in Tokyo. A panel session and symposium in Nagoya and a technical meeting in Tokyo are planned for the latter part of the year.

17th April fourth technical meeting and workshop (Yokosuka Research Park) together with the SDR Forum;
May panel session at VTC 2000 in Shinagawa (Tokyo);
21st July fifth technical meeting (Tokyo);
30th September panel session and symposium (Nagoya);
20th October sixth technical meeting (Tokyo).

So far, there have been 36 regular technical papers, 12 invited presentations and one panel session organised in the five technical meetings held since the Study Group's inception.

Table 2 is a list of the issues (to be) discussed at the technical meetings.

Table 2. A list of the issues to be discussed at the IEICE Software Radio Study Group in Japan

General	Applications of SDR that are suitable for Japanese wireless communications environment. Harmonization between the SDR Forum and Japanese SR activities. Introduction of SDR to Japanese researchers and industry. Backward compatibility between generations of cellular standards. Contribution to the Forum Standard (Prototypes). Contibution to the regulatory issues in MPT. Multiple wireless services for ITS application. Definition of wireless systems. Authentication and encryption. Stratospheric platform. Observation of illegal radio wave emission. Intellectual property aspects. Promotion of the SDR activities in Japan.
Protocols	Software download protocols. Reconfigurable protocols.
Architecture	Hardware architecture. Software architecture (APIs).
Devices	Antenna. RF, IF filter. A/D, D/A converter. DSP, FPGA. Battery.

2. Overview of the Prototype Software Radio Demodulators Developed in Japan

The software radio described in Table 3 was developed by NTT (Nippon Telegraph and Telephone). It has a software base station and a personal station system using a system with the Japanese standard Personal Handyphone System (PHS). An over-the-air software download function was implemented [4].

The software radio prototype described in Table 4 was developed by CRL with the purpose of multiple mode (PHS, GPS and Electronic Toll Collection (ETC)) communications targeted at vehicles [5].

A software radio using down conversion outlined in Table 5 was developed by Toshiba using the near-IF method [6].

The programmable real-time system [7] shown in Table 2 was developed to test various OFDM scenarios for the Digital Television Terrestrial Broadcast (DTTB) which is expected to start in Japan soon.

The prototype software radio demodulator shown in Table 2 was first developed in 1998, and is now being used to test automatic extraction and demodulation of unknown communications signals [9].

Table 3. Prototype software radio demodulators developed by NTT

RF frequency	2.45 GHz band
Bandwidth	13 MHz
Down conversion	Double super heterodyne
IF frequency (first/second)	215/39 MHz
IF input frequency	100 MHz maximum.
Access type	TDD
Sampling method	IF sampling (under sampling)
DSP	6400 MIPS
Features	3 RF inputs for an array antenna. Multi-mode operation, PHS-like protocol, modulation recognition capability

Table 4. Prototype software radio demodulator developed by CRL

RF frequency	1.6/1.9/5.8 GHz band
Bandwidth	8 MHz
Modulation scheme	BPSK, QPSK, π/4-QPSK, GMSK, ASK
Symbol rate	192 ksps (BPSK, QPSK, π/4-QPSK) 270.833 ksps (GMSK) 1024 ksps ETC
A/D converter	10 bit/60 MHz
D/A converter	14 bit/50 MHz
DSP	200 MHz × 4
Features	Triple mode (PHS/GPS/ETC), parameter-controlled software radio (filter, equalizer, detector, decoder)

Table 5. Prototype software radio demodulator developed by Toshiba

RF frequency	1.5/1.9 GHz band
Bandwidth	10 MHz
Down conversion	Direct conversion
Modulation scheme	n-PSK, π/4-QPSK, GMSK, MSK
Symbol rate	384 ksps (π/4-QPSK) 270.833 ksps (GMSK)
A/D converter	10 bit/64 MHz
DSP	166 MHz
Features	Direct conversion

Table 6. Prototype software radio demodulators developed by CRL and Sony/Tektronix

Modulation schemes	QPSK, 16QAM-OFDM
Number of carriers	108
Carrier interval	4 KHz
IFFT size	2048
Symbol length	250 μsec
Frame length	4.5 msec (16 symbol/frame)
Guard interval	1/8 (31.25 μsec)
Features	Digital TV test bed, channel measurement

Table 7. Prototype software radio demodulator developed by Toyo Communication Equipment Company and Tokyo Electric Power Company

Modulation schemes	FM, FSK, BPSK, QPSK, π/4-QPSK
RF frequency	370-380 MHz
IF frequency (Tx/Rx)	10-20 MHz/65-75 MHz (BW 10 MHz)
A/D converter	12 bits/40 MHz
D/A converter	12 bits/40 MHz
Sampling method	IF sampling
DSP	320 Mflops/board
Features	For an intelligent base station, software download capability

Table 8. Prototype software radio demodulator developed by Anritsu Company and NEC Company

Modulation schemes	AM, FM, FSK, MSK, GMSK, BPSK, QPSK, π/4-QPSK, 8-PSK, 16QAM
Access type	TDMA, SCPC
Symbol rate	10 M symbol/sec maximum
Speech codec	32kbit/sec ADPCM, CVSD-DM
Communication protocol	RCR-STD-28 (PHS)
Sampling method	IF sampling (under sampling)
IF input frequency	100 MHz
Software download	via LAN
Features	Modulation recognition capability, envelope analysis

3. Details of Technical Meeting of IEICE

3.1 Software Radio Study Group, 21st July 2000

The following are short summaries of some of the papers presented at the most recent Software Radio Study Group held at Keio University, Yokohama, Japan in July 2000:

H. Yoshioka *et al.* (NTT) presented research about automatic modulation recognition for software radio. They propose a recognition method based on the nearest neighbour rule. They changed the boundaries between different modulation methods when they added AWGN to the modulation prototype to improve recognition. They showed the effectiveness of their proposal by computer simulation [11].
S. Ishii (YNU) *et al.* proposed a space hopping scheme that can realise a low-cost smart antenna, since it requires only one RF circuit. In this paper, they discuss DOA estimation for their circuit and ways to overcome degradation of estimation accuracy [12].
K. Ikemoto (YNU) *et al.* proposed adaptive block coding and decoding for software-defined radio. Their system improves transmission efficiency. They achieve reduction in the number of control symbols by using encoder information. In order to reduce the Detection Selection Error (DSE) they propose a decoder scheme that estimates the encoder considering the encoder transition from the transmitter [13].
K. Umebayashi (YNU) *et al.* investigated blind modulation estimation with carrier frequency offset. Generally perfect carrier and symbol synchronisation is assumed, but in this paper the authors assume perfect symbol synchronisation but have examined carrier offset. Using each symbol's phase difference, it was possible to show that improvement in modulation estimation accuracy is achievable. Several algorithms are evaluated by computer simulation [14].

4. Summary

- Since Staring in 1999, 36 regular technical presentations and 12 invited talks have been given at 5 IEICE Software Radio Study Group technical meetings.
- Various prototypes of software radios have been developedby companies such as NTT, Toshiba, CRL, Sony/Tektronix, Toyo Communication and Anritsu.
- Theoretical research has been ongoing, particularly at Yokohama National University.
- Many invited speakers from the SDR forum and leading software radio companies from overseas have made presentations.

5. References

[1] J. Mitola "The Software Radio Architecture" *IEEE Communications Magazine*, Volume 33:5, pp26-38, May 1995.
[2] J.J. Patti, R.M. Husnay and J. Pintar "A Smart Software Radio: Concept Development and Demonstration" *IEEE Journal on Selected Areas in Communications*, Volume 17:4, pp631-649, April 1999.

[3] J. Mitola "Software Radio Architecture: A Mathematical Perspective" *IEEE Journal on Selected Areas in Communications*, Volume 17:4, pp514-538, April 1999.

[4] K. Uehara, M. Nakatsugawa, Y. Suzuki, U. Shirato and S. Kubota "Software Radio Base and Personal Station Prototypes (1)" *IEICE Software Radio Technical Meeting Proceedings SR99-10*, pp1-8, Nagoya, Japan, November 1999.

[5] H. Harada, Y. Kamio and M. Fujise "A New Multi-mode and Multi Service Software Radio Communication System for Future Intelligent Transport Systems" *IEICE Software Radio Technical Meeting Proceedings SR99-21*, pp81-88, Nagoya, Japan, November 1999.

[6] H. Yoshida, S. Otaka, T. Kato and H. Tsurumi "A Software Defined Radio Using Direct Conversion Principle" *IEICE Software Radio Technical Meeting Proceedings SR00-09*, pp59-63, YRP, Kanagawa, Japan, April 2000.

[7] H. Ohta, Y. Akita, M. Itami and A. Tsuzuku "A Study on the Programmable Real-time Signal Processing System" *IEICE Software Radio Technical Meeting Proceedings SR99-3*, pp17-22, YRP, Kanagawa, Japan, June 1999.

[8] T. Yokoi *et al.* "Software Receiver Technology and Its Applications" *IEICE Transactions on Communications*, pp1200-1209, Volume E83-B, No.6, June 2000.

[9] H. Ishii, T. Suzuki, T. Higuti, and T. Yamamoto "Development of Adaptive Receiver Based on Software Radio Techniques (1) – One Examination of Modulation Mode Identification by Using Envelope Analysis" *IEICE Software Radio Technical Meeting Proceedings SR00-14*, pp29-35, Nagoya, Japan, November 1999.

[10] M. Shinazawa and Y. Karasawa "Effect of Errors Due to Quantization and Clipping in Analog-to-Digital Conversion on Array Antenna Performance" *IEICE Software Radio Technical Meeting Proceedings SR00-13*, pp1-8, Yokohama, Japan, July 2000.

[11] H. Yoshioka, Y. Shirato, M. Nakatsugawa and S. Kubota "A Proposal of the Automatic Modulation Recognition Technique Employing Several Prototypes" *IEICE Software Radio Technical Meeting Proceedings SR00-14*, pp9-13, Yokohama, Japan, July 2000.

[12] S. Ishii, A. Hoshikuki and R. Kohno "MUSIC DOA Estimation Using Space Hopping Scheme – To Realize a Software Radio" *IEICE Software Radio Technical Meeting Proceedings SR00-15*, pp15-22, Yokohama, Japan, July 2000.

[13] K. Ikemoto, K. Umebayashi, S. Ishii and R. Kohno "A Study on Periodic Shift in Convolutional Coding and Viterbi Algorithm Based on a Concept of Software Defined Radio" *IEICE Software Radio Technical Meeting Proceedings SR00-16*, pp23-30, Yokohama, Japan, July 2000.

[14] K. Umebayashi, S. Ishii and R. Kohno "A Study on Blind Modulation Estimation with Carrier Frequency Offset Based on a Concept of Software Radio" *IEICE Software Radio Technical Meeting Proceedings SR00-17*, pp31-38, Yokohama, Japan, July 2000.

SDR Applications for the Next Generation Wireless Access: Prototype Implementation

Shuji Kubota, Kazuhiro Uehara and Masashi Nakatsugawa

NTT Network Innovation Laboratories, Wireless Systems Innovation Laboratory
1-1 Hikarinooka, Yokosuka-shi, Kanagawa 239-0847, Japan

Abstract: The advantages and application fields of software defined radio are described. Base and personal station prototypes created on the software defined radio technology are then introduced. The prototypes show acceptable performance and the remarkable advantage of flexibility for the next generation wireless access.

1. Introduction

The recent progress in digital signal processing techniques well supports the development of software defined radio (SDR) [1-5]. Intensive studies are underway to establish new system configurations for a universal platform for wireless communication systems. Moreover, the rapid growth in the multimedia and wireless communication services demands frequent upgrading of base station functions to offer a variety of new services. Migration from the existing systems to the next generation systems is also essential. SDR technologies are useful in reducing both upgrade and migration costs. In addition, the ability to handle different types of wireless services can improve the quality of service by allowing adaptive selection of the transmission scheme depending on the environmental. For these reasons, SDR technologies are essential in the next generation wireless communication systems.

This paper describes the advantages of software defined radio technologies, SDR application fields, a roadmap for the development for commercial products, and NTT's newly developed software defined radio base and personal station prototypes. Their flexibility in selecting modulation scheme, multiple access method, and bit rates, allows them to be used in both cellular and PCS communication systems. Functions and performance of the systems are also reported [6-7].

2. Advantages of SDR

The advantages of SDR can be categorized into three areas. They are "Multimode service capability", "System version-up capability" and "Adaptive transmission capability" as shown in Figure 1. The first one, "Multimode service capability" is the capability of realizing different systems or media. It offers wider cover areas and optimum QoS (quality of service) selection, for instance throughput (speed), bit error rate, voice clearness, cost and so on. The second one, "System version-up capability" is the capability of supporting version-up or the addition of new functions after installation; in other words, "future-proof". The third one, "Adaptive transmission capability", includes not only adaptive modulation, but also adaptive FEC (Forward Error Correction), variable bit rate, antenna characteristics control and so on. It offers better channel quality and an increase in system capacity. All these capabilities can be realized in terminals (personal stations), base stations, and relay stations for information bridging.

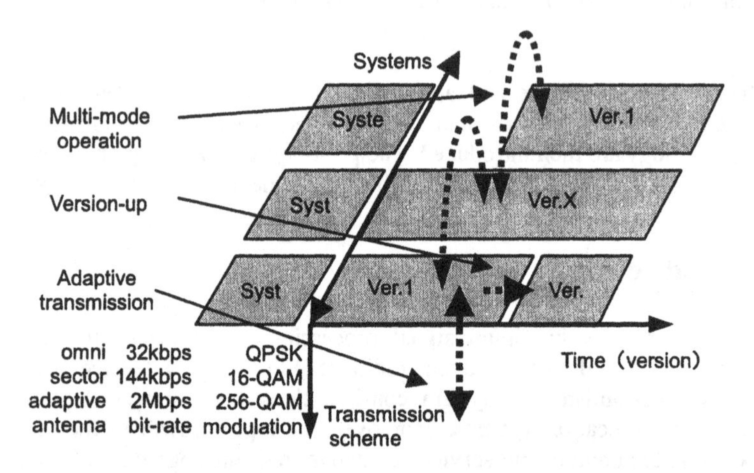

Figure 1. Advantages of software defined radio technologies

3. SDR Application Fields

The relationship between bit rate and the mobility of major wireless access systems in Japan is shown in Figure 2. The rapid growth of wireless communications has seen various wireless access systems put into commercial use. The number of PDC subscribers has reached about 50 million, and there are about 6 million PHS subscribers at the present time. The number of mobile subscribers now exceeds that of conventional analog fixed telephony subscribers. Following the 2nd generation wireless systems, IMT-2000 phase 1 is under development. In addition, AWA (Advanced Wireless Access) and EWA (Ethernet-Based Wireless Access) systems are being developed for even higher-bit-rate transmission, for instance 20 or 30Mbit/s to stationary or pedestrian users.

In order to provide the next generation wireless communication systems characterized by even higher-bit-rate service, we have started research and development activities. One approach is 4G (in other words, IMT-2000 phase 2), which will offer more than 2Mbit/s in high-speed mobile environments and 10 to 20Mbit/s to stationary users. Another approach, bit-rates above 100Mbit/s wireless access systems with limited mobility, will be required to realize multi-media support. This can be categorized as high-speed MMAC (Multimedia Mobile Access Communication systems). When installing such systems, the service will be limited to spot areas. Thus, we have to consider multimode operation between the new system and existing conventional systems to offer wide coverage. The multimode function of SDR terminals is one of the most promising solutions to this problem. The application range of SDR will extend from base station applications to terminal applications.

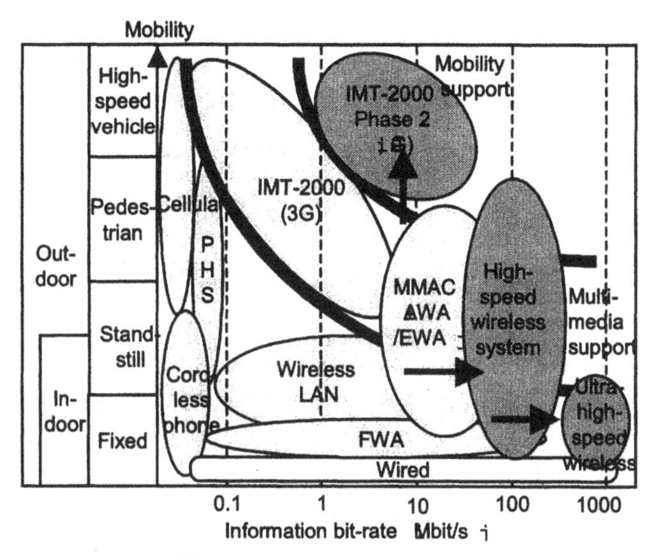

Figure 2. Mobility and information bit-rate of wireless systems

Its application in base stations will make it possible to use various types of existing air-interface terminals. Version-up and also bug fixes will become easier to achieve. Applications in home, SOHO, shops, cars, such space limited places are key areas for SDR applications. The bottleneck of the high power consumption of SDR is not fatal in base station applications. Its application to terminals will allow users to access multiple services, multiple systems and multimedia, not only communications but also broadcasting and ITS. It will be well accepted by mobile workers and so on. Also, it is useful for the realization of user-customized terminals, time sensitive services and less knowledgeable users, for instance the aged or the very young. From the common carrier's point of view, SDR technologies will be very effective in mobile networks, the last one hop of other communication networks and home and office applications. Global roaming is one of its most attractive features. Such seamless connection environments are the main targets of SDR implementation. In addition, multimode and multimedia operation with automatic system selection and seamless inter-system roaming during communication will be strongly required in the next generation wireless systems.

There are two strategies toward future wireless module development as shown in Figure 3. One is the specialized hardware approach. It targets performance, size, cost and so on. Second is generalized hardware, the SDR approach. It emphasizes functional flexibility. As dual mode terminals with two kinds of specific hardware already exist, for instance, PDC and PHS dual mode terminals, different multimode terminals will be created using specific hardware and generalized hardware, SDR. In such terminals, the specific hardware will realize high tier wireless functions that are difficult for SDR to implement, while SDR realizes the multiple low tier functions suitable for SDR implementation. This approach will realize high performance and flexibility at the same time. The level of SDR will be expanded step by step as shown in Table 1. Level 0 or 1 might use the same RF frequency, same air interface but variable functions. Next, the air interface including bit rate, and RF frequency will be extended. The ultimate SDR will have programmable devices connected directly to an antenna, and the functions of variable RF and variable bit rate will be realized.

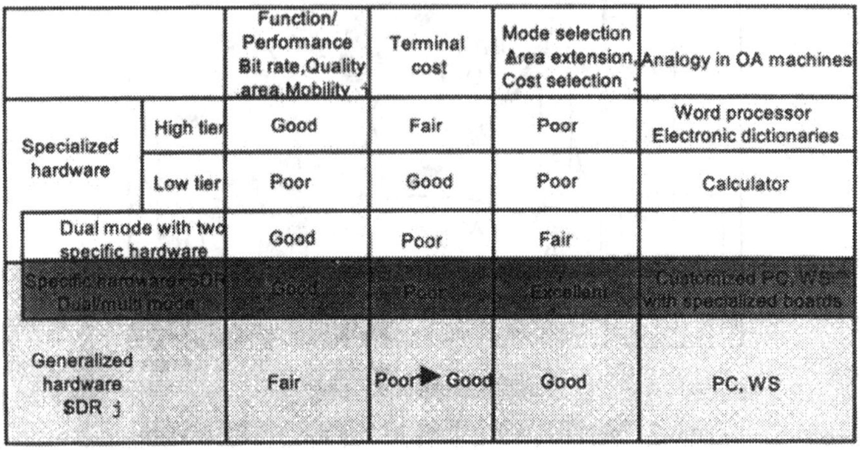

		Function/ Performance Bit rate, Quality area, Mobility	Terminal cost	Mode selection Area extension, Cost selection	Analogy in OA machines
Specialized hardware	High tier	Good	Fair	Poor	Word processor Electronic dictionaries
	Low tier	Poor	Good	Poor	Calculator
Dual mode with two specific hardware		Good	Poor	Fair	
Specific hardware+SDR Dual/multi mode		Good	Poor	Excellent	Customized PC, WS with specialized boards
Generalized hardware SDR		Fair	Poor ▶ Good	Good	PC, WS

Figure 3. Classification in wireless module development

Table 1. Level of SDR development

	RF freq.	Air interface	Remarks
Level 0	same	same	Mode switching
Level 1	same	same	Version-up
Level 2	same	multiple	PHS/DECT AMPS/IS95
Level 3	multiple neighboring	multiple	800MHz-band 2GHz-band 5GHz-band
Level 4	Multiple Different RF	multiple	800MH, Millimeter wave

4. SDR Prototypes

In order to clarify the feasibility of SDR, we developed a prototype SDR base station and personal stations as the first step in SDR application studies. Targets of our study are to verify the feasibility of the three major functions: Multimode operation, System version-up and Adaptive transmission. Implemented functions are, multiple communication services, over-the-air download capability and signal processing alteration. System parameters of the prototypes are summarized in Table 2.

Table 2. Major parameter of prototype SDR station

Parameter	Value
RF Band	2.45GHz
Bandwidth	13MHz
Freq. Conv. Method	Double Super Heterodyne
IF	215/39MHz
No. of RF/IF Systems	3
Duplex Scheme	Time Division Duplex (TDD)
External Interface	ISDN (Base Station) Voice, Bearer (Personal Station)
DSP Performance	6,400MIPS
AD Conversion	IF Under-sampling

4.1 Configuration

A block diagram of the system is shown in Figure 4. It is composed of RF/IF circuits, A/D and D/A converters, pre/post processors, a CPU, an I/F (interface) circuit and DSP parts. The CPU, DSPs, and I/F circuit are connected by a 32-bit VME bus. The base station offers an ISDN interface, while the personal stations have voice and bearer communication interfaces. The pre/post processors, which handle multiple access and waveform shaping, are employed to reduce DSP loads. In the receiving process, the 39±6.5 MHz IF signal is A/D converted using 12-bit quantization with 52MHz.resolution. The pre processor under-samples the IF signal, and a channel signal is digitally down-converted, filtered and fed to the DSPs. In the transmitting process, the reshaped and channel multiplexed base band signal is up-sampled with a 10-bit resolution D/A converter using 10-bit quantization at 104MHz. The signal is zero-stuffed before D/A conversion to improve the aperture effect due to the imperfection of the pulses. This enables us to utilize an undistorted image signal at the frequency of 39 MHz, which is used for the IF signal. This technique eases the IF filter requirements. Four commercially available 1,600 MIPS DSPs (offering 6,400 MIPS) are mounted on the same board. The loads of the transmission function are relatively simple compared to those of the receiving function. Thus, basically one DSP is assigned to the transmission side and three DSPs to the receiving side.

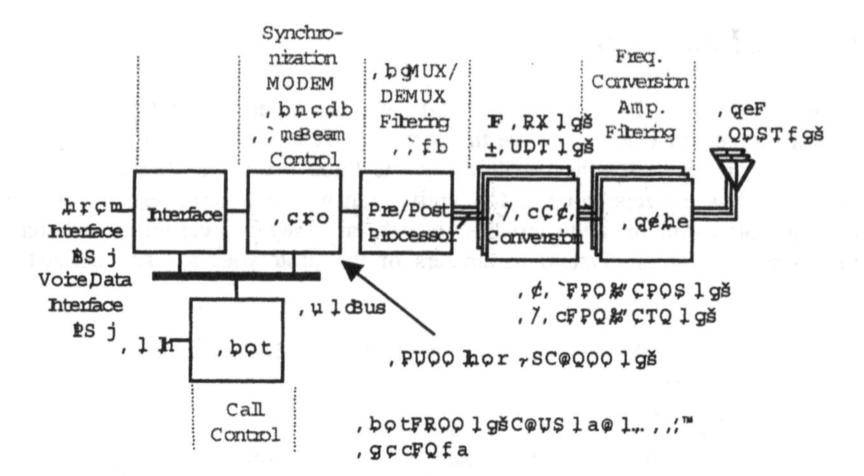

Figure 4. Configuration of prototype SDR station

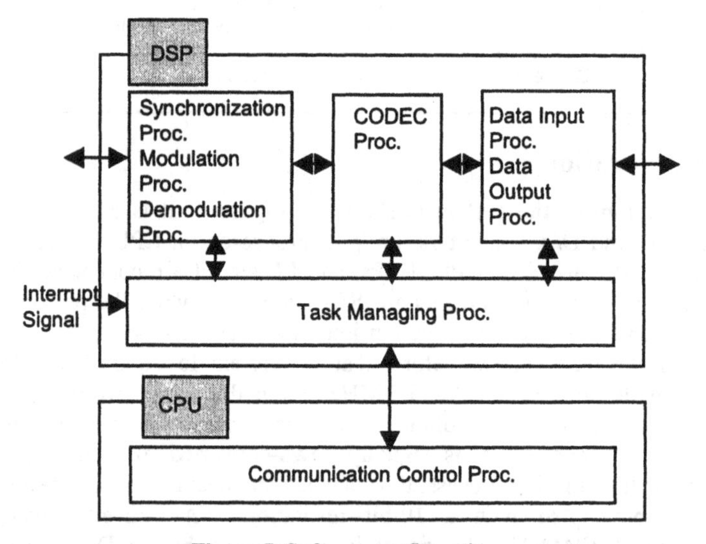

Figure 5. Software configuration

The software configuration of the developed system is shown in Figure 5. Task programs performing modulation and demodulation, CODEC, and adaptive array control are loaded onto the DSPs. Those programs communicate with the program on the CPU through the "macro interface". The CPU handles communication control and system management.

4.2 Performance

The developed prototype can operate in PHS, personal handy-phone system mode or/and lower bit rate mode comparable to the cellular system. Typical output signals state-space diagrams and eye patterns of the prototype SDR stations are shown in Figure 6. The average modulation vector error is less than 4%, which confirms acceptable performance. BER performance shown in Figure 7 well agrees with the theoretical values. The transmission spectrum is also acceptable as shown in Figure 8.

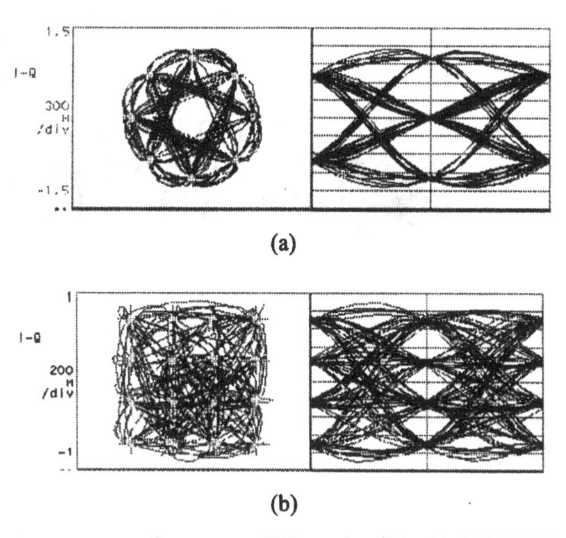

(a)

(b)

Figure 6. Vector error rate of prototype SDR station: (a) π/4-shift QPSK; (b) 16QAM

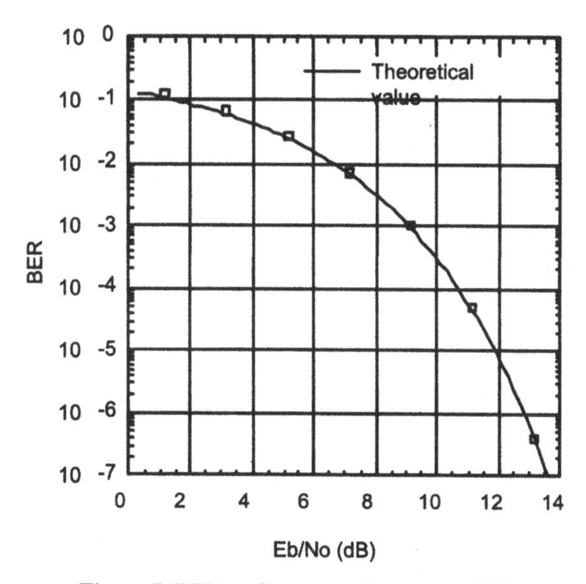

Figure 7. BER performance of prototype SDR

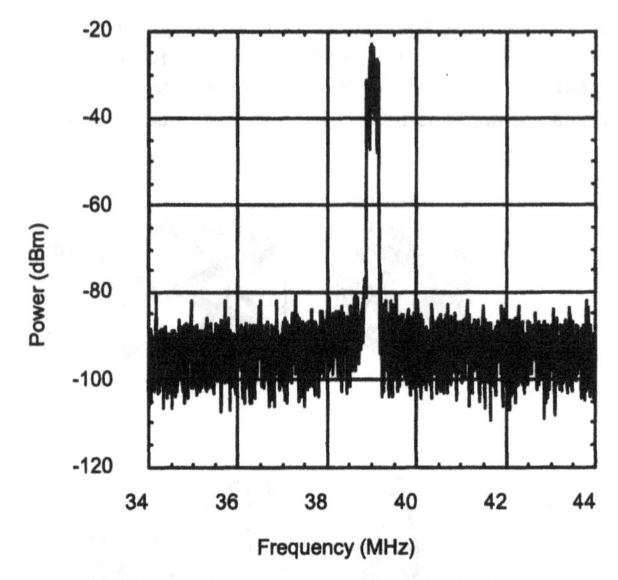

Figure 8. Transmission spectrum of prototype SDR station

5. Conclusion

By employing commercially available DSPs, CPU and other devices, software defined radio base and terminal stations have been successfully implemented. The software defined radio technology permits the modulation schemes, transmission rate and CODEC to be freely altered. These prototype SDR stations allowed us to confirm the feasibility of "Multimode operation", "System version-up capability" and "Adaptive transmission capability". The special advantages of this technology offer improved flexibility in designing the next generation wireless access systems.

6. References

[1] J. Mitola, "The Software Radio Architecture," *IEEE Commun. Mag.*, Vol.33, No.5, PP.26-38, May 1995.

[2] J. Mitola, "The Software Radio Architecture:A Mathemativcal Perspective," *IEEE J. Select. Areas Commun.*, Vol.17, No.4, pp.514-538, April 1999.

[3] R. Kohno, S. Haruyama, R. Miura, H. Harada, Y. Sanada, "Overview of Japanese Activities in Software Defined Radio," The 2nd SDR Workshop, Korea, pp.15-28, April 2000.

[4] H. Yoshida, "A Software Defined Radio using Direct Conversion Principle," The 2nd SDR Workshop, Korea, pp115-122, April 2000.

[5] H. Ishii, T. Suzuki, "Prototype of Software Defined Receiver and Its Application," The 2nd SDR Workshop, Korea, pp.123-134, April 2000.

[6] S. Kubota, K. Uehara, M. Nakatsugawa, “Common Carrier Activities in SDR: Prototype Implementation of SDR Systems,” The 2nd SDR Workshop, Korea, pp161-173, April 2000.

[7] Y. Suzuki, K. Uehara, M. Nakatsugawa, Y. Shirato and S. Kubota, "Software Radio Base and Personal Station Prototypes,” IEICE Trans. Commun., Vol.E83-B, No.6, pp.1261-1268, June 2000.

Mobile Multi-mode Terminal: Making Trade-offs Between Software and Fixed Digital Radio

Juha-Pekka Soininen, Anu Purhonen, Tapio Rautio [1] and Mika Kasslin [2]

[1] VTT Electronics, P.O. Box 1100, FIN-90571, Oulu, Finland
[2] Nokia Research Center, P.O. Box 407, FIN-00045 NOKIA GROUP, Finland

Abstract: The mobile multi-mode terminal has contradictory requirements on operational flexibility and physical characteristics. Optimisation of terminal kernel, *i.e.*, hardware resources and system software services, requires concurrent design of execution architecture and functionality. An approach for terminal architecture optimisation is presented. Functional optimisations and estimations are used to minimise the amount of basic functions and to identify resource sharing and configuration opportunities.

1. Introduction

Mobile phones have been a success and the number of mobile phone users has increased rapidly during last years. New radio interface standards have been developed and taken into use in order to provide the capacity needed. It has led into a situation, where mobile terminal manufacturers have had to develop dual- and multi-mode products that operate in several networks. Integration of PCs, PDAs and mobile phones will continue, and the future products will be multimedia terminals and personal communication centres instead of telephones.

The performance of digital signal processors and system chips has evolved rapidly [1]. It has been possible to push the interface of analog signal processing towards higher frequencies and to benefit from the advances of digital processing. A lot of the mobile terminal functionality can be implemented using software-defined parts. One definition for a software radio (SWR) is a multiband multi-mode radio with dynamic capability defined through software in all layers of protocol stack, including the physical layer [2]. The proposed SWR design methodologies and implementations are based on layered models [3, 4, 5]. A common property of the presented models is that the lowest layer contains typical architectural elements, either hardware or software, performing the actual operations required to achieve radio functionality and also the operations required for the architecture. The upper layers contain virtual machines that handle the configurations of the lowest layer and even the radio etiquette [6].

Traditional development of physical layer architecture starts from algorithms. The mathematical models are refined using simulations to floating point and bit-accurate data flow models, and the HW design and DSP core selection is assisted by benchmarks and reference designs. The role of software tools, techniques and architectures is becoming more important though [7]. Computer system design and application specific instruction set processor design relies on workload characterisation, trace generation and performance analyses that are often done using performance simulations [8]. Codesign has largely replaced the sequential design flows and introduced system level considerations. Several approaches that start from implementation independent functional specification and proceed to design space exploration, architecture design, and software/hardware partitioning have been presented [9]. The major challenges are still the specification languages, complexity of system synthesis, and verification capacity.

1.1 Research Problem

Future mobile terminals must be capable to handle multimedia data streams and to operate in most advanced wireless communication networks simultaneously. The search for optimal architecture and implementation requires that service, compression, data transfer and communication functions must be analysed, specified, designed and implemented in a joint design process. The process must be such that it can be studied how individual design decisions effect on the implementation of the complete product.

Our research hypothesis is that by modelling and analysing the essential functions of the product, it is possible to find commonalties that help in the optimisation. The optimisation must start already in the requirements phase, when selecting algorithms, services and operation configurations of the product. If the design process provides support for analyses and estimations, it is possible to make justified decisions earlier, to reduce the design exploration space, to speed-up the design and to improve the quality of the final product.

This paper presents an architecture design approach targeted for telecommunication kernels of mobile multi-mode terminals. Telecommunication kernel is an integrated circuit capable to execute both baseband and application functions. The main emphasis is in the functional specification, optimisation and decision support for architecture development. The approach is not limited to terminals but can be exploited in design of any multi-mode telecommunication device.

2. Mobile Multi-mode Terminal Requirements

The product scenario is a mobile terminal that is capable to connect to the next generation public mobile phone networks and future wireless local area networks [10, 11]. The service set of the terminal was chosen to demonstrate how the services effect on communication functions and on execution platform. It consists of following main applications:

- telephone services over GSM, WCDMA and WLAN;
- data base synchronised address book and calendar functions;
- office applications with protocol, security and encryption capabilities;
- Multimedia applications such as audio and video stream reception, sending, and presentation.

Naturally, the networks, the connection quality, the user interface, and temporal execution and memory capacity effect on the available service set.

Network support effects on terminal functionality. GSM and WCDMA networks are assumed to support the routing of telephone and data connections [10], while the WLAN is independent network. This means that the terminal has to monitor either GSM or WCDMA network constantly. The WLAN connection is different, since the standard does not consider how the WLAN co-operates with public networks, e.g., in use cases such as handovers and cost optimisation.

The main connections states of the mobile terminal are No Connection, GSM services, WCDMA services, WLAN services, Joint GSM and WLAN services and Joint WCDMA and WLAN service. The 18 transitions between them can be caused by changes in service set status, or by changes in connection quality.

Different services have different requirements for connections and handovers. Disconnecting handover aborts the active service. Seamless handovers are such that the user does not recognise them. However, the quality of service can improve, sustain or degrade because of the handover. Discontinuing handovers allow recognisable interrupts, but the connection will be established afterwards. The data can either be lost or buffered. The seamless handovers require simultaneous connections. Even if it is allowed to degrade the quality of service temporarily, the basic workload from communication exists. The acceptable handover type depends on the application, and it is one of the early important design decisions. Disconnecting handover must be accepted when target network does not support the service. Seamless handovers must be implemented for call and data stream services. Discontinuing handovers can be accepted in packet type transmission, such as Internet browsing and file transfer.

In addition to functional requirements, multi-mode terminal has also requirements concerning the three quality goals for the system architecture: performance, business and variability. Performance goal is characterised by requirements such as latency and throughput. Business goal includes restrictions to product development cost and development time. Variability portrays the architecture's ability to adapt itself to changing requirements.

In this scenario it is expected that typical ASIC can contain several DSP and RISC cores with the necessary embedded memories [12]. The initial architecture concept was therefore based on configurable DSP/RISC multicomputer platform that is supported by dedicated baseband-processor for WLAN and system software services that are optimised for application set. The analog and RF parts have been excluded from the platform.

3. Design Flow for Multi-mode Terminal

The objective of a design flow is to translate the user needs into a product that fulfils those needs, existing design constraints and that is feasible to produce. The proposed design flow for mobile multi-mode terminal is depicted in Figure 1. The methodology has conceptual, functional, architecture and implementation design layers. There are three types of activities to be performed in each layer: the actual modelling of the system, the analysis of whether we are designing the right system, and the estimation of the final system characteristics.

The conceptual design translates the informal and often obscure user needs into product concept and requirements. How to define the co-operation of networks and what kind of communication behaviour is acceptable for services are important multi-mode issues. The constraints are the standards, network support, protocols, and data formats of possible applications. Support from network operator is especially important when considering WLAN type of local networks. Data formats are essential in order to allow the terminal to communicate with existing infrastructure, such as multimedia providers and office applications.

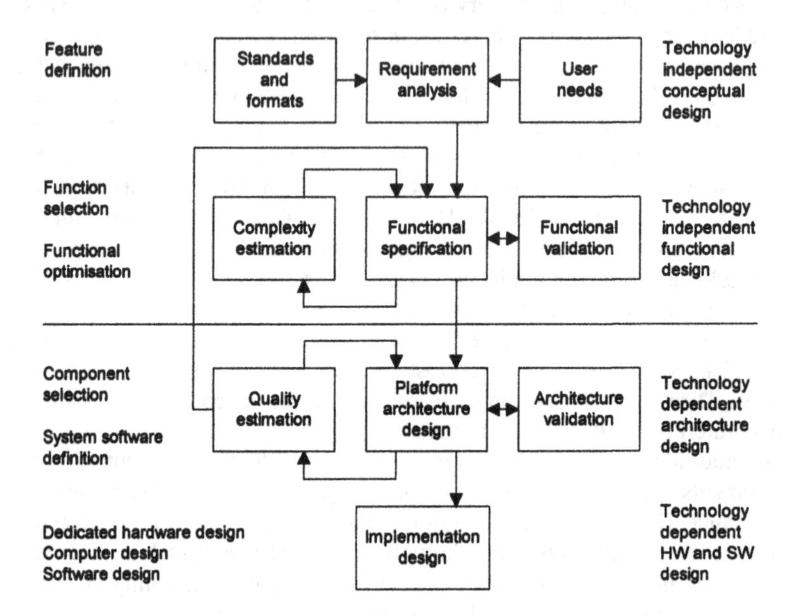

Figure 1. Terminal design flow

Functional specification translates the product concept into a formal and possibly executable specification. The behaviour must be decomposed into basic functions and operations. We have used UML extended with more detailed algorithm models. The objectives are to understand the operation in detail, to validate the requirements, to estimate what it requires to implement the product, and to optimise the complete product instead of optimisation of subsystems.

The quality factors are completeness, consistency and capability to support complexity estimations. The chosen set of algorithms and functions and operations effect on the characteristics of optimal processing units afterwards. In multistandard product both baseband and multimedia stream processing algorithms should be chosen such that they can utilise same resources as effectively as possible.

Terminal architecture consists of both hardware and software. Software architecture is defined as a structure or structures of the computing system, which comprise software components, the externally visible properties of those components, and the relationships among them [14]. Hardware platform architecture consists of processing units, memory units and communication channels.

The challenge in architecture design is to find the components and structures that minimise the costs while still maintaining the performance, configurability and flexibility requirements. The quality estimation of architecture must consist of both qualitative and quantitative measures. Qualitative measures are needed since in platform type of design there is always some degree of uncertainty. Quantitative measures are related to cost estimation.

In the implementation design, the software development needs hardware models for testing phases. Hardware design must combine both resource and functional design practices, since computer systems are generic execution resources, while dedicated hardware aims at peak performance for some specific function.

3.1 Specification Using UML

The structure of functional specification that is based on object-oriented approach and Unified Modelling Language (UML) [13] is shown in Figure 2. Use case diagram is a tool for requirements capture. Class models decompose the functionality into object classes. Scenarios describe co-operation of group of objects. Behaviour inside of an object is depicted with state diagrams. Data flow descriptions are needed for understanding co-operation of algorithm objects in data processing.

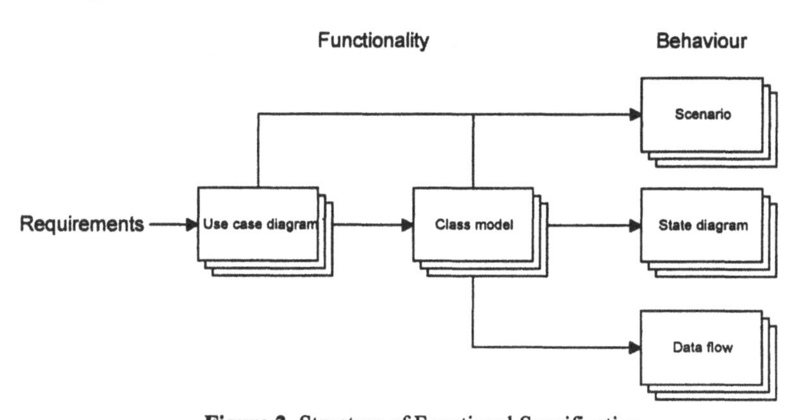

Figure 2. Structure of Functional Specification

Functional specification starts with modelling the requirements with use cases. A use case is a collection of possible sequences of interactions between the system and its users. These high-level use cases are further specified in lower-level use case diagrams. Figure 3 illustrates main use cases for the multi-mode terminal. The use case diagram has some of the internal functionality, because telecommunication systems are traditionally divided into layers. The Radiolink actor refers to physical connection whereas Network actor creates a logical connection to the terminal.

UML package diagram is used for describing the sub-subsystems. In the terminal model, there are five packages: Applications, Protocols, Devices, Transceiver and Power. Applications package contains the high-level functionality. Protocols package is reserved for communication with the network and control of the data transfer. Devices package includes displays and other equipment necessary for communication with the user. Power package is for managing battery. Transceiver package consists of signal processing functionality and the equipment necessary for communicating with the network.

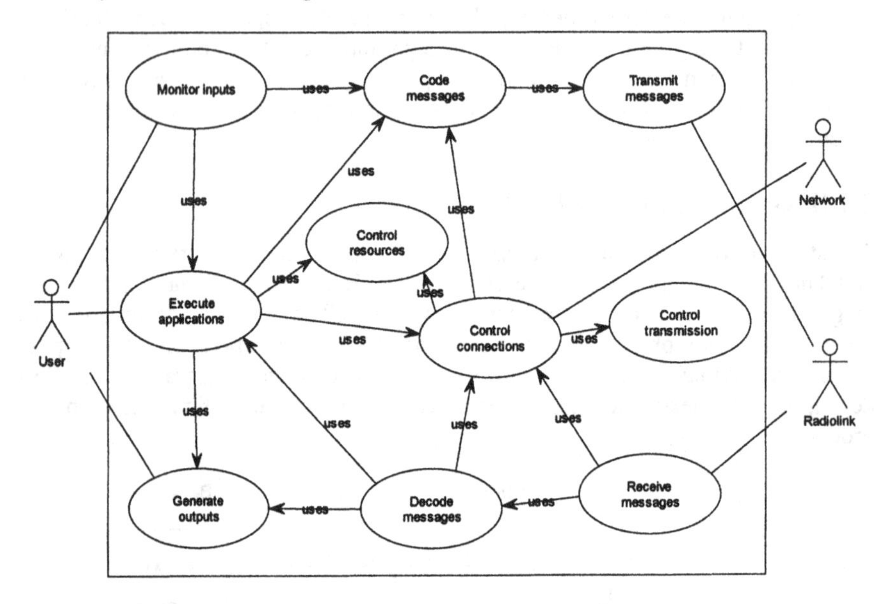

Figure 3. Multi-mode terminal use case diagram

The packages contain class models. A class model defines the types of objects that can be used in the system and the various kinds of static relationships between them. Figure 4 depicts the conceptual class model for the Transceiver. The class model allows dynamic reconfiguration of connections between terminal applications and the networks.

In Figure 4 the *Controller* is a high-level control for Transceiver. It communicates with the other packages and forwards the possible requests to the Connection. Controller knows what Connections exist at any time in the Transceiver. For each active radio link, there exists a *Connection*, which consist of Services. A connection knows what are its active Services at any time. *Service* is a base class for the actual services, which represent components in data flow models. Service is an interface to the actual functionality, which is implemented by some or group of Algorithms. *Algorithm* is a base class for the actual Algorithms that perform the actual functionality of Services. *Configurator* is a resource manager. It knows the status of each Algorithm implementation in a running system, and which ones are reserved and which ones can be used in a new mapping. Configurator maps the Services requested by a Connection to Algorithms so that Transceiver is always working optimally. *Configuration data* contains the instructions for creating Connections. *Mapping data* contains the instructions for mapping the Services to Algorithms.

The relationships between Service and Algorithm classes are further defined in separate class models. For example, Encoding-service is connected to one Channel encoding-algorithm from the five alternatives and to one or zero Puncturing-algorithms.

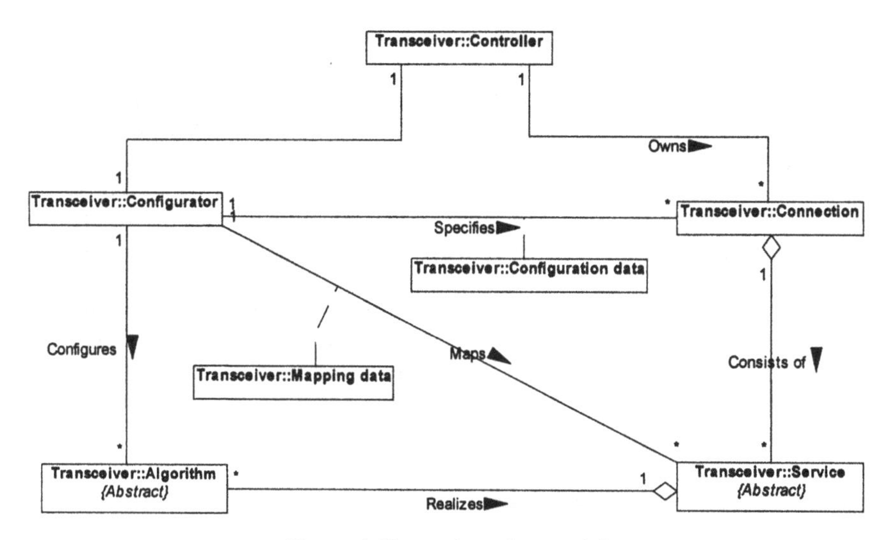

Figure 4. Transceiver class model

The internal behaviours of the classes are defined by state diagrams or algorithm descriptions. Collaboration of group of objects is illustrated with interaction diagrams. An interaction diagram captures the behaviour of a single use case. It shows the collaborating objects and the message flows between them.

Instances of Services and Algorithms create data flows that are defined in data flow descriptions, which include the definition of a Service chain and the selection of Algorithms. Service chain describes how connected Service instances create a data processing. Figure 5 describes a Service chain for WLAN transmitter. There can be several mappings of a Service to Algorithms and data flow descriptions, respectively.

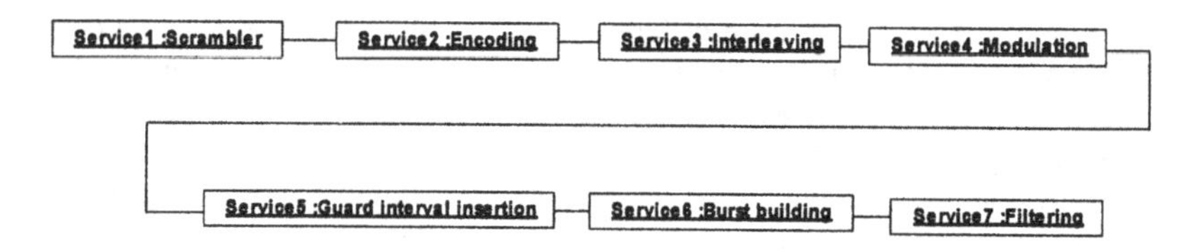

Figure 5. WLAN transmitter data flow diagram

Operation of the terminal consists of a sequence of configurations. Each configuration consists of a set of data flows diagrams that are implemented by system at the time when the configuration exists. Each data flow diagram consists of a set of Service instances, and each Service instance will be implemented with an Algorithm object. At the end, one Transceiver configuration is created by combining the objects in the selected data flows. The data flows included in a configuration depend on the multi-mode requirements and Transceiver configuration. Definition of possible configurations, or operation modes of the terminal, is done in mapping tables where the connections of the configurations, services and algorithms are done.

In the specification, the transceiver package has been of most interest. The chosen set of standards have been analysed and modelled. The Transceiver package had about 30 main configurations that consisted of about 35 Services and Algorithms. As shown, the UML based specification requires considerable effort to complete. The main reason to use UML is the system complexity. Even in a very restricted multi-mode case, we have lots of algorithms or functions used in several different configurations. Our main observations have been that UML force the designers to look into the essentials of the system and help to identify problems and to achieve common understanding. Identification of the right packages, classes and scenarios is not possible otherwise. It is also obvious that the model offers a seamless move from functional to the architecture design. Especially the use case diagrams and usage scenarios provide valuable information on co-operation of the objects of system. The class, state and algorithm models give a comprehensive view on the elementary complexity and execution requirements also.

3.2 Design of Terminal Architecture

The system architecture defines which components make up the system, how the components are connected to each other, and what functionality will be implemented by the components. The architecture design starts from the functional specification, design constraints, and existing building blocks. The results are the HW platform architecture and SW architecture.

The HW platform description consists of processors, memory organisations, buses, interfaces, intellectual property blocks, for example. The internal functionality of blocks and the structure changes by configuration are also defined.

The SW architecture model is divided into logical, process, physical and development views [14, 15]. The logical view is a static model of the system functionality. A physical view maps the functionality into hardware architecture. In process view, objects are mapped to processes and threads. The physical view and the process view together create a runtime view of the system. Because of dynamic configuration, a model for each different configuration is needed to get a complete runtime view. Finally in the development view, the software will be divided into modules that represent the actual code blocks.

Figure 6 illustrates the architecture design framework developed for multi-mode terminal design. It consists of activities in which the final architecture is detailed. Each activity is supported by analyses, and the design quality is validated using respective estimations. The process is preceded by requirement analysis, functional partitioning and complexity estimations. The design phases are the component selection, the function allocation, the resource mapping and the interface design.

Component selection phase results in technologies and processing units that will be used. It needs information on the operations and functions that are needed for the desired behaviours and on the capabilities of the possible processing units. The respective quality metric is the mappability of functions and processing units. In a case of a processor and an algorithm, mappability consists of expected performance, component and development costs, physical parameters of components, *etc.* In the mappability estimation, the target is to create a metric on processor suitability for given algorithm. The current techniques rely on benchmarks, algorithm characterisation, profiling, and performance simulations. However, in leading edge product development, it has to be done without compilation, execution and measurement, because the tools components are not available.

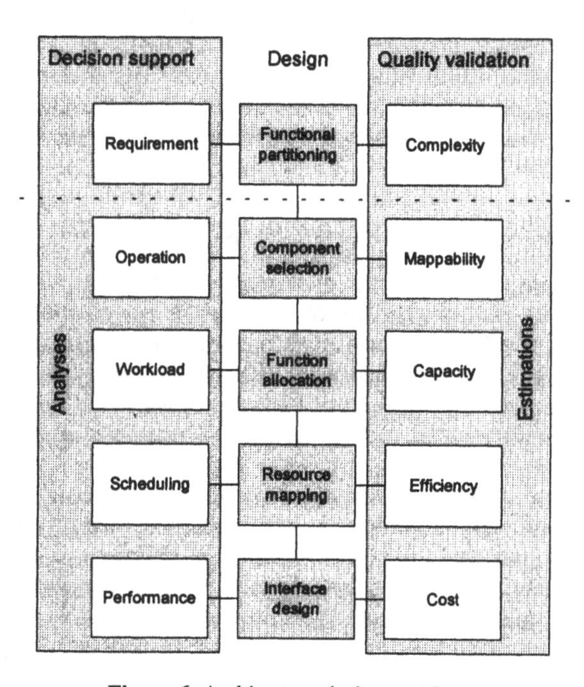

Figure 6. Architecture design matrix

In the function allocation phase it is decide which processing units execute the functions. This is an extension to software/hardware partitioning, since also the allocation to multiple processors has to be considered. Workload analysis and capacity estimation are respective activities. The objective is to analyse what kind of a workload a set of functionality represents for a processing unit, and what the allocation means to the whole platform. In capacity estimation, the objective is to figure out how to distribute the load. As and example one will to estimate utilisation of resources, need of system software support, and total power consumption.

Resource mapping phase attaches the operations to physical or logical resources. How many data paths or how many complex multipliers will be needed is a typical hardware problem. In case of software, the mapping of functions to processes and tasks is considered. Schedulability analysis is the most important decision support method. The efficiency of resource usage is the respective quality metric.

Interface and control design completes the architecture specification. The objective is to define how the hardware and software units communicate and how the execution is controlled at process and task level. In case of software architecture, this means that the operating system services must be defined. Performance analysis is the technique that can be used for guiding towards optimal solution. The final quality metric for architecture is the cost.

Multi-mode terminal imposes several challenges to the architecture development. The operations in chosen set of standards and applications are not very complex as such, but the total complexity and computational capacity requirement is beyond the current technology potential. Initial HW architecture would consist of a cluster of DSP cores for baseband and multimedia stream processing and additional RISC cores for the system control, the network control and the applications. Configurable on-chip memory organisation would be needed to ensure data transfer capacity.

3.3 Optimisation During Design Flow

The goals for optimisation are often diverse and contradictory. The best performing algorithms do not minimise the use of resources or power. When optimisation is applied during the design process, it is important to recognise how the design optimisations relate to the product goals, e.g., conformance to requirements and customer satisfaction. At each level of design, the measure of optimality should be expresses as a quality metric, which consists of both quantitative and qualitative measures.

The optimisation of mobile terminal kernel focuses on minimisation of cost and power consumption while maximising functionality and performance. Table 1 summarises the main activities. Several important observations were made during the terminal design. Firstly, the standards do not support multi-mode operations. It was difficult to identify common algorithms or functions that could be used in several data flows. Even the functions, which could have been reused by making them configurable or parameterised, were few. Configurable Viterbi decoder is one of the possible candidates. Secondly, techniques in the performance optimisation must be extended so that they take into account the power consumption, resource usage and portability. Possibility to use different resources depending on the current operation mode or configuration must to be considered. Thirdly, different configurations require different implementation characteristics and therefore parallel implementations may give better performance. For example, it may be feasible to replace conventional RAKE receiver with multi-user interference cancellation algorithm in some operation configuration.

In addition to traditional low-power design and hardware minimisation issues, there are several aspects worth studying, when considering the optimisation of hardware and the resource usage. Dynamic allocation of functionality to resources requires that the resources can execute the algorithms effectively. This can be achieved during the design by selecting the algorithms and functions so that the mapping to HW is known. Parameters to watch are the operation types, data dependencies, data accuracy, time budgeting, *etc.* Secondly, it is important to watch the resource usage as a function of time in addition to worst case scenarios. The amount of active logic must be kept minimum all the time. The resources that are needed must be used as effectively as possible, while the resources that are not needed must be switched off. Some algorithms are executed with processors that are not optimal for them. Thirdly, it is important to identify the system bottlenecks, and invest time and resources in them.

Table 1. Optimisation activities at different design levels

	Optimisation of functionality and performance	Optimisation of resources and power consumption
Conceptual level	Validations of requirements by studying use cases and their relationships to interfaces and constraints.	Estimation of requirement complexity and complexity of operation scenarios.
Functional level	Identification of minimal set of functions and the definition of operation configurations. Reuse of functionality and control schemes.	Using similar operations and data formats inside functions. Identification of common and configurable functions. Minimisation of parameter exchange between functions and definition of function interfaces.
Architecture level	Identification of platform software services. Identification of common hardware resources. Identification of performance critical parts.	Use of same processing resources for different functions. Studying possibilities for dynamic resource allocation and activation. Minimisation of communication between resources.
Implementation level	Design of communication channels. Optimisation of hardware speed and area. Optimisation of software.	Applying low power design practices. Management of both hardware and software configurations.

4. Future Work

There are several areas where future work is needed. The decision support and estimation solutions are inadequate for future needs. Feasible techniques that support core selection or function allocation to multiprocessor systems do not exist. The current codesign techniques are mostly oriented towards system synthesis, which is contradictory to intellectual property based design. It seems that approaches that integrate SW and HW architecture designs do not exist. In the traditional algorithm design, the techniques for dealing with contradictory requirements are scarce.

The work with terminal architecture will continue. Currently we have the functional specification and initial platform architecture completed. The final core selection is still in progress. The methodology development will continue in the integration of architecture design practices. The target is to create taxonomy for system architecture quality and techniques that support the quality estimations.

5. Conclusions

Generic software radio approach is not feasible for high-volume mobile terminal products. The heart of such product will be a telecommunication kernel that implements all the processing capacity. The optimisation of implementation must be done during the design and started already at the requirement specification.

The future telecom kernels integrate both hardware and software. The platform complexity exceeds the current capabilities and economical constraints. Therefore, the solutions must be based on extensive reuse and that they have to be reusable over longer time and different types of products. This requires intellectual property or core based design practices and concurrent development and optimisation of hardware and software architectures. Architecture development framework for such a kernel has been presented. Functional specification, decision support analyses and quality estimations are the basis of the process.

6. References

[1] Bursky, D., "Extreme Levels of Parallellism Escalate DSP Horsepower", *Electronic Design*, November, 22, 1999, p. 71-82

[2] Mitola, J., "Technical challenges in the Globalization of Software Radio", *IEEE Communications Magazine*, February, 1999, p. 84-89

[3] Mitola, J., "The Software Radio Architecture", *IEEE Communications Magazine*, May 1995, p. 26-38.

[4] Mitola, J., "Software Radio Architecture: A Mathematical Perspective", *IEEE Journal on Selected Areas in Communication"* Vol. 17, No. 4, 1999, p. 514-538.

[5] Srikanteswara, S. & al.: "A Soft Radio Architecture for Reconfigurable Platforms", *IEEE Communications Magazine*, February 2000, p.140-147.

[6] Mitola, J. & Maguire, Jr., G.Q., "Cognitive Radio: Making Software Radios More Personal", *IEEE Personal Communications*, August 1999, p. 13-18.

[7] Goosens, G. *et al.*, "Embedded Software in Real-Time Signal Processing Systems: Design Technologies", *Proceedings of IEEE*, 85(3), 1997 p. 436-454.

[8] Hennessy, J.L. & Patterson, D. A., Computer Architecture – A Quantitative Approach, 2nd ed. Morgan Kauffman Publishers Inc, 1996, 760 p.

[9] Ernst, R., "Codesign of Embedded Systems: Status and Trends". *IEEE Design and Test of Computers*, Vol. 15 No. 2, 1998 p. 45-53.

[10] 3rd Generation Partnership Project; Technical Specification Group: Services and System Aspects; Service aspects; Handover Requirements between UMTS and GSM or other Radio Systems (3G TS 22.129 version 3.2.0).

[11] Khun-Jush, J., *et al.*, "Structure and Performance of the HIPERLAN/2 Physical Layer". *Proceedings of Vehicular Technology Conference 5*, 19-22 September 1999, Amsterdam, Netherlands, p. 2667-2671.
[12] International Technology Roadmap for Semiconductors, Edition 1999, World Semiconductor Council.
[13] Fowler, M. & Scott, K., UML Distilled: Applying the Standard Object Modeling Language, Addison-Wesley, 1997, 179 p.
[14] Bass, L., Clements, P., & Kazman, R., Software Architecture in Practice, Addison-Wesley, 1998, 452 p.
[15] Kruchten, P. B., "The 4+1 View Model of Architecture", *IEEE Software*, November 1995, p. 42-50.

SDR Application for Mobile Appliance

Hiroshi Yoshida and Hiroshi Tsurumi

Corporate Research & Development Center, Toshiba Corp., 1 Komukai Toshiba-cho, Saiwai-ku, Kawasaki 210-8582, Japan

Abstract: A software defined radio (SDR) is one of the solutions for realizing a multi-mode terminal for various mobile communication standards. The SDR can be changed by replacement of the software, while the conventional radio is equipped with one receiver chain for each standard. In this paper, analog stage architecture suitable for the SDR implementation is described. And a multi-band and broadband RF-stage configuration, which introduces the direct conversion and low-IF principle, is presented for realizing the multi-mode SDR.

1. Introduction

Various mobile communication standards are expected to continue to exist in the future, and so it is important to develop a mobile terminal that can be used as a multi-mode transceiver. Such a terminal would not only enable regional and global roaming, but also reduce the costs of introducing new technology superimposed on antiquated legacy technology.

An SDR is one of the solutions for realizing a multi-mode terminal for various mobile communication standards [1]. The SDR has the flexibility to be a multi-standard connection with replacement of software, such as modem, filter, equalizer, cancel codec, and synchronization. The multi-mode SDR hardware should consist of a reconfigurable baseband digital signal processing stage and a multi-band and broadband analog RF stage.

In terms of the analog stage, the conventional super-heterodyne transceiver architecture requires some analog filters fitted to the carrier frequency and channel bandwidth of each of the communication standards. The disadvantage of this architecture is that it is difficult to make it multi-band and broadband, because these filters are fixed narrowband passive components unsuitable for a multi-band and broadband system with multi-mode operation.

The direct conversion and the low-IF architecture is suitable for such multi-band and broadband requirements, since unlike the heterodyne principle, passive filters and band-limited components in RF and IF stages are not required [2] [3]. Recently, the direct conversion and the low-IF principle has attracted widespread attention due to its compatibility with multi-standards [4].

This paper first describes the analog stage architecture for a handheld SDR terminal and then presents the multi-band and broadband analog stage configuration introducing the direct conversion and the low-IF principle. The discussion focuses on receiver hardware implementation.

2. Analog Stage Architecture for Handheld SDR

The ideal SDR consists of a combination of an analog-to-digital converter (ADC) connected directly to an antenna and a digital signal processor [5]. However it is as yet difficult to realize high-resolution ADC or high-speed DSP which will be needed for the ideal SDR for the current mobile communication application. The first step toward realization of a practicable SDR is to distribute the minimum analog components for constructing an analog stage where received RF signal is down-converted to the baseband before ADC, and thus to realize quadrature demodulation and channel selection in the digital signal processing stage, which have been included in the analog stage in the conventional receiver. The analog stage for the practicable SDR is required to equip multi-band and broadband characteristics, and thus to feed the received signal, which is modulated according to various mobile standards, to the digital stage.

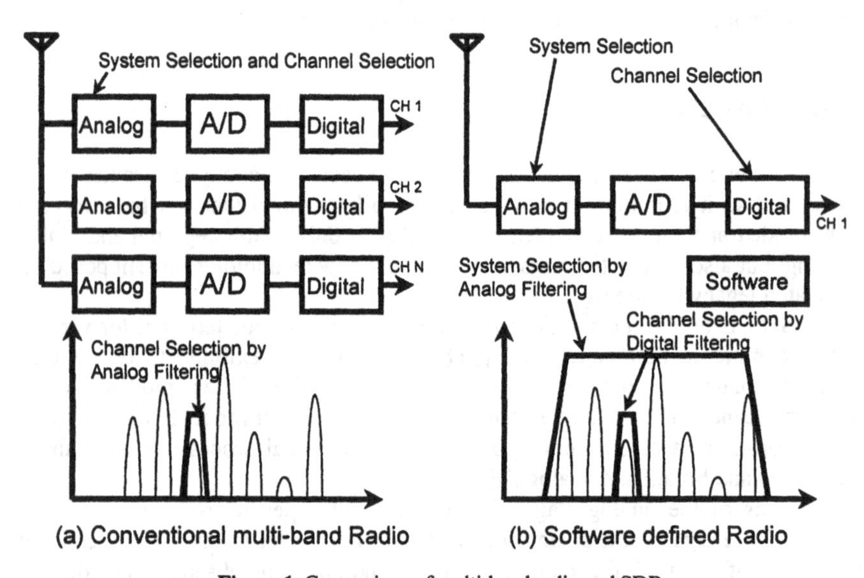

Figure 1. Comparison of multi-band radio and SDR

Figure 1 shows the schematic view of the multi-mode terminal using (a) a conventional multi-band receiver with multiple chains and (b) a multi-band receiver using the SDR. For multi-mode applications, the multi-band receiver in Figure 1 (a) is equipped with one receiver chain for each standard. It selects the channel according to the different carrier frequency and the different channel bandwidth using a fixed analog-defined channel selection filter. On the other hand, in the SDR in Figure 1 (b), all signals over the objective system bandwidth are filtered in the analog stage and fed to the broadband high-speed ADC: then the desired channel is selected from the digitized multi-channels with the software-defined channel selection section in the digital stage. It means that the system bandwidth of one mobile communication standard is entirely analog-to-digital converted to select the channel bandwidth by programmable digital filter. In this latter architecture shown in Figure 1 (b), the required bandwidth and roll-off characteristic of the digital filter can be reconfigured for every desired system by changing the coefficient of the digital filter programmatically. Therefore, this concept (analog system-selection and digital channel-selection) is appropriate for multi-standard receiver architecture for the practicable SDR. The SDR in Figure 1 (b) requires the multi-band and broadband analog stage which ideally feeds the signals over the entire system bandwidth to the ADC for each standard.

The system specifications such as frequency band, system bandwidth, channel bandwidth, and modulation scheme of significant mobile communication standards are shown in Table 1. The analog stage for the SDR is required to receive various frequency bands (*i.e.*, 800 MHz – 2 GHz) and broadband (*i.e.*, 20 – 60 MHz) signal, while the digital stage selects the desired channel from the "band-selected" signal and detect the "channel-selected" signal modulated in various modulation schemes programatically.

Table 1. Significant mobile communication standards

	Frequency Band [Hz]	System Bandwidth [Hz]	Channel Bandwidth [Hz]	Modulation Scheme
AMPS	800M	25M	12.5k	FM
PDC	800M/1.5G	20M	21k	π/4-QPSK
PHS	1.9G	23M	192k	π/4-QPSK
IS-136	800M	25M	24.3k	π/4-QPSK
GSM	900M/1.8G	25M	200k	GMSK
cdmaOne	800M	25M	1.25M	QPSK
W-CDMA	2G	60M	3.84M	QPSK
cdma2000	800M/2G	25/60M	1.25/3.75M	QPSK

3. Multi-band and Broadband Analog stage Configuration for SDR

In this section, first, the problems respecting realization of the multi-band and broadband receiver by super-heterodyne principle are discussed in Section 3.1, and then, the direct conversion and low-IF principle are described in Sections 3.2 and 3.3, respectively.

3.1 Super-heterodyne Principle

Figure 2 shows an example of a receiver chain of the super-heterodyne architecture. This architecture requires frequency-dependent passive components including dielectric filters in the RF stage and surface acoustic wave (SAW) filters in the 1st stage for image suppression. Ceramic or crystal filter is also needed in the 2nd IF stage for channel selection with another crystal for the 2nd local oscillator. In general, the center frequency and bandwidth of these filters cannot be variable, but must be fixed according to the radio standard specification. Furthermore, the bandwidth of the passive filter is restricted by the Q value and usually narrower than the entire system bandwidth. Therefore, the super-heterodyne approach is unattractive for the multi-band and broadband analog stage configuration. It is possible to overcome to the functional limitation on multi-band/broadband and programmability by using switched capacitor filter banks and direct signal synthesis. The size and weight penalties incurred may be acceptable for base stations, but are unacceptable for mobile terminals.

One of the candidates for realizing such a multi-band and broadband analog stage for the SDR is the direct conversion receiver.

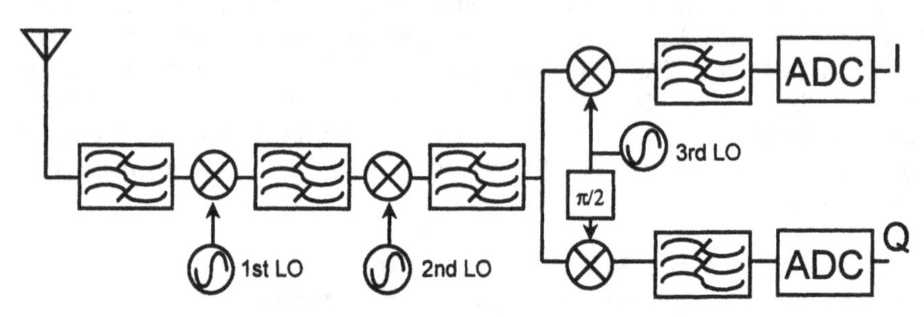

Figure 2. Super-heterodyne configuration

3.2 Direct Conversion Principle

The direct conversion architecture is inherently equipped with the multi-band and broadband characteristics because the direct conversion principle dispenses with the need for the passive filters and band-limited components.

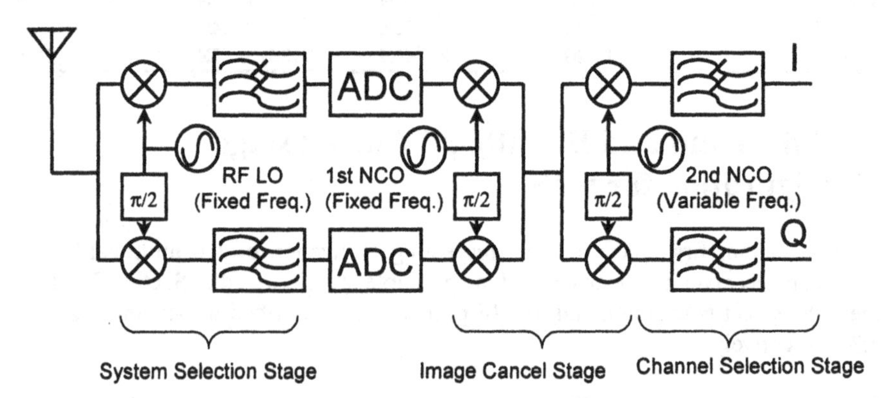

Figure 3. Direct conversion configuration

Figure 3 shows one of the configurations of the direct conversion receiver employing the analog system-selection and digital channel-selection concept, where the entire channel bandwidth including desired channel is considered to be one broadband signal to be down-converted to the baseband by the front-end quadrature demodulator with frequency-fixed local signal prepared for each standard. According to this principle, the down-converted channels exist from DC to 10MHz for 20MHz bandwidth system. The baseband IQ signals which contain entire system bandwidth are "system-selected" by bandwidth-variable analog low-pass filters for an anti-alias purpose and then converted into digital signals by IQ ADCs. The digitized baseband signals are re-converted to the programmable IF with image rejection configuration. The IF signals are digitally down-converted to the baseband with programmable down converters for "channel-selection". In this architecture, the entire system channels are down-converted to the baseband with fixed local signal whose frequency is the center of the desired system. However, in this case, care needs to be exercised, as the corresponding image signal exists inside the system band. The image signal is canceled in image cancel stage before digital channel selection stage.

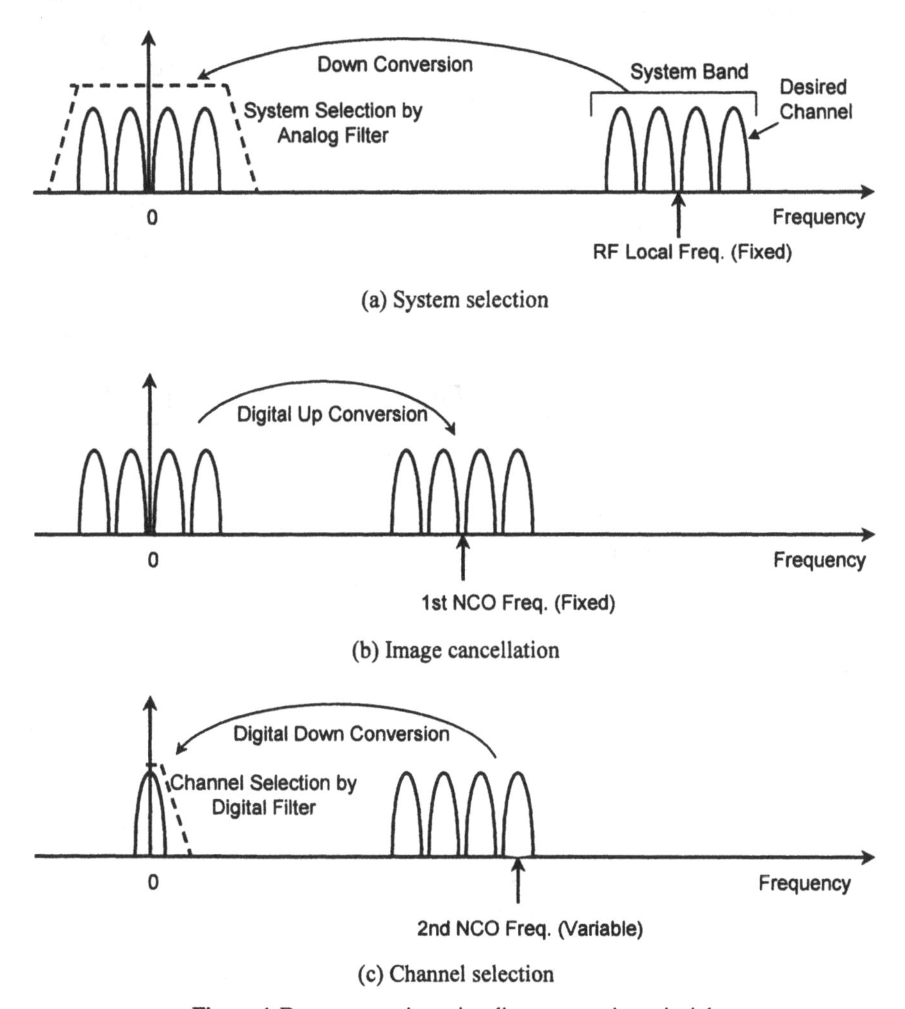

(a) System selection

(b) Image cancellation

(c) Channel selection

Figure 4. Down conversion using direct conversion principle

Figure 4 illustrates the principle upon which the configuration in Figure 3 is based with respect to the frequency domain. The entire system bandwidth shown in Figure 4 (a) is down-converted to the baseband and system-selected by IQ analog active filters. The baseband signal including entire system channels is digitized in IQ ADCs, and then re-converted to the digital IF for image cancellation (Figure 4 (b)). The IF signal is down-converted again in the channel selection stage where the desired signal is finally channel-selected by the programmable digital filter (Figure 4 (c)).

3.3 Low-IF Principle

The direct conversion principle is suitable for the SDR. However, imperfections in the RF quadrature mixer such as I-Q gain imbalance or phase error cause residual or latent image responses. Even if IQ gain and π/2 phase shift are carefully designed, image suppression level cannot be more than 40 dB; however, the image suppression of more than 70 dB is required to meet the specification of the receiver chain for mobile standards.

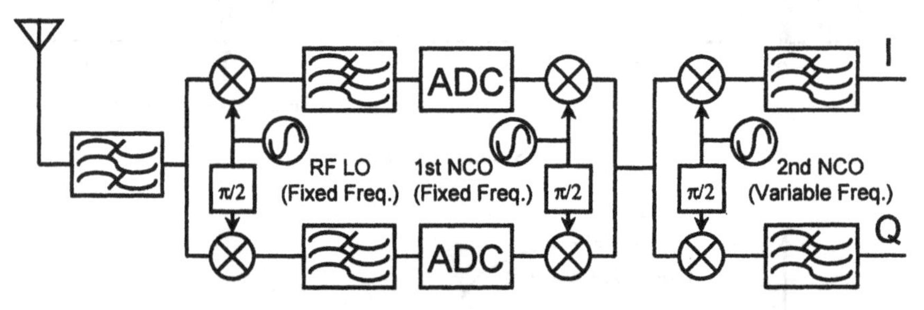

Figure 5. Low-IF configuration

The configuration shown in Figure 5 is an effective solution for improving image characteristics where an RF band-pass filter, which is not required for the direct conversion, is inserted at the top of the receiver front-end for broadband signal selection. The first local frequency is fixed at a few megahertz lower (or few megahertz higher) than the lowest (or highest) frequency of the system band of interest so as to eliminate image signals with the band-pass filter in the order of 30 to 40 dB. In this case, the image signals which should be taken into consideration are located outside the system band. When the first local frequency is set to 10 MHz lower than the lowest channel frequency of the desired system, the image signals, which appear 20 MHz separate from the lowest channel frequency, can be suppressed to 30 to 40 dB by the RF filter. And the bandwidth of interest in the low-IF band exists in the range of 10 MHz to 30 MHz for 20 MHz bandwidth system. Adopting the configuration shown in Figure 5, overall image rejection ratio of 70 to 80 dB can be achieved, which is acceptable for the mobile communication application. This architecture still retains the advantage of enabling low-pass filter and ADC activating at lower frequency than in the case of conventional super-heterodyne architecture.

As shown in Figure 6 (a), the RF local frequency is not set to the center of the system bandwidth of interest; therefore, each channel in the system is down-converted to not zero frequency but to very low-IF. The low-IF signals are then system-selected by IQ analog filters and re-converted to higher frequency with the 1st NCO for image rejection (Figure 6 (b)). The desired signal is finally down-converted to the baseband with the 2nd NCO and channel selected by the programmable digital filter (Figure 6 (c)). This mechanism is considered to resemble the well-known "low-IF" topology [6]. However, this architecture is basically different from the well-known "low-IF" architecture where the desired signal is detected by frequency-fixed analog filter.

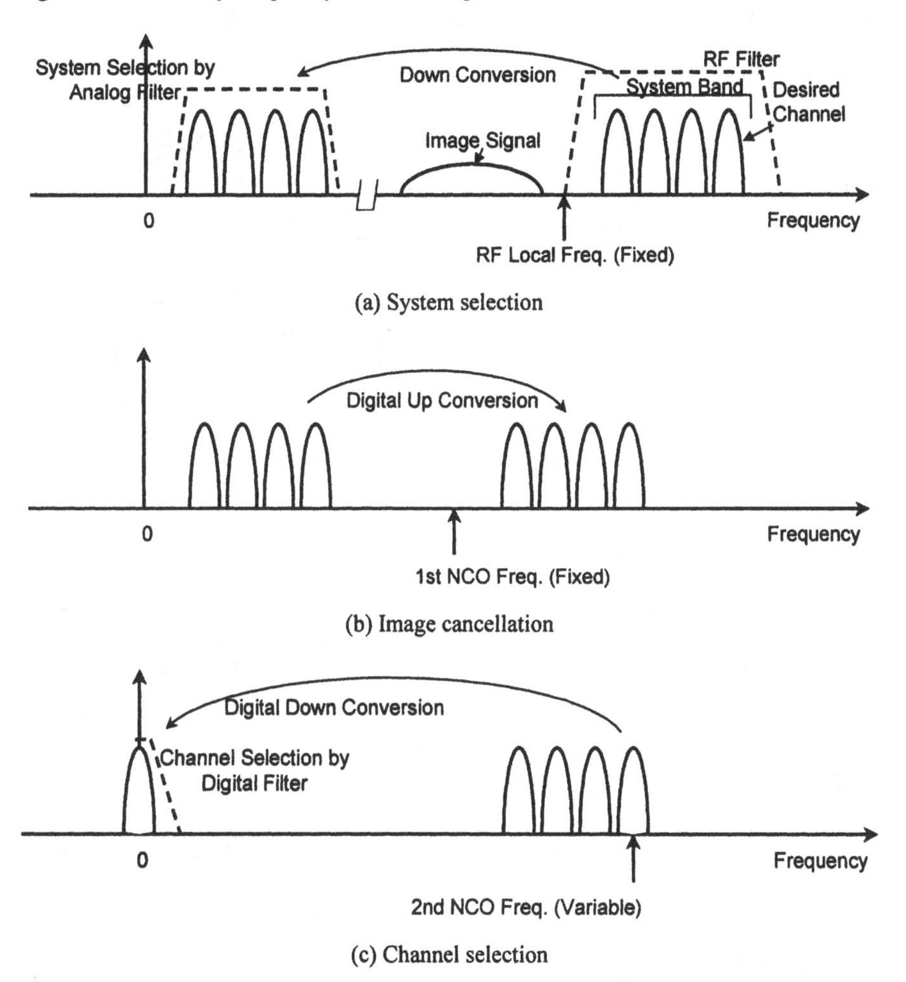

Figure 6. Down conversion using low-IF principle

Figure 7 shows another configuration with two digital quadrature demodulators and a variable frequency NCO for channel selection. These two quadrature demodulators function as a complex multiplier with image rejection characteristic. Consequently, it operates identically to the configuration shown in Figure 5, where "system-selected" digitized signals for I and Q channels (I_{IF} and Q_{IF} in Figure 7) are fed into two quadrature demodulators, and are down-converted into the baseband signals by the NCO with the center frequency of the desired signal. Four baseband signals from the quadrature demodulators, "II", "IQ" for the I_{IF} and "QI",

"QQ" for the Q_{IF}, are summed or subtracted and then filtered with the digital low-pass filters for "channel-selection". In this configuration, the image signal is canceled using complex operation in low-IF domain. For the handheld SDR terminal application, this configuration is more suitable than the configuration shown in Figure 5, since the operating frequency in the digital stage is approximately a quarter that of configuration shown in Figure 5.

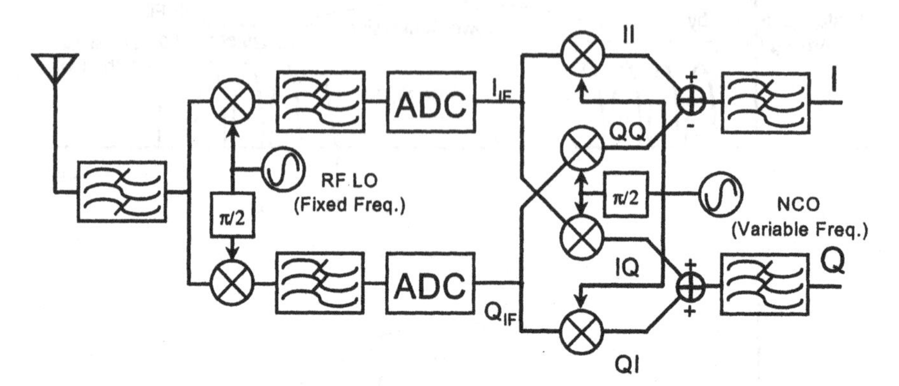

Figure 7. Low-IF configuration with complex multiplier

4. Conclusions

Multi-band and broadband RF analog front-end architecture for handheld SDR terminal applying the direct conversion and low-IF principle was presented. The architecture can enable the SDR to be a multi-mode handheld terminal which is reconfigurable only by exchanging the software for various mobile standards.

5. References

[1] J. Mitola, "The Software Radio Architecture", IEEE Communications Magazine, vol. 33 No. 5, pp. 26—38, 1995.
[2] H. Yoshida, H. Tsurumi, and Y. Suzuki, "Broadband RF Front-end and Software Execution Procedure in Software-defined Radio", Proc. IEEE VTC 1999-Fall, vol. 4, pp. 2133—2137, 1999.
[3] H. Tsurumi, H. Yoshida, S. Otaka, H. Tanimoto, and Y. Suzuki, "Broadband and Flexible Receiver Architecture for Software Defined Radio Terminal Using Direct Conversion and Low-IF Principle", Special Issue on Software Radio IEICE Trans. Commun., vol. E83-B, no. 6, pp. 1246—1253, 2000.
[4] A. Fernandez-Duran, T. Sanjuan, J. Sevenhans, J. Dulongpont, J. M. Parez, J. Casajus, J. Barrett, T. Roste, and G. Fletcher, "Zero-IF Receiver Architecture for Multistandard Compatible Radio Systems. Girafe Project", Proc. 46th IEEE VTC, vol. 2, pp. 1052—1056, 1996.
[5] J. Mitola, "Software Radios Survey, Critical Evaluation and Future Directions", IEEE AES Systems Magazine, vol. 8 No. 4, pp. 25—36, 1993.

[6] P. R. Gray and R. G. Meyer, "Future Directions of Silicon IC's for RF Personal Communications", CICC '95, pp. 368—369, 1995.

SDR Application for Intelligent Transport Systems

Hiroshi Harada and Masayuki Fujise

Communications Research Laboratory, MPT, Japan
3-4 Hikarino-oka Yokosuka, Kanagawa, 239-0847 Japan

Abstract: In this paper, the multi-mode & multi-service software radio communication system (MMSR) is proposed as an application of software defined radio (SDR) technology suitable for the Intelligent Transport Systems (ITS). Moreover, the parameter-controlled-type SDR technology is introduced as one of key technologies to develop MMSR system. Then, the transmission performances of the proposed system are evaluated using experimental prototype.

1. Introduction

Recently, demands for mobile radio communication systems have been increased, and several proposals and examinations on the broadband multimedia mobile communication systems which include not only voice communication but also data, picture and moving-image communication have been carried out.

In the environment where various radio communication systems intermingle, we must utilize the different mobile communication systems by using one integrated terminal.

A representative method is multimode terminal based on software defined radio (SDR) communication technology [1][2], in which user can utilize their favorite mobile communication systems by downloading the digital signal processing software (DSPS) which describe the elemental components for realizing specific communication systems to the digital signal processing hardware (DSPH) like DSP and/or FPGA of the terminal. By using the SDR technology, it is possible to operate the manifold mobile communication systems by one multimode terminal, to correspond to the updated various modulation systems easily and to construct flexible communication systems for the variance of the propagation environment. In addition, from the viewpoint of the manufacturer, there are some merits in only having to concentrate on the research and development of the software instead of hardware when SDR technology is widely spread. Moreover, from the viewpoint of environmental conservation, we can reduce the amount of industrial waste [3].

However, in order to realize SDR technology, the following two items must be discussed: (a) The necessary elemental technology and (b) The scenario for the introduction of SDR technology. As for item (a), it is possible to categorize into three parts: RF(Radio frequency) part related to up- and down-converter, IF(Intermediate frequency) part related to orthogonal modulator and demodulator, analog-to-digital (A/D) and Digital-to-analog (D/A) converter, and baseband part related to DSPH.

Table 1. Communication systems utilized in the ITS system

Category	Communication			
System	PDC	PHS	IS-95	IMT-2000
Frequency	800MHz 1.5GHz band	1.9GHz band	800MHz band	U/L:1.9GHz band D/L:2.1GHz band
Modulation	π/4QPSK TDMA-FDD	π/4DQPSK TDMA-TDD	U/L:QPSK-CDMA D/L:OQPSK-CDMA	BPSK, QPSK CDMA
Data Rate	42kbit/s	384kbit/s	9.6, 4.8, 2.4, 1.2kbit/s 14.4, 7.2, 3.6, 1.8kbit/s chip rate: 1.2288Mcps	CDMA2000 64-384kbit/s chip rate:1.2288Mcps chip rate:3.6864Mcps W-CDMA2000 64-384kbit/s chip rate:3.84Mcps
Bandwidth	50kHz/1ch	300kHz/1ch	1.25MHz	less than 5MHz
Priority	3	3	3	3

Category	Control			Broadcasting			
System	GPS	VICS	ETC	Radio Broadcasting		TV	Satellite
Frequency	1.5GHz band	FM band	5.8GHz	0.5-1.6MHz	76-90MHz	90-222, 470-770 MHz	11.7-12.0 GHz
Modulation	BPSK+ DS-SS	LMSK	ASK	AM	FM	VSB-AM FM	FM PSK
Data Rate	50bps	16kbps	1024kbps	Analogue	Analogue	Analogue	Analgoue
Bandwidth	1.023 MHz	100kHz	<8MHz	15kHz/ch	200kHz	6MHz	27MHz
Priority	1	3	3	3	3	2	2

As for item (b), two points must be discussed: (1) “Which services must be integrated in one multimode terminal?” and (2) “How and in which field is the SDR technology socially recognized?”. The question (1) is quite important question. This is because we can realize multimode terminal easily by developing specified LSI for each communication system and arranging the LSI on the one circuit board if we don’t use any SDR technology. Therefore, it is necessary to introduce the SDR technology to the application field in which many communication systems must be required. Moreover, as for the question (2), we must promote the SDR technology from the user terminal in order to spread the concept of the SDR technology widely.

As one of the application fields which can fulfill the answer of (1) and (2), the author noticed intelligent transport systems (ITS) [4][5], because a number of services must be utilized in the car. For example, broadcasting service, mobile radio communication service, the Electronic Toll Collection System (ETC), the

Vehicle Information and Communication System, (VICS), and Global Positioning Service (GPS) are the representative examples. However, the space for such communication terminals is limited in the car. Therefore, we must integrate all terminals for all communication service utilized in the car into one multimode terminal.

In this paper, the multi-mode and multi-service software radio communication system (MMSR) is proposed as an application of software defined radio (SDR) technology suitable for the ITS. Moreover, the parameter-controlled-type SDR technology is introduced as one of key technologies to develop MMSR system. Then, the transmission performances of the proposed system are evaluated using experimental prototype which can realize GPS and ETC and PHS (Personal Handy Phone System) services by two new SDR technologies.

2. The Necessity of SDR Technology in ITS

2.1 Affinity between SDR and ITS

At present, the communication services utilized in the ITS are arranged in Table 1. As shown in Table 1, the communication services can be divided into 3 systems: communication-based system, control-based system, and broadcasting-based system. The representative examples of communication-based system are PDC(Personal Digital Cellular) system, PHS system, and the digital cellular communication system using CDMA (Code Division Multiple Access). Moreover, the third-generation mobile communication system based on CDMA : IMT-2000 will be realized in the near future. And, GPS system, VICS system and radar system are representative examples of control-based system. Moreover, the Advanced cruise-assist Highway System (AHS) for the driving-support based on roads information such as traffic accident, fall object, and meteorological phenomenon, and ETC system will be utilized in the car. Then, the representative examples of the broadcasting-based system are radio broadcasting system, analog-television broadcasting system and satellite broadcasting system, and the digital television service will be utilized in future. As mentioned in the above, the number of communication systems utilized in future ITS system will increase. However, there is the limit on the space of the car to equip these systems, and space saving and high integration of the system component must be desired in order to realize the such manifold services smoothly and flexible.

2.2 The Proposal of the Multimode and Multiservice Software Radio Communications System

In the ITS in which many systems intermingle, multimode software radio communication (MSR) system using the SDR technology is suitable to realize several systems by using only one terminal efficiently. Fundamental system configuration is shown in Figure 1. This MSR system consists of three units: RF unit (RFU), IF unit (IFU) and baseband unit (BBU). The RFU consists of the antenna, up-converter (U/C) and down-converter (D/C) blocks. The IFU consists of orthogonal modulator and demodulator block, digital-to-analog converter (D/A) block, analog-to-digital converter (A/D) block. The BBU is composed of several baseband digital signal processing hardware (DSPH) like DSP and FPGA. These three units also have two parts: TX module and RX module. TX module has relation to the transmitter. On the other hand, RX module is related to the receiver.

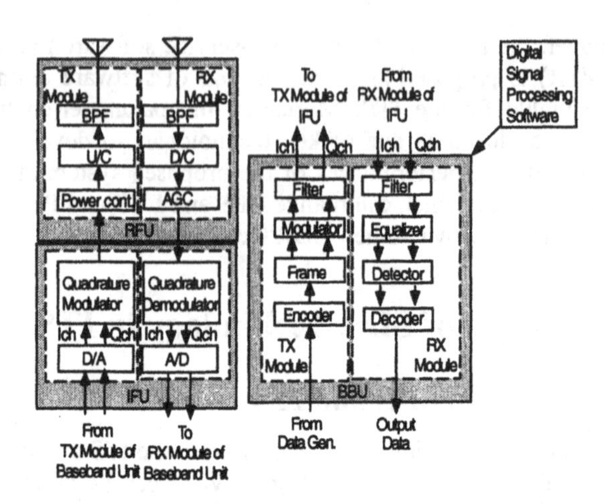

Figure 1. The configuration of full-download-type software radio system

When we realize a telecommunication system by MSR system, first of all, all digital signal processing software (DSPS) which configure a required telecommunications system are downloaded to BBU. After finishing the download of software, the configuration check program is executed. Finally, BBU configures the required baseband modulation and demodulation circuits. Then, transmission data are fed into BBU of TX module. If we change all the software whenever user changes the required communication system, the SDR technology can be called as full-download-type SDR technology.

At present, many researches on individual component of U/C and D/C blocks, A/D block, and D/A blocks have been progressed, and many prototypes which can realize this MSR system have been developed [5][6]. However, it is difficult to apply this MSR system for the ITS. This is because there are several demands for MSR system from the viewpoint of users. In the ITS system, the services can be divided into three priorities:

1. the services which are always necessary during driving a car;
2. the services which are absolutely necessary in any specific time;
3. the services which are sometimes required by the feeling of the user.

The priority is given for the systems shown in Table 1 by using the classification. If we realize MSR system on the ITS, the demands by which users utilize several services simultaneously may come out. However, research and development in which users select only one system from several communications systems was carried out in MSR system. For the future ITS system, not only MSR system but also multiservice software radio communication system must be realized. Then, we newly proposed the combination system between multimode and multiservice software radio communication system as multimode and multiservice software radio communication (MMSR) system [7].

The configuration for realizing this MMSR system can be arranged into the following four systems [4]. The signal processing method is performed in the condition that the signal of each system in MMSR system is converted to a digital signal 0 or 1:

(a) the method in which DSPH of the BBU perform digital signal process for the manifold systems simultaneously by installing plural hardware components for the software radio communication system;
(b) the method in which DSPH of the BBU perform digital signal process for the manifold systems simultaneously by time division multiplex or time division multiple access;
(c) the method in which DSPH of the BBU perform digital signal process for the manifold systems simultaneously by code division multiplex or code division multiple access;
(d) the method in which DSPH of the BB unit perform digital signal process for the manifold systems simultaneously by random access protocol.

Configuration examples of transmitter and receiver described in (a) are shown in Figures 2(a) and 3(a), respectively. In this configuration, several DSPHs, which realize desired communication systems are prepared separately. Then, a demand of the user is transmitted to each DSPH from the outside controller, and manifold services are utilized. As a similar configuration, the plural DSPHs can be integrated in one DSPH, and it is also controlled from outside controller [7]. Since each flow of digital signal processing for each system is separated individually, this MMSR system does not cause any interference between services. However, plural DSPHs must be prepared, and the redundant signal processing are performed in some DSPHs. As a result, the scale of hardware may increase with the increase of the number of the integrated systems.

Configuration examples of transmitter and receiver described in (b) are shown in Figures 2(b) and 3(b), respectively. In this configuration, one DSPH is prepared, and the hardware carries out the signal processing for plural communication services by time division based multiple access technique in order to eliminate the redundant signal processing, which becomes a problem by preparing plural DSPHs in MMSR system (a). As for the method of time division based multiple access technique, two methods are considered: the time division multiplexing (TDM) method, which allocates the fixed time slot to signal processing for specified communication services, and the time division multiple access (TDMA) method, which allocates the time slot to signal processing for specified communication services on demand.

In the MMSR system used TDM method, efficient MMSR system can be realized if the slot timing is decided. However, if the number of integrated systems increases, we must prepare many time slots for digital signal processing. Moreover, if we allocate one time slot to a communication service with priority (3), the time slot is sometimes used and mostly it is not used. By using TDMA technique, we can solve the problem, but we need to research various management techniques such as time-domain interruption technique and so on.

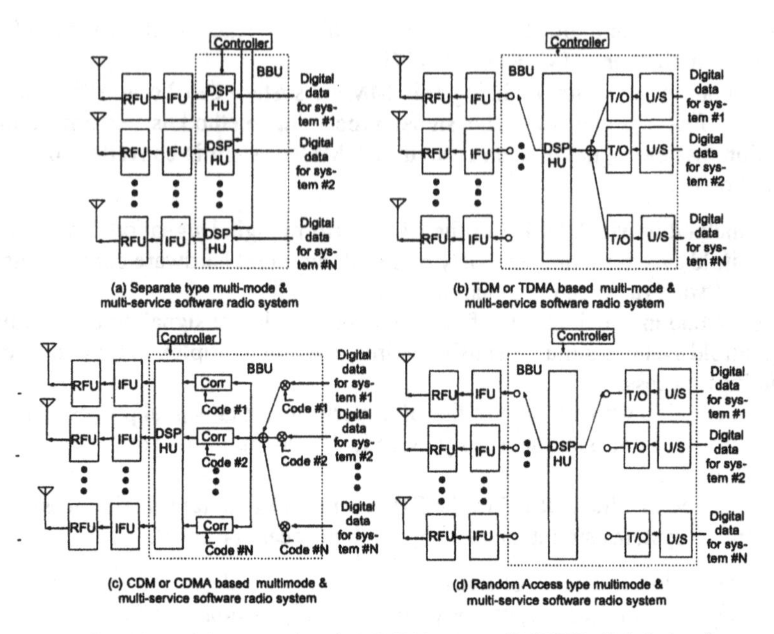

Figure 2. Configuration of the transmitter in MMSR system (DSPHU: digital signal processing hardware unit, T/O: time offset, Coor: correlator)

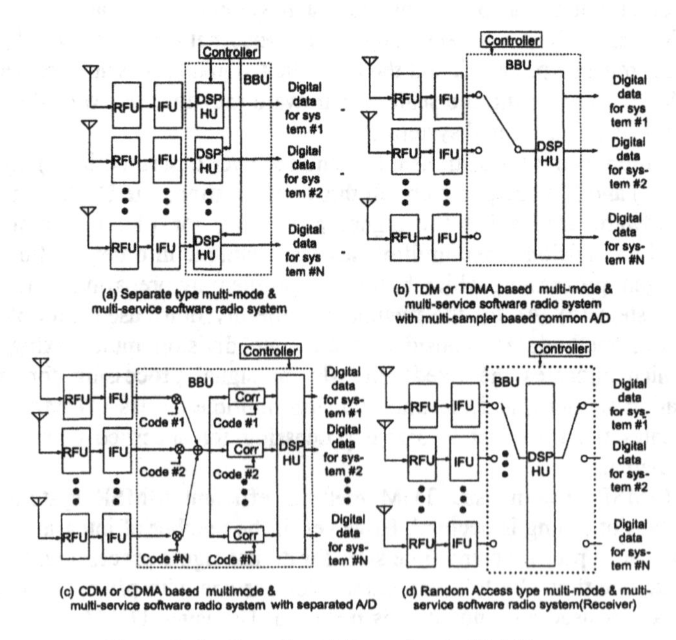

Figure 3. Configuration of the receiver in MMSR system

Configuration examples of transmitter and receiver described in (c) are shown in Figures 2(c) and 3(c), respectively. In this configuration, in order to eliminate the redundant signal processing by preparing plural DSPHs in MMSR system (a) and to avoid synchronous signal processing like TDM or TDMA technique in MMSR system (b), The code is superposed for digital data of each system which is obtained before D/A conversion in TX module of BBU and after A/D conversion in BBU of RX module and the superposed digital data is multiplexed by Code Division Multiplexing (CDM) technique. Then, when the request which needs a specified service occurred, user obtains the information data of the required service by using spread code for the specified service. As for the codes used in this case, it is possible to utilize the cryptogram used in the cryptology theory as well as spread code for spread spectrum communication system. The merit of this system is to utilize the required service by using the code whenever user needs to utilize the system. However, the superposion of the code means that the clock speed of digital signal processing increases. We, therefore, need the high-speed digital signal processor.

Configuration examples of transmitter and receiver described in (d) are shown in Figures 2(d) and 3(d), respectively. This configuration is fundamentally one of the application systems of system (b). Instead of TDMA and TDM technique, random access protocol must be adopted. Moreover, except for the MMSR system (a)-(d), we can utilize hybrid systems, in which system (b) is adopted for the configuration of transmitter and system (c) is adopted for the configuration of receiver.

The MMSR is efficient for the integration of plural systems. However, as for the SDR technology, we can make it more efficient by a new concept for the configuration of BBU.

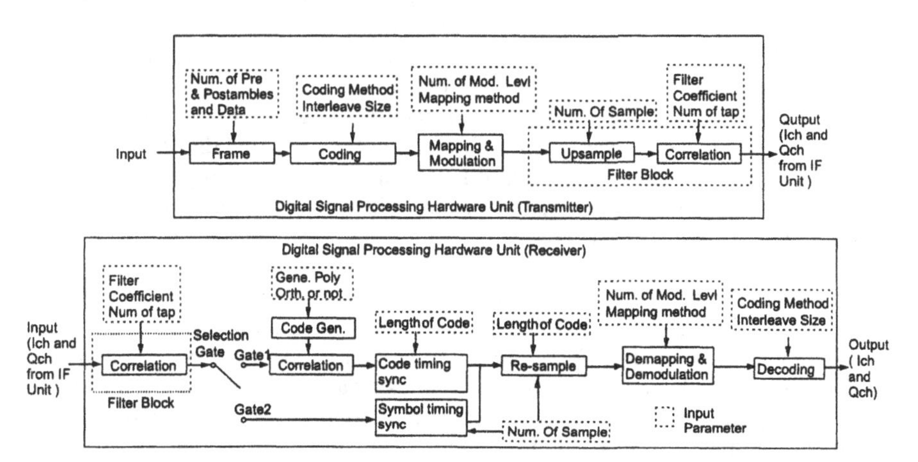

Figure 4. The configuration of BBU in the proposed SDR technology

2.3 Problems on Conventional SDR Technology

In the conventional full-download-type SDR technology, we can change the system configuration in accordance with our request. However, the following problems are involved:

1. the volume of software downloaded into the DSPH increases, as the contents of the required telecommunication component blocks become more complicated;
2. the period for configuration check of the DSPH also increases, as the contents of the required telecommunication component blocks become more complicated;
3. In the download software, there are often several component blocks which are related to the know-how of manufacturer, e.g., the optimization method or calculation algorithm for some special blocks, *etc*. The know-how may leak out by the download of software. And, there is the possibility altered from the other people.

In this paper, we introduce a new SDR technology to overcome these problems. In next subsection, we explain the configuration in detail.

2.4 Concept of the Parameter-controlled-type SDR Technology

The proposed SDR technology [3]-[5] is only related to the BBU of both TX and RX modules. For the other units, RFU and IFU, we may prepare these units for each system individually or we may prepare an integrated RFU and IFU for all systems. The BBU configuration of the proposed system is shown in Figure 4. In the BBU, basic telecommunication component blocks like encoder, frame, modulator, filter blocks of transmitter and equalizer, detector and decoder blocks of receiver have already been programmed and implemented in the DSPH in advance. And the functions of the telecommunication blocks are not fixed but programmable and changeable easily by downloading external parameters. Namely, if we would like to change the configuration of digital filter of the transmitter, we send only coefficient information of the required digital filter to the filter block of BBU. Then, the proposed BBU unit becomes one of general-purpose transmitter and receiver by external parameter. The software radio system by a new configuration method of BBU is called as parameter-controlled-type software radio system.

The proposed system is realized by downloading external parameters, which is small and important information for the realization of a required telecommunication system. Therefore the important blocks have been configured in the DSPH in advance. As a result, it is expected that we can obtain stable performance and reduce the period for configuration check in comparison with full-download-type SDR technology. Moreover, the information which have relation to the know-how of manufacturer such as the optimization methods for the specified telecommunication components never leak out, because we only download a general and small volume software. Moreover, since the volume of download software to the DSPH is quite small, we can utilize several strong coding methods to the download software. Consequently, the proposed software radio communication system can keep high concealment.

3. Performance Evaluation of the Proposed MMSR System by Developed Prototype

In order to show the effectiveness of proposed software radio system, an experimental prototype was developed and its transmission performance was evaluated. The following are shown in Figures 5 and 6: system configuration of the experimental prototype and the appearance. Moreover, the system parameters are summarized in Table 2. The experimental prototype can make use of three real telecommunication systems: PHS and GPS and ETC systems as Service mode for ITS. And as for PHS and ETC or GPS and ETC, we can utilize two services simultaneously by MMSR system. This is the first prototype of MMSR system for ITS in the world [4][5]. Moreover, as User mode, it is possible that the user freely conducts several modulation schemes of GMSK, π/4QPSK, BPSK and QPSK. As for PHS and GPS, we integrate the antennas of two systems into one antenna because the frequency utilized in GPS (1.5GHz band) is quite similar to PHS (1.9GHz band). And, the external parameters, which need to change the system, are supplied from a notebook type computer connected with experimental prototype by the 10Base-T Ethernet cable. In the notebook computer, management software of the experimental prototype has been installed.

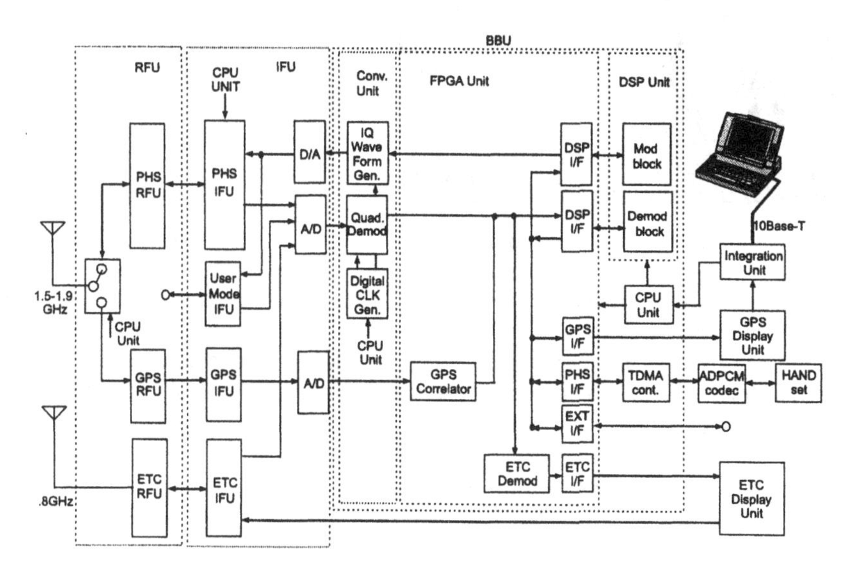

Figure 5. System configuration of the experimental prototype

When user would like to realize PHS system by using the configuration shown in Figure 5, the external parameters which is necessary for constructing the PHS system, is stored at the CPU unit through 10Base-T Ethernet cable from a notebook type computer. The CPU unit gives the parameters to BBU, IFU, and RFU. Then, this experimental prototype becomes the PHS mode. The similar technique can be utilized for the GPS and ETC systems.

In order to evaluate proposed software radio system, the following comparisons are carried out between full-download-type software radio system and parameter-controlled-type SDR technologies: Volume of download program and length of download time. As for the volume of download program, the detail is described in [5]. In this paper, the length of download time is evaluated in Table 3 by using User mode. In User mode, we can realize GMSK, $\pi/4$QPSK, BPSK or QPSK by changing parameters. As shown in Table 3, the average download time of the proposed software radio system becomes around 1/100 in comparison with full-download-type software. In addition, the average download time for all modulation schemes for the realization of User mode is independent on the modulation schemes and almost same by parameter-controlled-type software radio system. This is because the download software is almost similar in all modulation schemes of User mode, and the volume is almost same for all modulation scheme for the realization of User mode.

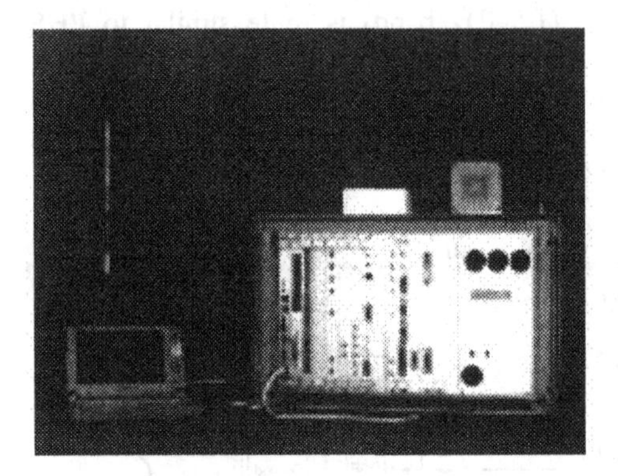

Figure 6. Appearance of the experimental prototype

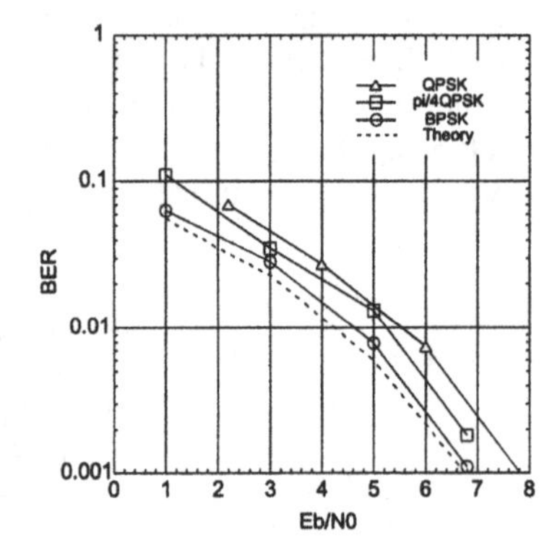

Figure 7. BER performance

Finally, the BER performance of this experimental prototype in the case of user mode is shown in Figure 7. In Figure 7, we also include a theoretical BER value vs. Eb/N0 of all system. In this case, since we adopt the coherent detection for all systems, the BER performance vs. Eb/N0 become same value in BPSK, QPSK and $\pi/4$ QPSK. As shown in Figure 7, it is clear that the experimental prototype obtained quite similar BER performance to theoretical BER performance within 1 dB in all modulation schemes.

4. Conclusions

In this paper, the necessity of SDR technology into the ITS was described. Then, the MMSR system and the parameter-controlled-type SDR technology were introduced in order to configure several radio communication systems effectively and speedy by software download. Moreover, the effectiveness of the proposed configuration method is evaluated by using computer simulation and developed experimental prototype. And it is confirmed that our proposed system becomes one of good candidates for the realization of future flexible software radio system. In order to realize MMSR systems, we must consider many subjects. Especially, the Communications Research Laboratory (CRL), M.P.T researches digital signal processing method to process several communication systems simultaneously by using common sampling rate [8]. In this case, the common sampling rate is sometime non-integer times of symbol rates of integrated systems and the synchronization method must be considered. Paper [8] proposed a method and constructed TDM or TDMA based MMSR system by using the method. Moreover, by using the parameter-controlled-type SDR technology, we can reduce the download-time of software. The author, therefore, consider the download protocol to suit for the ITS by not only download from wired network but also wireless network as one of further studies. Especially, as one of download method by wireless network, the usage of FM broadcasting band is one of candidates, because all cars have radio broadcasting receiver and the coverage area is quite large. Moreover, the receiver unit is quite cheap. As for the topic, the author starts to publish a few papers [9]. Moreover, in Japan, a national institute: TAO also started to research SDR technology for ITS as well as CRL, MPT. In the near future, the application of SDR technology will be widely spread to the world.

Table 2. System parameters of the developed experimental prototype

Services	Frequency
(Service mode) PHS, ETC,GPS (User mode) BPSK, QPSK, π/4QPSK, GMSK	PHS-1.9GHz, ETC-5.8GHz, GPS-1.5GHz, User Mode-IF band
Modulation	**Data Rate**
PHS---π/4DQPSK ETC---ASK GPS---BPSK+Spread Spectrum User Mode--- BPSK, QPSK, π/4QPSK, GMSK	PHS---384kbps (carrier bit rate) ETC---1024kbps:2048kbaud (data rate) GPS---50bps (data rate):1.023Mcps (chip rate) User Mode BPSK, QPSK, π/4 QPSK—384kbps (carrier bit rate) GMSK-270.822kbps (carrier bit rate)

Table 3. The average configuration time in User mode

BPSK -> QPSK	37.5ms	BPSK -> GMSK	30.5ms
QPSK -> BPSK	32.5ms	GMSK -> BPSK	35.25ms
BPSK -> π/4QPSK	32.5ms	QPSK -> GMSK	30ms
π/4QPSK -> BPSK	35ms	GMSK ->QPSK	40.5ms
QPSK -> π/4QPSK	30ms	π/4QPSK -> GMSK	30ms
π/4QPSK -> QPSK	35ms	GMSK -> π/4QPSK	35ms
Full Download and User mode	3410ms		

5. References

[1] European Commission DG XIII-B : " Software Radio Workshop," May 1997.

[2] Special Issue on Globalization of Software Radio, IEEE Commun. Mag., Feb 1999.

[3] H.Harada, M.Fujise, "Multimode Software Radio System by Parameter Controlled and Telecommunication Toolbox Embedded Digital Signal Processing Chipset", 1998 ACTS Mobile Communications Summit , pp.115-120, Jun. 1998.

[4] H.Harada, Y.Kamio and M.Fujise, "A New Multi-mode & Multi-service Software Radio Communication System for Future Intelligent Transport Systems," Technical Report of IEICE, SR99-21, pp.81-88 , Nov.1999 (in Japanese).

[5] H.Harada, Y.Kamio and M.Fujise, "Multimode Software Radio System by Parameter Controlled and Telecommunication Component Block Embedded Digital Signal Processing Hardware", IEICE Trans. Commun., vol.E83-B, No.6, Jun. 2000.

[6] H.Ishii, T.Suzuki, T.Yamamoto, "Development of Adaptive Receiver Based on Software Radio Techniques", IEEE PIMRC'99, pp.16-20, Sept. 1999.

[7] H.Harada, "A Proposal of Multi-mode & Multi-service Software Radio Communication Systems for Future Intelligent Telecommunication Systems," Proc. of International Symposium on Wireless Personal Multimedia Communications (WPMC'99), pp.301-304, Sept. 1999.

[8] R. Sawai, H. Harada, H. Shirai, M. Fujise, "A Study on an adaptive symbol timing synchronization method for multi-mode & multi-service software radio communication system," Technical Report of IEICE, SR00-04, pp.25-32, Apr. 2000 .

[9] H.Harada and M.Fujise, "A Study on Software Download Method for Software Defined Radio Communication Systems", The 2000 Communications Society Conference of IEICE, Sept. 2000 (in Japanese).

Software Radio for ATC Application

Andrea Berti, Pierluigi Fantappiè, Paolo Maltese, Fortunata R. Sorace
OTE-Marconi, Mobile Radio Business, ATC Engineering

V. Barsanti 8, 50127 Firenze - Italy
Contact author e-mail: andrea.berti@marconi.com

Abstract: OTE – Marconi has a long tradition in terms of designing, producing and installing ground-based radios for the communications involved in the Air Traffic Control (ATC). If these radios are, up to now, analogue AM (amplitude modulation) radios, adhering to an old standard lasting since 1923, OTE-Marconi is now involved in designing the new generation of radios who will support the introduction of digital transmissions in the ATC world.

1. Introduction

The new generation of radio transceivers that is being developed by OTE-Marconi has been thought and devised to adhere to all the new standards for digital transmissions in ATC applications, *i.e.*, ICAO VDL Mode 2, VDL Mode 3 and VDL Mode 4, as well as to the old analogue transmission one, based on amplitude modulation.

An analysis of state of the art of the current technology has been done at the very beginning of the design for verifying which new architecture could be introduced in the framework of costs, component availability and time-scale.

The characteristics and standards requirements of the various modes of operation for the new generation of transceivers are summarised in Table 1.

2. Modulation Standards for Air Traffic Control

Table 1. Analogue and digital modes of operation – technical characteristics

	AM-25 kHz [1]	AM-8.33 kHz [1]	ACARS [1]	VDL 2 [2], [3]	VDL 3 4], [5]	VDL 4 [6]
Modulation type	Amplitude modulation	Amplitude modulation	AM-MSK @ 2400 bps	D8PSK @ 31500 bps	D8PSK @ 31500 bps	GFSK @ 19200 bps
Frequency band (MHz)	118 – 137 118 – 156	118 – 137	118 – 137	118 - 137	112-137	108 - 137
Channel spacing (kHz)	25	8.33	25	25	25	25

Table 1. Continued

	AM-25 kHz [1]	AM-8.33 kHz [1]	ACARS [1]	VDL 2 [2], [3]	VDL 3 4], [5]	VDL 4 [6]
Offered services	Voice	Voice	Point to point data	Point to point data	Coded voice, point to point data and broadcast data	Point to point and broadcast data
Offered connectio ns	Ground to air	Ground to air	Ground to air	Ground to air	Ground to air	Ground to air Air to air
Channel coding	N.A.	N.A.	NO	RS (249, 255) octet coding	Various block codes according to the burst type (Voice/ Data / Mgmt)	NO
Frame structure	N.A.	N.A.	NONE	NONE	4-slot TDMA or : 3-slot TDMA, with or without synchronisati on to UTC second	STDMA structure synchronise d to UTC minute. 4500 slots/minute
RX dynamic range (dB)	100	100	100	100	100	100
Co-channel protection ratio (dB)	N.A.	N.A.	N.A.	20	20	12
Adjacent channel protection (dB)	60	60	60	44	44	44
Blocking	>= 80 dB at ± 1 MHz from tuned frequency	>= 80 dB at ± 1 MHz from tuned frequency	>= 80 dB at ± 1 MHz from tuned frequency	-33 dBm, starting from the 2nd adjacent channel	-4 dBm at ± 3 MHz from tuned frequency, with 10 dB of allowed degradation	-33 dBm, starting from the 2nd adjacent channel
Intermod ulation protection	> 70 dB	> 70 dB	> 70 dB	>= 43 dB (*)	>= 70 dB	>= 43 dB (*)
Protection against impulse noise	Required	Required	N.A.	N.A.	N.A.	N.A.

NOTES :
N.A.: Not Applicable
(*): DRAFT requirement

3. Comparison with Cellular Systems.

With respect to cellular systems, ATC communications have some peculiarities:

- specific frequency band: 108÷156 MHz;
- simplex one-frequency operation;
- maximum reliability (high redundancy);
- operation in high RF density environment (airports);
- direction of the communications : ground to air and *vice versa*, or air to air (VDL 4 only).

The hardest performances come out from the simplex frequency operation: both uplink (in this case from ground towards the mobile – the aircraft) and downlink operates on the same frequency band. The same radio channel is occupied by the PTT (push to talk operation) by the pilot or by the controller.
This means that no duplexer filters can be used in the branching system and transmitting and receiving channels are separate by only few hundreds of kHz. For this reason, the dynamic of transmitters and receivers must be as high as possible. Key parameters are front-end noise figure but also linearity, oscillator noise and power amplifier linearity (to avoid spectral regrowth).

AM modulation has some peculiarities like a poor co-channel protection level and a high sensitivity to any kind of distortion: not only intermodulation but even crossmodulation must be taken into account.

Performance must be maintained over the different operating modes: different modulation schemes, channel arrangement, signal bandwidth.

4. DTR100 Functional Architecture

The DTR100 is designed in order to be a flexible radio transceiver, capable of implementing either the *analogue* mode of operation, based on the traditional AM DSB modulation technique, either the newly devised VDL digital modes of operation, which use both Differential 8-ary Phase Shift Keying and GFSK modulation techniques.

In particular, the radio is designed to be capable of operating with the following "modes":

- analogue AM modulation with 25 kHz (with or without offsets)/ 8.33 kHz channel spacing, with the possibility of transmitting either voice or data (according to the ACARS 2.4 Kbps MSK tone coding by an external modem);
- ICAO VDL Mode 2, with D8PSK modulation, 31.5 Kbps on 25 kHz channel spacing and CSMA channel access policy;
- ICAO VDL Mode 3, with D8PSK modulation, 31.5 Kbps on 25 kHz channel spacing and TDMA channel access policy;
- ICAO VDL Mode 4, with GFSK modulation, 19.2 Kbps on 25 kHz channel spacing for ADS-B applications, AND/OR D8PSK modulation,31.5 Kbps

on 25 kHz channel spacing, for VDL 2 – like operations. In both cases, the channel access is based on a STDMA channel access policy.

The operational mode is entirely software configurable.

The transceiver has a modular structure: TX (Transmitter) and RX (Receiver) modules perform the RF functions rather than the BB (Base Band) module performs the digital processing. Other modules of the radio perform power supply, interface and control (configuration and diagnostics).

Much of the processing power of the radio is concentrated on the Base Band, which is responsible for the implementation of all the functions of the physical layer and the MAC layer's tasks, in all digital modes of operation, besides the AM modulation-demodulation related functions in the analogue mode.

The modulation and demodulation of the received signals is performed through *in phase and quadrature* components. Since the goal of our design was to develop a device that were as close as possible to the concept of "Software Radio", we adopted the *Quadrature Undersampling* digital processing technique to derive the I and Q base band channels. This technique includes the advantages of *undersampling* and the capability of extracting quadrature components from a single sampled sequence. The theory is based on the following equations [7],

$$Fq = \frac{4Fc}{2m+1} \tag{1.1}$$

being Fc the carrier's frequency, m an integer number and Fq the AD converter sampling frequency,

$$Fs = \frac{2Fc}{n} \tag{1.2}$$

being Fs the final sampling frequency for I and Q, n an integer number,

$$Fq = qFs \tag{1.3}$$

being q an integer factor.

If the sequence output from the AD converter is at the sample rate of Fq, then, by taking the first two samples of every q-ary group and discarding the others, we can construct two decimated sequences which are at the sample rate of Fs and relative to signals with a phase shift of 90°, as shown in Fig 1.

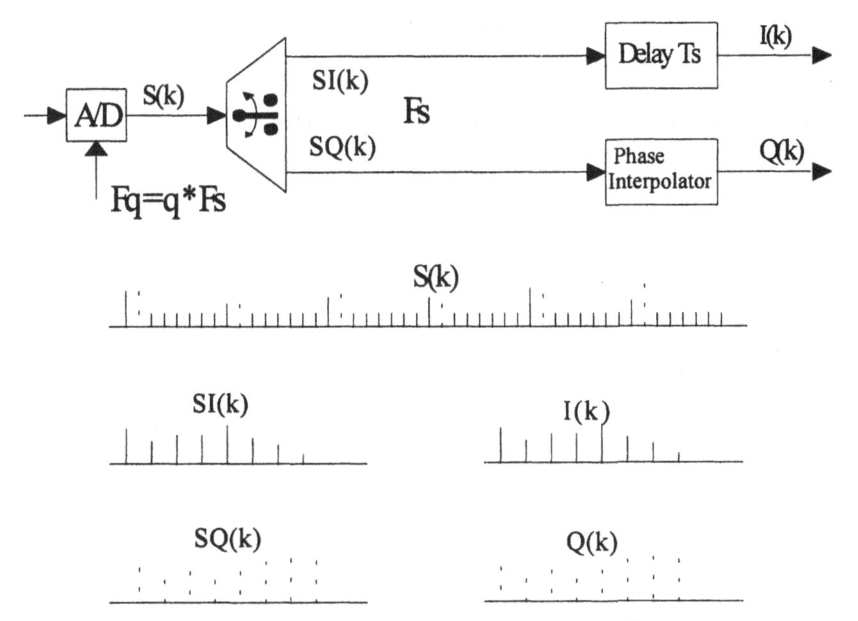

Figure 1. Quadrature undersampling architecture

The base band I and Q signals are then obtained by digital low pass filtering, since one replica of the IF spectrum is centred around DC. Since both channels are obtained from a single original sequence, there is the need to compensate for the relative delay between each I sample and its corresponding Q one (this delay is exactly $Tq = 1/Fq$). A digital Lagrange's interpolator filter has been found suitable for this task.

In Figure 2, the receiver's architecture is shown, up to the Analogue to Digital Conversion.

The highly extended receiver's input dynamic range, -107 dBm to +7 dBm, is covered thanks to an Automatic Gain Control technique which involves a mixed analogue and digital solution. In fact, two reception chains with different overall gains are implemented and two AD converters are used so that their dynamic ranges overlap. Further processing in the base band module enables to select which of the two chains' samples are to be actually demodulated. The AM mode of operation requires an even better AGC circuit, capable of making the audio output signal independent of the received RF signal's power. Also this feature is implemented in the digital processing of the Base Band module.

In the beginning of the DTR100 Receiver design process, we decided to perform only one down conversion to an Intermediate Frequency of 21.4 MHz and then sample the signal according to the quadrature undersampling theory. This solution appeared particularly attractive, since the goal was to develop a "software radio" with most of the radio-channel functions implemented in digital algorithms rather than analogue circuits. Unfortunately, the tight and demanding specifications regarding the receiver's sensitivity required outstanding performances for the AD converter in terms of dynamic range (SNR) and distortion, if the input frequency were to be 21.4 MHz. The major mixed analog-digital vendors were selected, including Maxim, Analog Devices, Intersil, Burr Brown and others, but it has not been possible to find a device with the needed cost-to-performances ratio. This technological limit has led us to introduce a second down conversion to the IF of 455 kHz, thus defining the receiver's architecture illustrated in Figure 2.

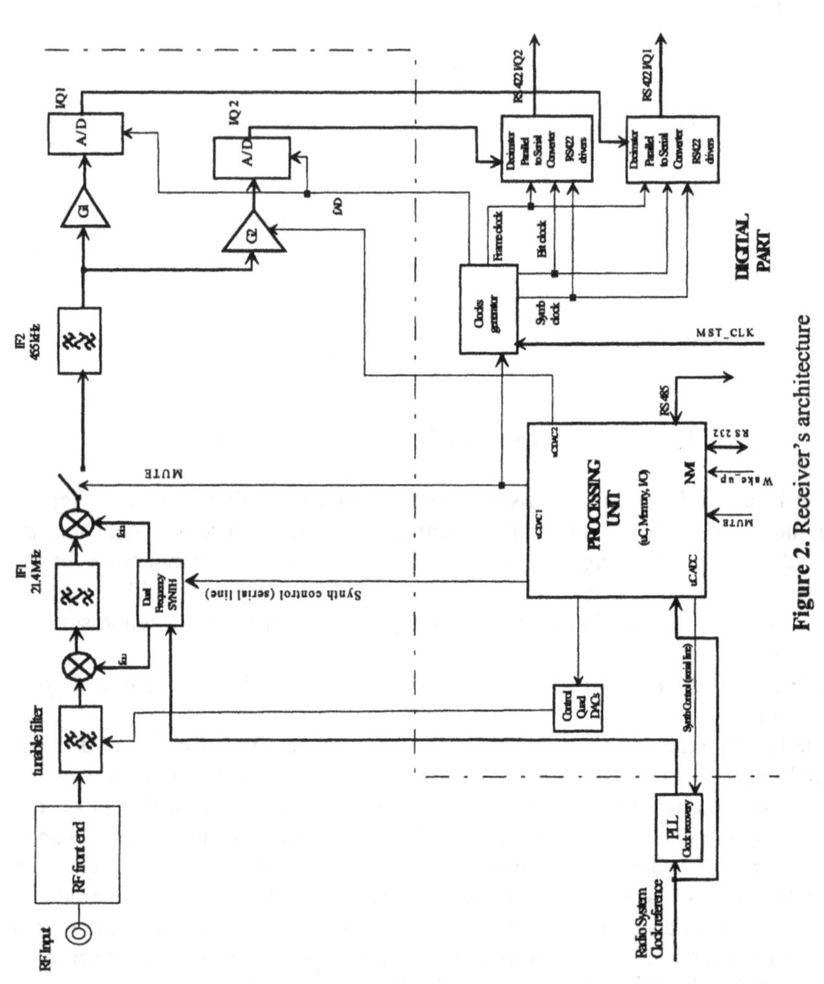

Figure 2. Receiver's architecture

Concerning the channel selection, again, a mixed solution of analogue and digital filtering has been chosen. In fact, the channel bandwidth is set to be 25 kHz for all digital modes and for today AM mode of operation, while in the close future the AM channel will be reduced to 8.33 kHz. The analogue filtering before the AD conversion has the fundamental role of avoiding *aliasing* of spectral replicas and also of attenuating the adjacent channels, while the base band digital filtering is very sharp and gives the most of the required adjacent channel rejection. Avoiding the analogue filtering before the AD conversion and implementing the channel selection all in digital would deteriorate the received signal quality and, most important consideration, would endanger the AGC algorithm which is based on the measurement of the RSSI before the channel filtering. The presence of adjacent channels highly above the signal level would cause the selection of the wrong reception chain, thus leading to the loss of the useful signal.

5. The Future

As stated earlier, the need to keep the DTR100 transceiver's cost contained, while designing a very flexible and "intelligent" radio, has led to an architecture which does not fully implement the radio-channel interface in digital processing. Nevertheless, this target is not far from being achieved, as soon as the needed technology will become available at a lower cost.

Having this target in mind, the primary objective of a future revision of the project is the achieving of the single 21.4 MHz down conversion in the receiver (conditioned to the AD converters performances) while keeping the base band architecture as stable and unchanged as possible. This solution would in fact reduce the "analogue" presence in the reception chain, with all the advantages concerning immunity to temperature and components' parameters drifts, thus improving the performances of the radio.

6. References

[1] " Electromagnetic compatibility and Radio spectrum Matters (ERM); Ground-based VHF hand-held, mobile and fixed radio transmitters, receivers and transceivers for the VHF aeronautical mobile service using amplitude modulation; Technical characteristics and methods of measurement", ETSI ETS 300 676 , V1.2.1, Dec. 1999

[2] "EMC and Radio Matters (ERM) – Very High Frequency (VHF) radio equipment operating in VDL Mode 2 Technical characteristics and method of measurement", ETSI EN xx.xxx v. 1.1.1, March 2000

[3] "VHF DIGITAL LINK STANDARD AND RECOMMENDED PRACTICES" (CSMA mode) - vers. 2.9 – ICAO, April 1996

[4] "NEXCOM subsystem Segment Specification for the Multi Mode Digital Radio (MDR)", FAA specification – sept. 1999

[5] "VHF DIGITAL LINK MODE 4 STANDARD AND RECOMMENDED PRACTICES" – ICAO, Sept. 1999

[6] "EMC and Radio Matters (ERM) – Very High Frequency (VHF) radio equipment operating in VDL Mode 4 Technical characteristics and method of measurement", ETSI EN xx.xxx v. 1.1.1, March 2000

[7] "Quadrature sampling of FM-Bandpass signals – Implementation and error analysis", Werner Rosenkranz, DIGITAL SIGNAL PROCESSING N. 87, 1987

Part 5

Network Architecture: Protocols and Services

Software Defined Radio: What Do We Do with It?

Gerhard Fettweis

Dresden University of Technology, Germany

Abstract: A fragmented cellular market has been the main driver for the call for a software defined radio technology development so far. As the hardware architecture technology is maturing, the question needs to be addressed as where do we see a fit for developed solutions:

- will the cellular terminal market be the driver?
- will the cellular base stations be the point of market entry?
- will home networking be the driving force as the first and biggest market?

To be able to point out answers to these questions the current technology, service, and the markets need to be analyzed to understand opportunities and fallacies of software defined radio. The ultimate question is: who will pay the ticket to ride the technology into the market?

1. Motivation

We define the customer need for a new quality of service, e.g., higher rate data, and then standardize everything down from higher layer protocols to the nuts and bolts of the physical layer. Even applications, e.g., voice coders, are specified and standardized from algorithm down to the quantization of arithmetic implementation. This leaves us with the result that every new standard leads to huge investments in equipment which is limited to the predefined standardized technology and services. By perfecting the details we as communications engineers therefore block recurring democratic innovation and define dictatorial standards as road block for future innovation. The effect of this can be seen in industry. Over decades the PC world was driven by providing more and more processing power enabled by microelectronic technology evolution, leading to increasing prices per microprocessors over time. In contrast, telecommunications microelectronical technology innovation is exploited in telecommunications typically only to decrease the cost per unit sold, see Figure 1.

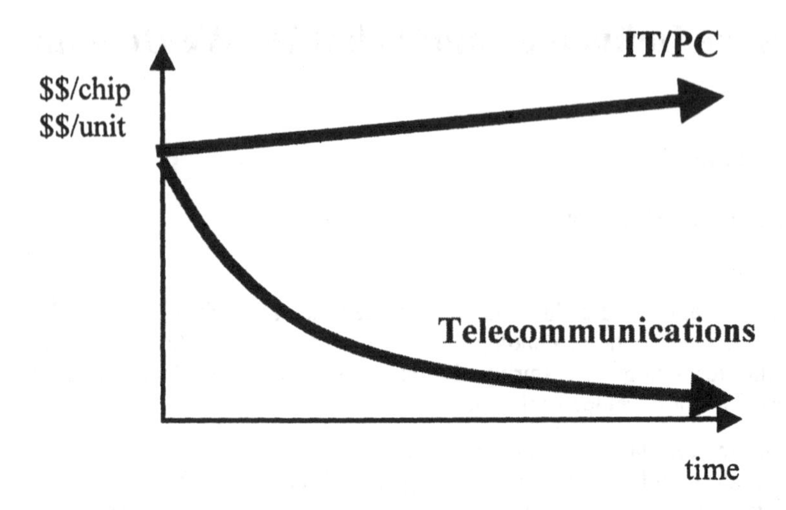

Figure 1.

How can we change this, *i.e.*, how can we ensure that microelectronics technical innovation is continuously feed back to ensure continued innovation in wireless communications?

One answer is: By allowing the user to download/ define whatever air interface is used, a network centric view of defining everything in standards is replaced by a user-driven open technology innovation platform (Figure 2).

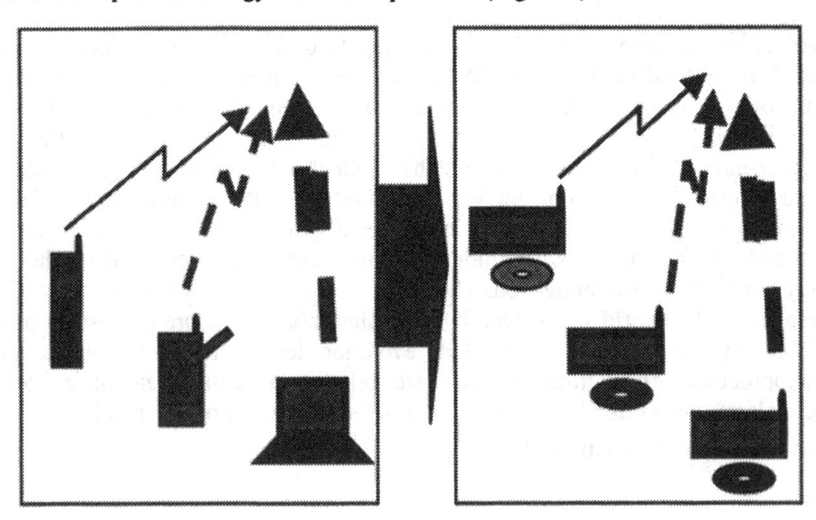

Figure 2.

Hence, in principal every engineer coming up with a new improved modem software can offer this freely to provide higher quality, rate, or network user density. Clearly this is the business model of information technology industry which via SDR can enter (wireless) telecommunications as well.

Of course we have to ask the question, is there still innovation potential e.g., at the physical layer that can be the motor for benefiting from new technology? Space-time processing has just emerged as a clear opportunity of how no end is in sight today for additional signal processing power provided by microelectronic technology advances.

Thus, SDR is a viable opportunity to drive the telecommunications curve of Figure 3 to the information technology curve of the PC industry!

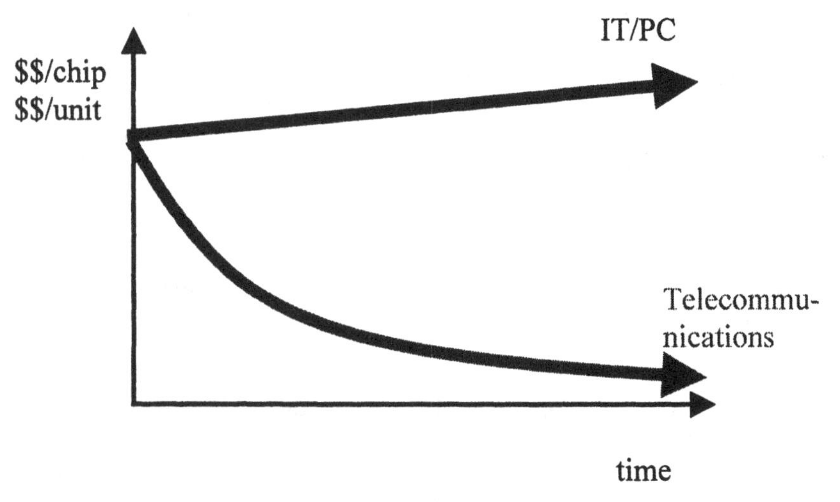

Figure 3.

2. How Far Is SDR Technology to Date?

After seeing the great potential SDR can provide, the question remains to see when SDR hardware technology is ready to provide a cost-effective low-power solution. The signal processing pipeline of a pragmatic SDR-receiver can be viewed in Figure 4.

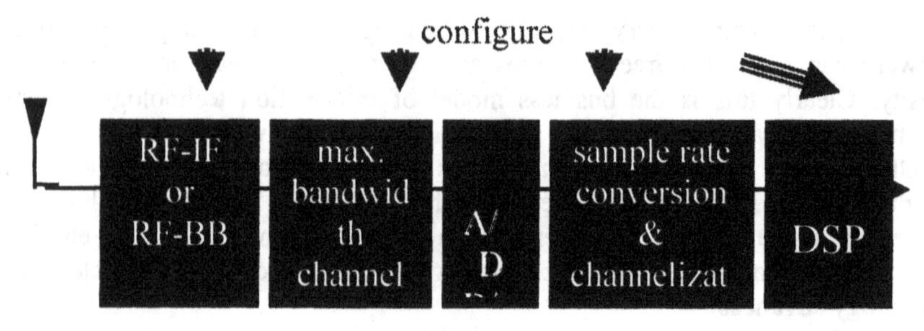

Figure 4.

First the RF part transforms the RF signal either to IF or directly to baseband.

Then a wide analog channel filter with the maximal channel bandwidth of interest is implemented followed by the A/D converter.

Subsequently the sample rate conversion is carried out due to likely incommensurate clocks of the realized hardware master crystal and the implemented wireless system. This is combined with the channelization filtering. Finally, the baseband demodulation is carried out by a DSP. Is this technology ready?

RF

Triple-band GSM phones are built today at competitive pricing in large volumes. Hence, multiband RF with high dynamic range is feasible.

Channel Filter

The channel filter, if designed for 16 MHz bandwidth, includes all of today's important wireless systems starting from HiperLAN/2 over 3GPP to IS-95 and GSM. If designed for 5 MHz bandwidth, all current cellular systems can be addressed. [1]

A/D

A/D conversion typically would be preferable at 4 times oversampling, *i.e.*, around 80 MHz. Higher rates make $\Sigma\Delta$ approaches better adapted as well as simplify the channelization and rate conversion problem. [2], [3]

Channelization & Rate Conversion

Channelization and sample rate conversion can be solved with low complexity by a time-variant polynomial FIR filter implemented by a transposed Farrow-structure. The implementation complexity and power consumption is marginal. [4], [5]

DSP

Today's standard DSPs do not deliver the required processing power, and in particular also not at the desired price point and power consumption levels. [6]

However, with the introduction of Systemonic's OnDSP™ platform, 100 MIPS order-of –magnitude complete baseband modem implementations are feasible at small die-size and power consumption. Hence, a DSP processor platform as the basis for SDR is available.

3. What Do We Do?

SDR hardware technology has come to a point such that a pragmatic implementation as described above is feasible within the next 2 years. However, now a protocol integration needs to be provided to enable logging into the network, supporting mobility, authentication, and encryption. This "network access and connectivity channel (NACCH)" has been proposed to be the protocol glue to make SDR happen. Where do we find it?

Work is underway at many SDR related forums. At the same time UMTS/3GPP licenses have been paid and the network roll-out is happening at a large scale. Does this leave room for innovation by introducing SDR, or is this the ultimate road-block?

It turns out that 3GPP/UMTS today allow for different radio access networks (RANs) to be integrated, *i.e.*, today we see within UMTS:

- FDD;
- TDD;
- GSM/GPRS/EDGE;
- HiperLAN/2;

and research & development is on the way to integrate:

- Bluetooth;
- DVB-T.

Therefore it is easy to see that others can be integrated as well. In addition, Service Access Points (SAP) have been defined which enable the access to the core network without knowledge of the physical air interface. Hence, we can see that within UMTS/3GPP the protocol functionality has already and further is being developed that is needed for SDR to take off soon.

4. References

[1] H. Tsurumi, Y. Suzuki, "Broadband RF Stage Architecture for Software-Defined Radio in Handheld Terminal Applications", IEEE Commun. Mag., Feb. 1999, 90-95

[2] R.H. Walden, "Performance Trends for Analog-to-Digital Converters", IEEE Commun. Mag., Feb.1999, pp.96-101

[3] T. Hentschel, G. Fettweis, „Software Radio Receivers", CDMA Techniques for Third Generation Mobile Systems, ed. by Francis Swarts et. all, Kluwer Academic Publishers 1999, Chapter 10, p. 257-283

[4] T. Hentschel, M. Henker, G. Fettweis, „The Digital Front-End of Software Radio Terminals", IEEE Personal Communications Aug.1999, 40-46

[5] T. Hentschel, G. Fettweis, Continuous-Time digital Filters for Sample Rate Converstion in Reconfigurable Radio Terminals", Proceeding of European Wireless 2000, Sep. 12.-14., 2000, Dresden, VDE-Verlag, p. 55-59

[6] D. Murotake, J. Oates, A. Fuchs, "Real-Time Implementation of a Reconfigurable IMT-2000 Base Station Channel Modem", IEEE Commun. Mag., Feb. 2000, 148-152

OPtIMA: An Open Protocol Programming Interface Model and Architecture for Reconfiguration in Soft-radios

K.Moessner, S.Vahid, R.Tafazolli

Centre for Communication Systems Research (CCSR)
University of Surrey
Guildford, Surrey GU2 7XH.
E-mail : k.moessner@eim.surrey.ac.uk
s.vahid@eim.surrey.ac.uk

Abstract: Despite increasing numbers of software-radio implementations reported recently, there remains a lack of complete formal design methodology for implementation of these radios. Most implementations concentrate efforts on reconfiguration of physical hardware via proprietary application programming interfaces. At the networking level, projects such as the Radio API (GloMo) and the IEEE 1520 are the prime examples of the efforts to open up the protocol stacks and to introduce open platforms and service programmability. Software reconfigurable radios will also benefit from open implementations and access to protocol functionalities based on standardised programming interfaces within a complete framework. This article introduces an object-oriented software framework that splits classical layers (protocols) into two distinct parts: 'pro-layers' and active 'pro-interfaces' to achieve software reconfiguration. Benefits of this approach are the facilitation of the exchange of single protocols (pro-layers) during run-time, run-time adaptation between protocols as well as management of protocol extensions. An example implementation with application specific protocol extensions for QoS signalling incorporated within a stock-trading multimedia communication application, is presented in this article.

Index Terms: Open Programming, API, Class, signalling, software-radio, re-configuration, Java/RMI, QoS.

1. Introduction

Software-reconfigurable radios or soft-radios are evolving towards all-purpose radios which can implement a variety of different standards or protocols through re-programming [1,2]. Soft-radios are emerging as viable alternatives to multimode terminals without the "Velcro approach" of including each possible/existing standard. Next generation mobile terminals and network nodes will require significantly richer capabilities in the control plane due to the need to support large numbers of diverse applications with different QoS requirements and traffic characteristics. Deployment of active nodes and interfaces [3,4,5] will certainly facilitate introduction of differentiated or integrated services to support new multimedia applications as well as provide for smoother interworking functionality between media protocols (internet and IN) or different signalling systems (SS7, H323 *etc.*). Also network management will become more intelligent and network capabilities will evolve rapidly through software changes without the need to upgrade the network infrastructure. It is therefore obvious that future reconfigurable mobile networks will benefit greatly from the deployment of active nodes and service interfaces.

Soft-radios terminals will also need to be re-programmable and this reconfiguration capability will not be confined to the physical layer alone. To achieve reconfigurability of the complete protocol stack calls for introduction of flexible interfaces between protocol layers *i.e.*, Protocol Programming Interfaces (PPIs) to replace the rather static service access points. Re-configuration through introduction of programming interfaces between protocol strata will provide the possibility to write both single protocols or even whole protocol stacks in a manner similar to the way applications are written in high level programming languages (e.g., Java applications use different APIs which are part of the class libraries with binding at runtime - this means that the functionality is out-sourced to the API and the application simply defines the sequence and determines the parameters passed to methods within the APIs). Other important design issues in software radio design concern proper reconfiguration of the radio *i.e.*, reconfiguration management, which is responsible for runtime reconfiguration management [6], over-the-air down load protocol [7] and security related aspects.

Protocol stacks are aggregations of several single protocols, each of which has certain functionality and serves a certain task. Traditionally, protocol frameworks use stratification as a composition mechanism. Protocols in one layer of the stack are impervious to the properties of the layers below. Each layer is treated as a black box and there exists no mechanism to identify/bypass any functional redundancies, which may occur in the stack. A well-documented example for protocol stacks is the OSI_RM (Open Systems Interconnection Reference Model), which consists of seven layers ranging from application, presentation, session, transport, network and data link layers to the physical layer. Each of these layers represents a complete protocol that offers its services to the next upper layer or expects services from the layer immediately below [8]. The same principle applies for mobile and fixed line telecommunication networks, and hence interworking functionality is defined in separate protocols. Communication between layers is accomplished via Service Access Points (SAPs). Through these SAPs sets of primitives in a given layer become available to the next layer up the hierarchy. Services define operations to be performed within the layer and can be requested by upper layers. SAPs are used to encapsulate the layers, to hide their complexity and to uniquely describe the functionality that a layer provides and what upper layer users may request from them. However, SAPs are static and lack any flexibility, they do not support flexible changes in the protocol stack and need to be re-standardised in case any change becomes necessary.

In this article we introduce a novel concept that redefines the interfaces between protocol layers, classifies interactions between different layers within the protocol stack and provides an architecture supporting protocol reconfiguration. This is achieved by implementing active programming interfaces as objects within the protocol stack and by using object oriented design methods to define this new protocol stack architecture. Standardised active programming interfaces will introduce the additional degree of freedom necessary for standard reconfiguration of protocol stacks in both terminal and network. This paper is organised as follows. Section II presents a brief review of related approaches within the area of open platforms and protocols highlighting overlapping concepts. This is followed in section III, by introduction of the architecture and protocol specification of the proposed approach and a description of how OOD principles are used to implement the proposed active protocol interfaces. In section VI composition of protocol layers and interfaces is outlined. The implementation model and an example application are presented in section V while section VI outlines the benefits of the described architecture. Finally the section on conclusions delivers an outlook on future research directions in this area that can build up on this novel approach.

2. Related Work

2.1 State of the Art

The need for common open interfaces on the application layer has been widely acknowledged in industry and research community. Research projects pursued in Europe and USA address this area and describe introduction and implementation of open application programming platforms for mobile terminals. These interfaces reside on top of standard traffic channels and provide access to (usually) legacy, and in case of UMTS, to future communication systems traffic channels. Platforms like the MASE [9,10] or MExE [11] are relying on a platform within the Mobile Station and additionally on control elements (proxies) within the network, which represent the mobile stations and adjust data streams to the resources and capabilities provided by the mobile access network and mobile station, respectively.

A further step in this development is then to address the interfaces between the lower layers of protocol stacks and to define them in a future proof manner. Protocol stack implementations are rather static and re-standardisation procedures are necessary to introduce the slightest improvements or to add new services to existing protocols and protocol stacks. Therefore it is necessary to introduce methods, comparable to those used in the aforementioned higher layer projects, however applied to the lower layers of protocols. This has been tackled in the Radio API [12] project by Rooftop, the main efforts in that project has been to open up the service connections between protocols within a protocol stack for cellular systems. The Radio API project deals with the renunciation of static SAPs to access the services within a protocol, in favour of open APIs (or for lower layers PPIs), which seems a logical step towards freely programmable and flexibly re-configurable protocol stacks.

The IEEE P1520 [13] initiative represents an approach that aims to provide programmable interfaces to abstractions of network entities in support of service programmability, and whose definitions will be based on encapsulation of features of numerous protocols within generic interfaces. The MASE, MExE, Radio API

(API Framework-GloMo [12]) and IEEE P1520 have certain main features in common, and all four are primarily concerned with defining APIs at different levels of abstraction in the networking area. Aiming at protocol stack reconfigurability means to derive generic classes of protocol services and to offer flexible, object oriented/component based, structures to implement protocol stacks [14]. Hence, the need to provide APIs (Application Programming Protocols) and PPIs (Protocol Programming Interfaces) that consist of sets of exchangeable and extensible classes, defining protocol services and functionality. The concept of active programmable protocol interfaces, proposed in this article delivers an implementation approach that supports these requirements.

3. The OPtIMA Approach

Software reconfiguration has been identified as a crucial technology to facilitate the realisation of software-defined radios. The main functionality of interest is the ability to exchange protocol software 'on the fly' and to reconfigure complete protocol stacks, which may become necessary in various circumstances.

The approach presented here implements a framework for protocol stack reconfiguration. Protocol stacks are split into a number of functional entities described in terms of generic "classes" organised in class libraries, with dynamic binding at runtime to implement reconfigurable protocol stacks. The framework is also capable of supporting composible protocols. A single protocol layer may be replaced by a collection of components each of which implements a particular function. Inside a soft-radio terminal for instance, a "Reconfiguration Management" unit would control component selection, deletion/upgrade and communication with the PPIs. Within the OPtIMA specification, we have also provided guidelines to build/implement standardised and proprietary protocol stacks using the defined APIs and PPIs.

3.1 Application Programming Interfaces and Object-oriented Design Principles

Application Programming Interfaces (APIs), in general, rely on the paradigm that interfaces hide the complexity of how functionality is actually realised. This enables programmers to simply apply the guidelines of how an interface has to be used and which parameters are to be passed for each single function call. Applications become sequences of calls to functions pre-defined within these API implementations. Much of the complexity within such applications is therefore moved down, below these programming interfaces. One of the advantages of APIs is their extensibility and partial or even complete exchangeability without necessarily requiring the complete re-writing of the application code. Other benefits of programming interfaces include the scalability and the simple use of such structures.

The basic idea of Object Orientation is to put a systems behaviour and its properties into discrete entities [16] this requires that functionally different parts of a system have to be identified. Once inter-relationships between entities and state or behaviour (functionality) of these entities are analysed, a formalisation and description as classes has to be done. Object Orientation relies on a number of basic principles of which the class is the major one; further concepts include Objects, Abstraction, Attributes, Operations/ Methods/ Services, Messages, Encapsulation, Inheritance, Polymorphism and Reuse [17]. Examples in which OOD and APIs are used together are manifold. For example most parts of the JFC (Java Foundation Classes) are implemented in classes, which are derived (via one to several levels of hierarchy) from the base class 'object'. The same applies for the MFC (Microsoft Foundation Classes), which contain the base implementation classes for application programming for the Windows platform.

3.2 System Architecture

The intention of the OPtIMA model is to introduce a framework, which enables the exchange of protocols during run-time as well as the active involvement of the interfaces that enable direct signalling communication between interfaces of different protocol stack layers without accessing the layer implementation. This requires a somewhat different system view; in this approach protocols become split into 'pro-interfaces' and 'pro-layers' as shown in Figure 1.

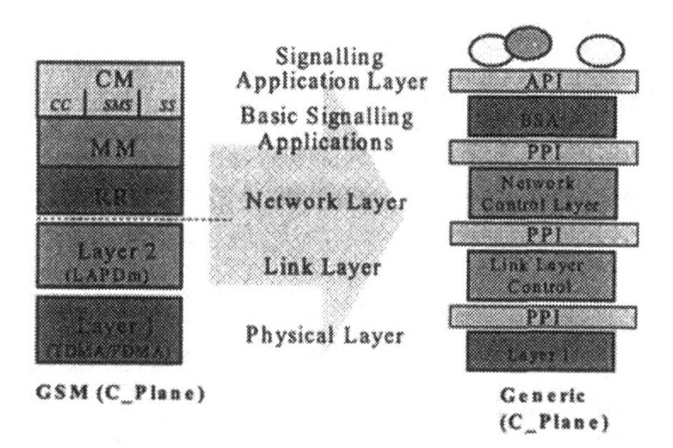

Figure 1. Introducing active interfaces

Pro-interfaces are active implementations defined within classes[1], which are derived from a base class. This base class defines the generic functionality of all possible pro-interfaces. Further specialisation can then be achieved within the derived pro-interface classes. When implemented as objects, interfaces (*i.e.*, pro-interfaces) deliver both additional functionality and additional architectural complexity. *Pro-layers* are the actual protocol implementations, which obtain data through pro-interfaces, manipulate these data and export them through the pro-interfaces. Execution of functions/methods within pro-interface and pro-layer classes is controlled by thread-objects. Such a construction delivers a highly flexible platform to replace the classical protocol stacks and to enable flexible

[1] Entity definitions are according to their position in the stack (*i.e.*, pro-interfaces are between two pro-layers).

stack re-configuration during run-time. In other words thread objects are implementing classes, which control message transport and manipulation throughout all pro-layers and pro-interface.

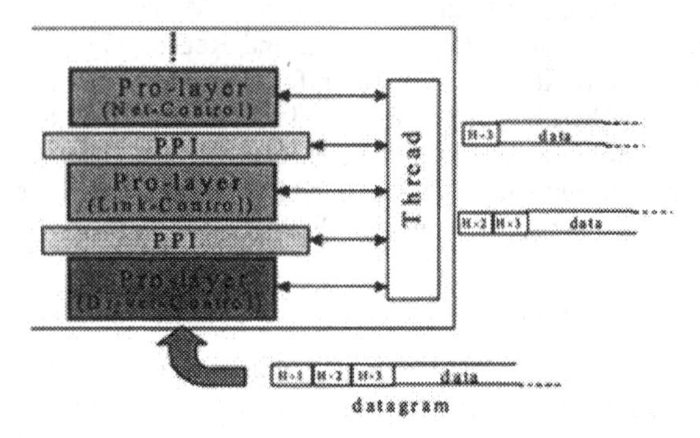

Figure 2. Thread controlled message handling

Upon arrival, a thread takes over responsibility for the messages and their processing within the complete protocol stack and then passes the references of these messages to their destinations (see Figure 2).

3.3 Class Architecture

A 'functional decomposition' of protocol stacks represents the composition of the basic entities that open protocol implementations should consist of, is presented in Figure 3.

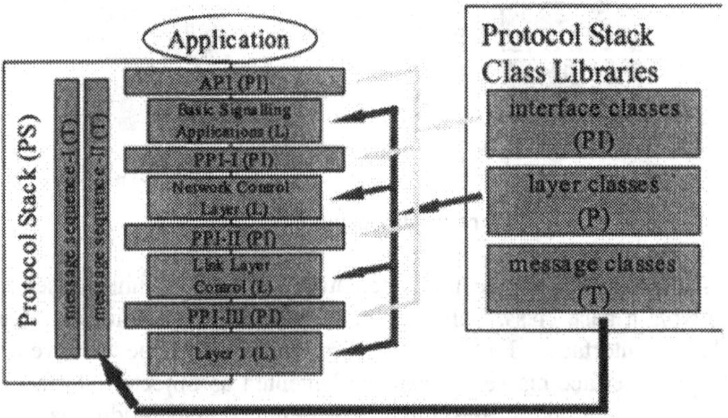

Figure 3. Protocol stack structure and class libraries

Four different class types have been identified which are:

- *Layer classes* (L-classes defining pro-layers) represent the functionality of single protocols within the protocol stack (*i.e.*, physical-, link, network- and signalling application functionality). Layer classes (pro-layers) inherit from a generic (pro-) layer class, as shown in Figure 4.

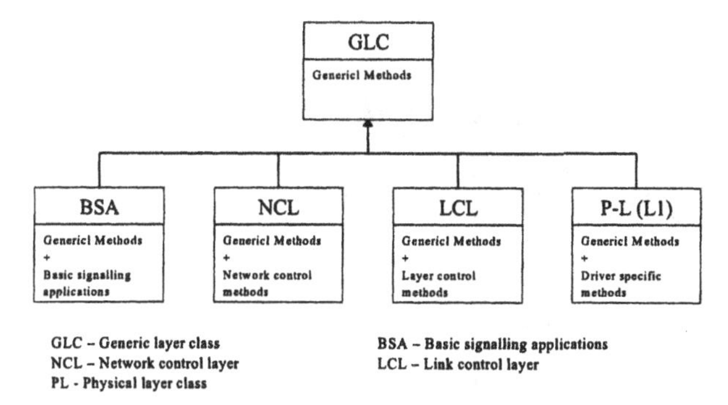

Figure 4. Introducing pro-layer classes

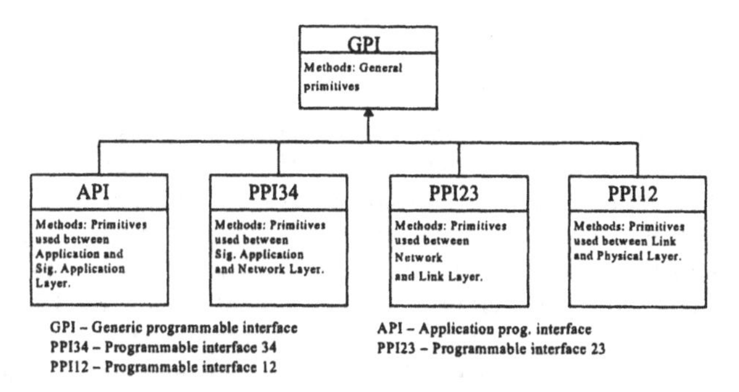

Figure 5. Interface class hierarchy

- *Programming Interface classes* (PI-classes defining pro-interfaces) represent the various programming interfaces between the protocols (L-classes). PI-classes consist of methods, which further specify and implement the generic primitives defined in a base class (GPI class in Figure 5). Figure 6 depicts the protocol stack emphasising the primitive description of this base class. PI-classes, in general) detect events (*i.e.*, messages from some protocol) and trigger the execution of the appropriate thread object.

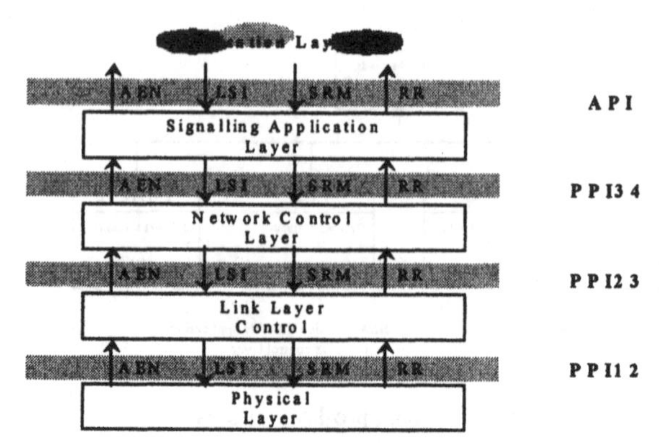

Figure 6. Interface primitives

- *Thread classes* (T-classes) implement pre-defined procedures (e.g., Connection-, Mobility-, Radio Resource- or QoS-Management - signalling). Other, non-defined and non-standardised, signalling procedures may be implemented within customised thread classes (see Figure 7). In addition to the implementation of message sequences, threads also enhance the flexibility of the complete protocol stack. The use of threads, to execute signalling procedures, enables programmers to implement several signalling procedures in parallel. This feature may be advantageous in regard to point to multipoint communications. Thread classes incorporate and use the methods defined in PI- and L-classes.

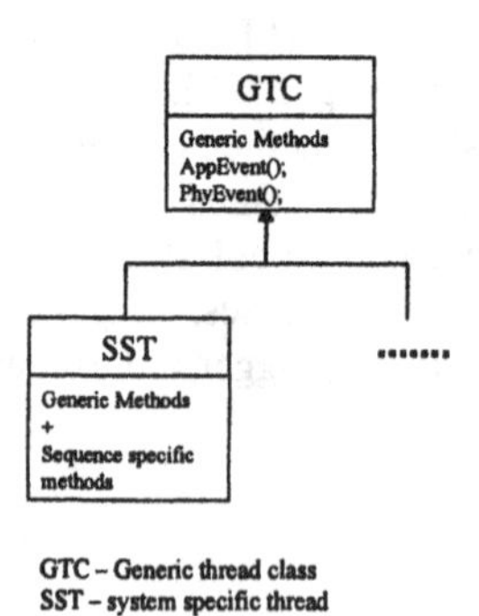

Figure 7. Thread classes hierarchy

- *Protocol Stack class* (PS-class) defines and represents the implementation of a complete protocol stack. Different legacy stacks (e.g., the GSM stack) may use their specified constructors to include the appropriate classes to implement the standard whilst other stacks may be freely defined. The PS-class is the only class publicly accessible within the protocol class library. Instances of this class implement any chosen protocol stack functionality, furthermore, the PS-class manages protocol re-configurability when L-, PI- or T-classes are exchanged during run-time (see Figure 8).

L, PI and T classes are defining the capabilities, methods and properties of a protocol stack, the PS-class implements those classes and contains also the protocol stack structure.

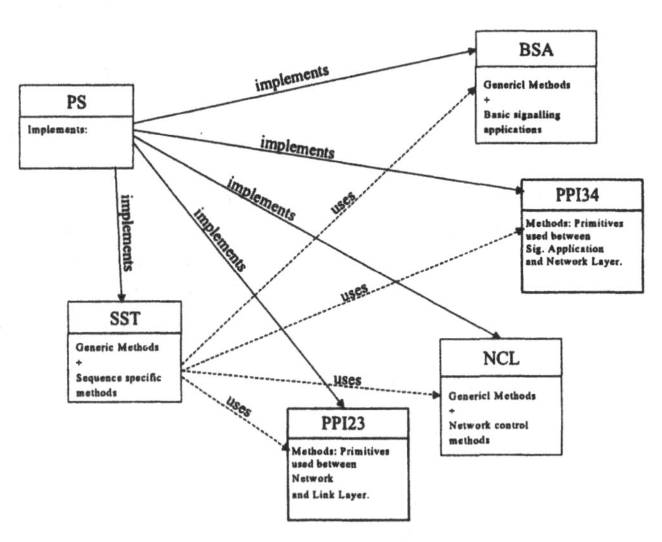

Figure 8. PS-class relations

3.4 Design Guidelines

OPtIMA relies on the aforementioned classes and a set of design rules, which define how classes and the complete protocol stack ought to be implemented. Basis for the protocol stack implementation is a code-skeleton (see Figures 12a and 12b section V) in which the different interface and layer classes become implemented. The 'Protocol class', within one protocol stack, is the only class exporting public interfaces; it implements L-, PI- and T-objects and defines the structure of the protocol stack. The architecture as a whole uses inheritance to define a hierarchy of both different interface and different layer classes; therefore it uses a generic interface class and a basic protocol class to derive pro-interfaces and pro-layers, respectively.

3.5 Protocol Stack and Thread classes

The three groups of classes (PI,L,T) are instantiated, implemented and controlled by an instance of the *Protocol Stack Class*; this PS-object defines therefore the complete protocol stack. Classes within one of the functional groups (threads, interfaces and layers) are located within class libraries, each of which provides the functionality for their particular group of specialised (derived) classes. Figure 3 shows the dependencies and the 'logical location' of classes within the protocol stack and their class libraries. *Thread classes* are those entities that actually manage the message passing throughout the protocol stack; they are used to handle single message sequences (*i.e.*, each thread controls one sequence).

4. Composition of Protocol Layers and Interfaces

Distribution of protocol stack functionality into pro-layers and pro-interfaces forms a new paradigm for protocol stack development where dissemination of functionality is facilitated through open protocol programming interfaces. The OPtIMA architecture consists of five layers each with their own tasks and functionality. All entities within the 'decomposed' protocol stack (including pro-layers, pro-interfaces and Threads) are implemented as separate classes. Pro-layer classes (*L – classes*) contain the attributes of this protocols (layers) and are used to store information obtained from (and related to) corresponding pro-layers. L-class methods are used to process incoming information and to produce an adequate response or initiate follow up sequences. The Layer classes (*i.e.*, within the scope of this architecture) and their main functions are:

- *Physical Layer,* (modem, channel access, FEC, ciphering, *etc.*);
- *Link Layer Control,* (link control);
- *Network Control Layer*, (controls message routing);
- *Signalling Application Layer*, (Connection Management, Resource and Mobility Management signalling);
- *Application Layer*, applications need to be compliant to the API specification.

4.1 Active Programming Interfaces

Pro-layers are isolated by pro-Interfaces (PI), which ensure exchangeability and extensibility of protocol[2] implementations during runtime. Appropriate PIs are defined (see class architecture) and introduced between the protocol implementations (pro-layers). Pro-interfaces provide the open protocol-programming platform; they enable interchange of signalling messages and deliver access to pro-layer classes. Four pro-interface classes are defined to implement a complete protocol stack (as depicted in Figure 6):

- *PPI12* (between Physical Layer and Link Layer Control);
- *PPI23* (between Link Layer Control and Network Control Layer);
- *PPI34* (between Network Control and Signalling Application Layer);
- *API* (Application Programming Interface).

Pro-Interfaces are derived from a generic interface class which defines four basic types of control messages known as *primitives*, whose task is to inform layers about events, pass new values for variables between the pro-layers and trigger methods implemented in these pro-layer classes. These four 'primitive' types are:

[2] Protocol in this context refers to a legacy protocol, in contrast to the pro-layer which is the implementation of legacy protocol functionality in the scope of OPtIMA

- *Service Request Messages (SRM)*: Asynchronous primitive to perform immediate, typically non-persistent actions. The flow direction for a SRM is from a higher layer to a lower one;
- *Request Responses (RR)*: Persistent state or long-term measurement primitive. The flow direction for a RR is from a lower layer to a higher one;
- *Layer State Information (LSI)*: Persistent state primitive. The flow direction for a LSI is from a higher layer to a lower layer;
- *Asynchronous Event Notification (AEN)*: Asynchronous primitive to report recent, typically non-persistent events. The flow direction for an AEN is always from a lower layer to a higher one.

Exploiting object oriented programming technology, each of the pro-interface classes inherits a generic functionality from a super class called GPI. This 'parent' class is not part of the implemented protocol stack as a separate entity; but it exclusively defines the minimum access interface (*i.e.*, the primitives) for all the (derived) PIs, as depicted in Figure 9. The object-oriented structure of active interface objects in conjunction with the overall architectural framework allows implementation of an additional feature: (active) pro-interfaces offer an alternative medium to pass messages and facilitate therein a common signalling and data API for application development.

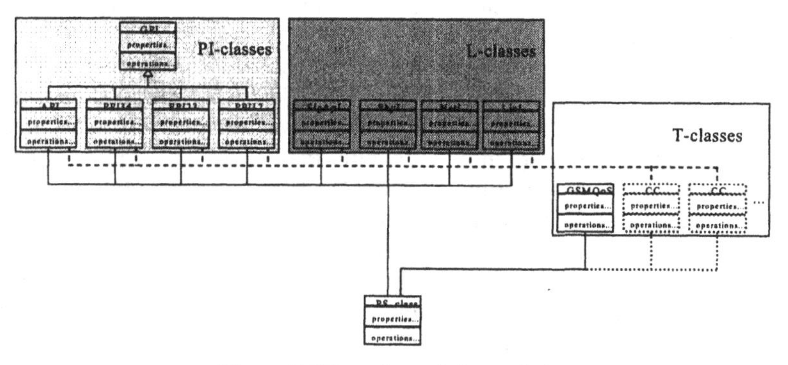

Figure 9. Hierarchy and class relations

As depicted in Figure 10, active interfaces are to be used to provide access to attributes within the pro-layers in two ways: sequentially (series A) and non-sequentially (series B).

The advantages of this architecture become apparent when OPtIMA is used in different communication networks; the signalling interface for applications then remains unchanged independent of the implemented protocol stack and the means of transport in the system. Adaptation to the target signalling system only takes place in the appropriate protocol (pro-layer) and that can flexibly be exchanged during run-time.

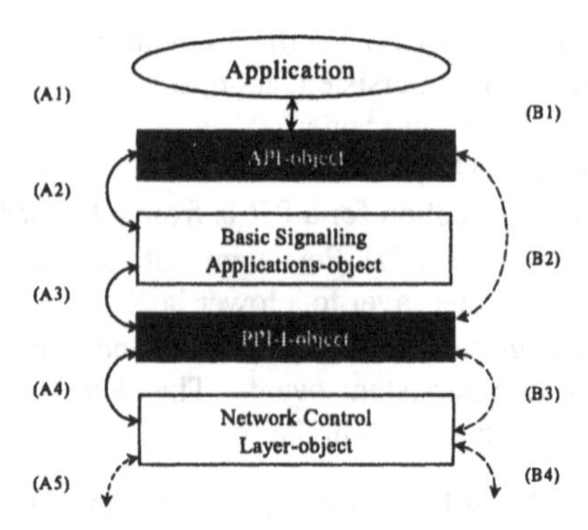

Figure 10. Active interface objects

4.2 Implementation of the Architecture

L– and PI–classes form the two major families/groups of classes that implement a protocol stack based on open programmable protocol interfaces. However, the functionality necessary to control signalling sequences, to deal with periodic tasks (e.g., channel measurements in the lowest layer) and to respond to external triggers has to be provided as well.

To fulfil these tasks, the family of Thread classes (*T–Classes*) has been introduced; they use the multithreading features of Java[3] to implement concurrent message processing. But T–classes do not define any other functionality, rather they access methods defined in L (and also PI)–classes and call appropriate functions in those classes. Priority of threads (and therefore execution priority of the message sequence) can be defined during thread instantiation.

Protocol Stack classes (PS–Class) represent the whole of a protocol stack; attributes of this class are instances of all other (previously explained) classes. Thus, a PS–class exploits the functionality and the information of L–classes, uses the primitives defined in PI–classes and controls the execution of the tasks of the instantiated T–objects. This denotes that instances of the PS–class deliver the desired protocol stack functionality by implementing all required objects. Appropriate PS-class constructors can be used to implement any standard protocol stack (provided the layer classes for these standards are available). Using this structure, a protocol stack can be dynamically adapted to any particular set of requirements by exchanging the appropriate (pro-layer and thread) class(es).
The following class relationships represent the complexity of the architecture (see also Figure 11):

- *L–Classes depend on the PI–Class definitions:* L–objects can access primitives of a PI–class to communicate with higher or lower layers, whilst

[3] Java has been chosen as implementation platform because of features such as dynamic binding and platform independence

the PI–class provides access to the information stored in a L–object or trigger any method implemented within this (L-) object.

- *L and PI-classes depend on T–classes:* T–objects can use the appropriate attributes and methods of L- and PI-objects to carry out its pre-defined task.
- *PS–class has L, PI and T–objects as attributes:* PS–objects can use the functionality and the information defined in these objects directly.
- *T–classes depend on the PS class:* PS-objects control starts and stops of various threads in its main method, but T–objects are solely responsible for the task execution (see Figure 11).

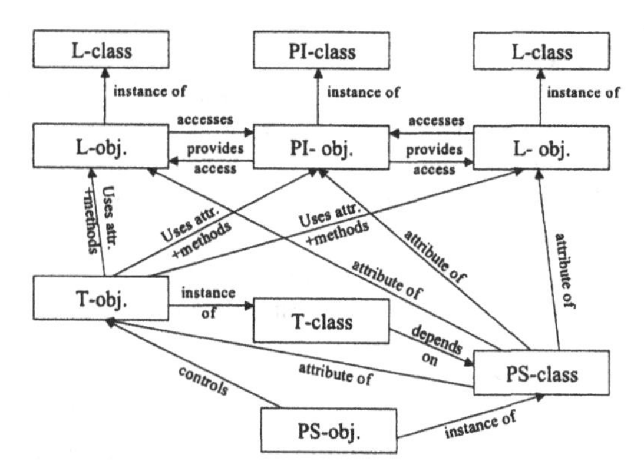

Figure 11. Relations within the protocol stack

All classes are structured in a class library, which provide the required signalling functionality to any network entity. The reason why classes are grouped in packages is to provide a base level of security. Declaring only the PS-class public[4] ensures that the other objects become (theoretically) invisible and inaccessible to any other non-related object.

Although T–objects are directly used in the main method of a PS–object, L, PI and T–objects are used as attributes within the PS–class, and these attributes are the instances which define the appropriate objects as members of one protocol stack (e.g., the reference of the PS–object is mostly used as parameter in the methods of an L–class. In this way, if a method of an L–object is called, the object can identify the PI–objects belonging to the same PS–object and use their attributes and methods (*i.e.*, messages), this is also depicted in Figure 9. Two operational modes for Protocol Stack objects trigger signalling/message sequences. These are either: internal (*i.e.*, the PS–object initiates or terminates thread execution without external triggers but caused by a time-out) or external (*i.e.*, events occurring at the application or physical layer). Both of these operational modes lead eventually to the instantiation or termination of the appropriate thread.

The flexibility of this architectural approach is based on the introduction of the generic PIs and is further consolidated by the possibility of using several

---

[4] The 'public' object is allowed to access the 'private' or 'protected' methods within the scope of its own attributes.

concurrent threads. Generic PIs deliver the means necessary to access L-objects and to support T–objects, they provide the flexible structure necessary to support and implement different protocol stack standards. The generic set of primitives between the functional entities of the PS remains unaltered (even if new pro-layer implementations are introduced, as long as the messages comply to the four primitive types); this denotes that the PI–classes can access any pro-layer class, without being subject to changes.

5. Implementation Models

To assess proper functionality of programming interfaces, a set of QoS related messages, methods and classes has been defined and organised in a class library. The motivation for choosing QoS signalling as a test case has been the increasing complexity of interactions and variety of QoS demands expected in future generations of mobile networks, when users will be able to request mixed services (*i.e.*, voice/video/data and other Internet services). Aurrecoechea et. al. [18] compare in a number of different QoS architectures and discuss their particular features, it is stated that most QoS architectures merely consider single architectural levels rather than the end-to-end QoS support for multimedia communications. Furthermore, it is pointed out that QoS management features should ideally be employed within every layer of the protocol stack and QoS control and management mechanisms should be in place on every architectural layer.

We therefore implemented two models , both incorporating support for end-to-end and level controlled QoS negotiation as examples to show both functionality and to test proper operation of communication signalling within the OPtIMA architectural framework. The remainder of this part contains the definition of a generic QoS signalling message specification, a description of our protocol stack and the QoS negotiation implementation. Model 1 implementation is based on RMI platform (running on Sun Solaris workstations) where two applets are implemented and used as representations of signalling (application) end points. Model 2 implementation is a standalone stock trading application running over a CORBA ORB (OrbixWeb platform) within windows NT environment.

A set of messages and parameters are specified and defined for a generic QoS signalling structure built on the OPtIMA protocol framework. Messages and their types (*i.e.*, primitives) for the API – pro-interface were presented in Figure 12b. Naming of methods/function calls follows a general rule. The message format and message name specification is defined as:

```
[Name of Programming Interface][Type of Primitive][Description](arguments/parameters) { }
```

Choosing the following message as example,

```
APISRMSerReq(String typeOfService) { }
```

This message is part of the API (Interface between Signalling Application Layer and Application Layer), it is derived from the primitive type 'Service

Request Message' (SRM) and contains a string as argument (*i.e.*, typeOfService). All messages (*i.e.*, SRM, LSI, RR and AEN) are unidirectional in downward or upward direction, whilst SRM and LSI are downwards (from upper layer to lower layer), RR and AEN are active in upwards direction (*i.e.*, lower to higher layers). The flow direction from higher to lower level or from lower to higher level is also indicated in the numbering system used:

PPI23**RR**TypeOfVariable(arguments) { } → Message from Layer 2 to Layer 3.
PPI32LSITypeOfVariable(arguments) { } → Message from Layer 3 to Layer 2.
API_PPILSITypeOfVariable(arguments) { } → Message from Application Layer to L4.
PPI_API**AEN**TypeOfVariable(arguments) { } → Message from Layer 4 to Application Layer

Some of the messages may use the active feature of the OPtIMA and may have to access other than subsequent layers (e.g., application sending a message directly to the link layer), this as well is defined in the command naming:

API_2SRMTypeOfSRM(arguments) {}

Table 1 contains the complete list of messages (SRMs, LSIs, RRs and AENs) defined for the QoS signalling API and PPIs.

Table 1. Set of QoS signalling messages

Service Request Messages (SRM)

SRM	Parameters	Description
APISRMSerReq	p, index	Service Request
APISRMSerChange	p, index	Service Modification Request
APISRMDisconnect	p	Disconnection
APISRMSerContd	p	Service Continues

Layer State Information (LSI)

LSI	Parameters	Description
API_PPILSIQoSProvided	p, qosAccept	QoS Provided Accept/Deny

Asynchronous Event Notification (AEN)

AEN	Parameters	Description
APIAENSerReqOK	p	Service Request OK
APIAENSerReqFail	p	Service Request Fail
APIAENQoSProvided	p	Provided QoS for Service Request
APIAENSerChangeOK	p	Service Modification Request OK
APIAENSerChangeFail	p	Service Modification Request Fail

Request Reply (RR) not applicable

Both implementation models (see following section) have enabled verification and validation of both the architecture and the specification of programming interfaces. To prove the validity of OPtIMA and proper functioning of the pro-interface implementations, the QoS messaging part of the API has been taken as example.

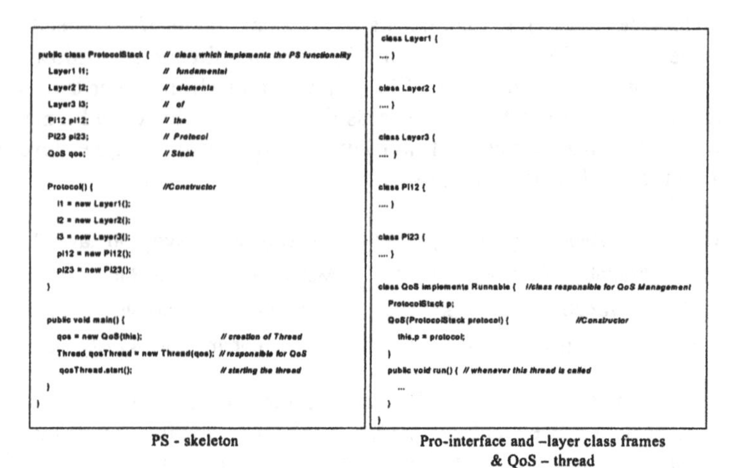

Figure 12. a. and b. sample skeleton code

Figures 12a and 12b, depict skeleton code for QoS negotiation and the set of messages for the API pro-interface respectively. Subsequent sections contain descriptions of the implemented models and class structure for QoS negotiation and management at the API and also a QoS specification/monitoring tool is also included within a stock trading application.

5.1 Model Description

In model 1 the implementation consists of a server applet and a number of client applets, with both applet classes displaying the QoS negotiation. QoS classes defined in OPtIMA facilitate messaging between client and server using Java RMI as transport media. The applets are used as signalling end-points to negotiate and display the QoS settings. Communication between these end-points takes place via the API using an underlying layer class (pro-layer), which implements the RMI connection via interfaces (Stub and Skeleton) to the Java Object Broker (RMI). Client-Stub and Server-Skeleton implementations in this model, are representative of the pro-layer classes.

The protocol stack (used in this test platform) consists of an application layer (L-class), an API (PI-class) and a general layer class (L-class) representing the remainder of the protocol stack[5]. Application layer classes are implemented as

[5] The functionality of this model relies on the basic client/server principle, where objects are distributed across the network and communicate via a middleware layer (*object broker*). Clients request services, via the request broker, from remote server objects. The (RMI) broker uses interfaces bound to the implementations of clients and servers called stubs and skeletons, respectively. Stubs and skeletons hide the complexity of the communication between client and server, they control serialisation and de-marshalling of parameters and they establish, maintain and terminate connections between the remote entities.

applets (*Client Applet* and *Server Applet*), they provide the graphical user interface (GUI) of the signalling end-points. Java AWT (abstract windowing toolkit) is used to implement the GUI and Solaris on Sun workstations as computing platforms.

The general layer class (RMI Class Client and RMI Class Server) represents the test-platform-version of the protocol stack. This class is derived from the signalling application layer class, all QoS information available in lower layers is accessible via this class. Moreover, the general layer class accesses the Java's RMI Stub and Skeleton classes. RMI-Client and -Server classes are implemented as separate classes they provide the means for (RMI) distributed object computing endpoints, see depiction 13.

Client and Server applets are depicted in Figures 14 and 15, respectively. Their data fields represent requested and provided QoS parameters.

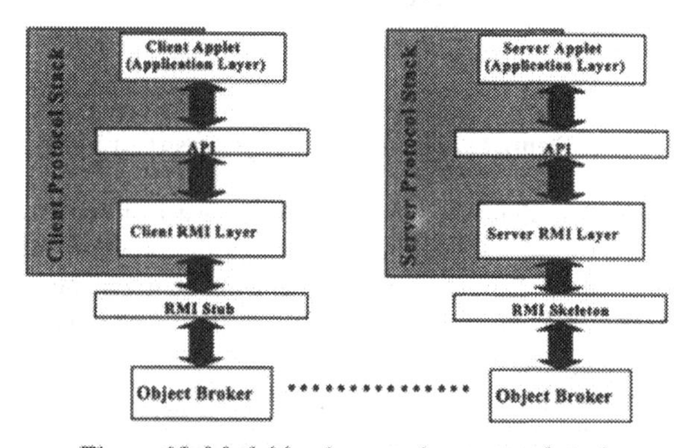

Figure 13. Model implementation protocol stack

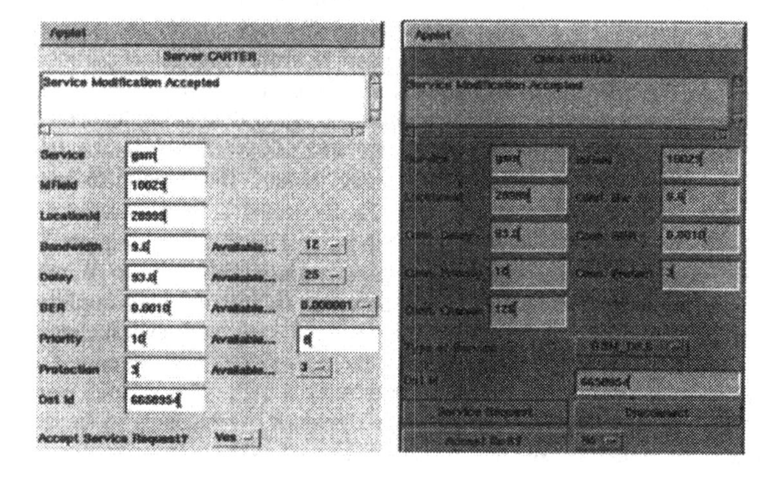

Figure 14. Server applet **Figure 15.** Client applet

The experimental set-up consisted of the server applet running on a SUN Solaris server and a number of client applets running on distributed x-terminals, which were connected to the server machine via an ATM hub (over 10/100 Mbits/s Ethernet connections). As an example of operation of model 1 implementation, we briefly describe the operation of the simplified signalling messages sequence for QoS re-negotiation presented in Figure 16. The client requests a modification of the already provided service from the server (Service Modification). It passes to the server all the required QoS parameters and information related to the identity of the client. The server processes this request and examines the required QoS, the availability of local resources *i.e.*, current loading due to all clients, and the resources already assigned to this client.

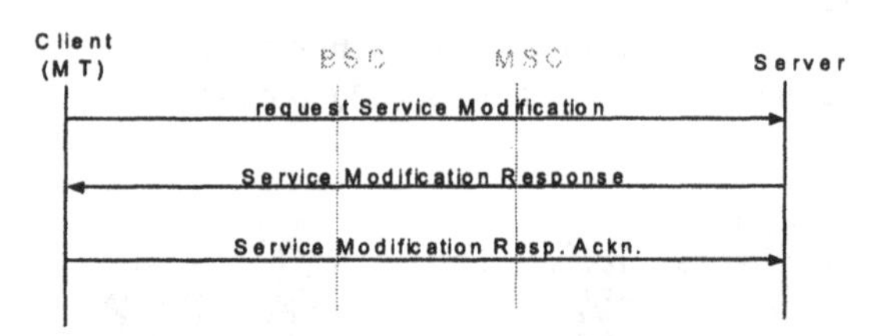

Figure 16. QoS modification message sequence

Then it informs the client whether the new request is accepted or not (Service Modification Response). The client acknowledges the previous message (Service Modification Response ack.). In case the request is not accepted, the client can either disconnect or continue the current session with the previous QoS values.

5.2 The OPtIMA _Trader Application

Although the interfaces in model 1 and 2 prototypes make use of the same underlying QoS signalling (API) structure there are however a number of major additions in model 2 to enable the additional functionality of the user interface. Model 2 prototype offers an interface to users to search and select particular share symbol, specify their QoS requirements and renegotiate QoS during a session. It has been implemented as standalone application rather than as applet. The interface presented to the user incorporates username and password registration fields, text messaging capability, an online share index 'ticker' and a point-to-point video conferencing tool as depicted in Figure 17. In model 2 RMI is replaced by the CORBA based object broker OrbixWeb (used for QoS messaging part) and the complete model has been ported from Sun Workstations over to Windows NT environment on networked PCs. Java however is still used as implementation language. All OPtIMA classes remain unchanged, only stub and skeleton classes are exchanged (to enable connection to the new ORB) and of course the new graphic interface used to access the OPtIMA QoS API. The various parts of the *OPTIMA_Trader* application use different transport mechanisms to transmit their information: user and password registration, as well as the text messaging

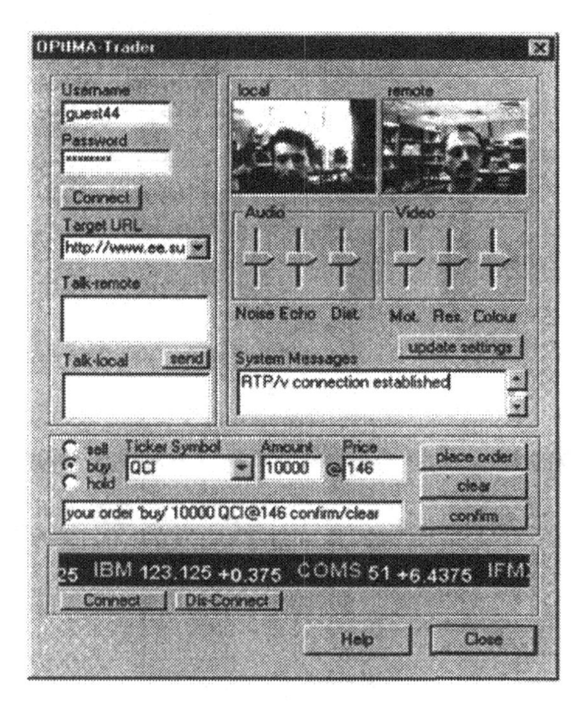

Figure 17. OPtIMA_Trader GUI

service are using the RMI facilities to transmit alphanumerical symbols and strings from client to server. A RTP connection is used for the video conferencing tool, whilst an IIOP (CORBA) connection carries the QoS signalling mentioned in the first part of this section (V). The 'share price ticker' is a java applet constantly updated through connection to a remote Web server [19], it uses the http protocol for the information transport (see Figure 18a).

The RTP session manager is the 'signalling endpoint' for RTP connections (real-time video/audio), it establishes, maintains and terminates RTP connections and reacts to QoS changes requested by the application (via the OPtIMA QoS API). QoS signalling related fields in the GUI, show capabilities such as requests of changes in colour, motion and resolution as desired setting (useful for fast re-negotiation of QoS during an active session). The *Update Settings* button however brings up a separate profile window (Figure 18b) where desired, minimum and offered values for frame rate, colour, brightness and resolution for video and noise, distortion *etc.* for audio can be set by the user (via sliding scale bars). Each custom profile can then be saved for later use or discarded as required. Moreover the RTP session manager provides additional network information. Whilst the profile window contains subjective settings as: required, minimum and offered QoS, the RTP Statistics window displays numerical information about the packet transmission, obtained from the RTP session manager.

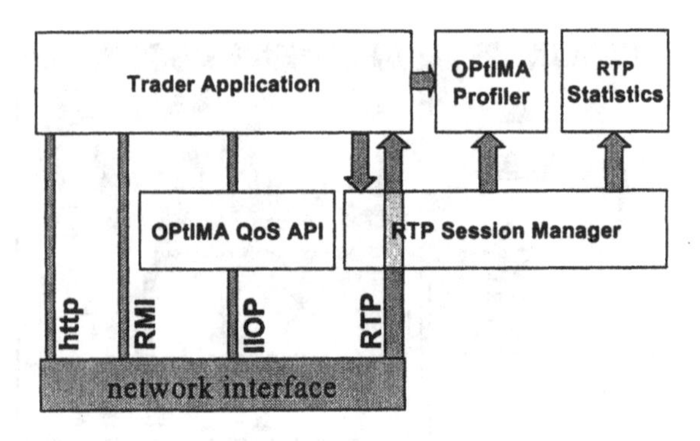

a

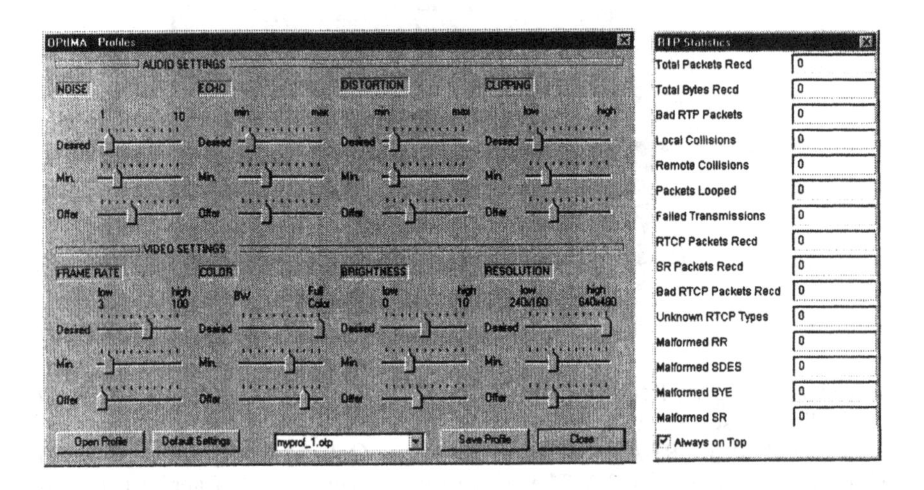

b

Figure 18. a. OPtIMA_Trader data/signalling model; b. Video/audio profile and RTP Statistics windows

The model application is complex in several regards; a number of different software packages and various transport systems are used alongside. Java Swing has been applied to implement the control components (*i.e.*, buttons, sliders, text-fields, *etc.*) within the GUI whilst the JMF (Java Media Framework) implements the video conferencing part. Using these two packages together within one application has proved non-trivial: swing uses 'light weight' whilst JMF on 'heavy weight' components which means that swing components do not require as much of the system resources as JMF components do as they rely heavily on JVM (Java Virtual Machine) resources. Due to these different computing resource requirements, using light and heavy weight components alongside may cause 'jumping' effects when the GUI is being dragged across the screen. Swing is part of the Java 2 SDK, JMF a collection of class libraries that provide multi-media capabilities. The third software package is OrbixWeb 3.0 from Iona and contains a Java implementation of the CORBA 2 standard. Transport of signalling messages and content uses IIOP (Internet-Inter Operation Protocol), http (hyper text transport protocol) and RTP (Real Time Protocol) connections are on top of TCP and UDP, respectively over IP. Although initial results are encouraging, proper evaluation of the proposed protocol framework, will require an extensive suite of tests on aspects such as tolerable delays and response times, packet loss and jitter and server loading, delay/throughput/loading at the ORB level (similar to [20]) and subjective QoS assessments which are currently in progress. Future work is envisaged to demonstrate software download of protocol code (new L2/3) and application components (based on protocol in [7]) and reconfiguration capabilities using the trader application.

Figure 19 depicts the testbed used for both parts of the QoS – API model implementation, whilst in model 1 Solaris workstations were used as signalling endpoints, two PCs operating Windows NT were employed to host model 2.

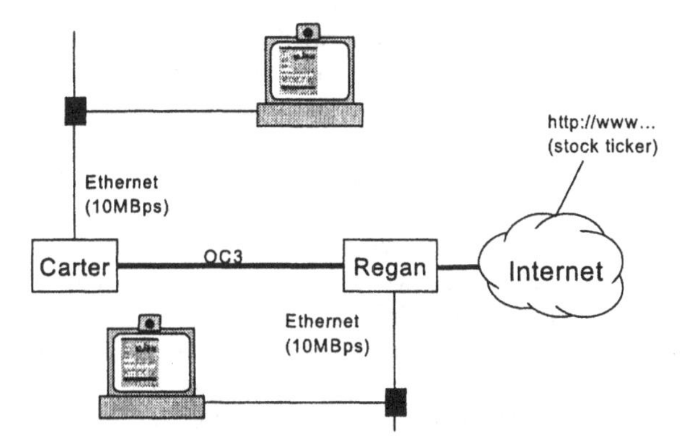

Figure 19. OPtIMA testbed

The trader application has a complex structure and is using a multitude of different transport mechanisms on top of both UDP and TCP over IP. Despite the overload in transport techniques used within the testbed, it serves a multitude of different tasks; it not only shows the use of a QoS negotiation API defined and implemented using the OPtIMA framework, but it also provides a platform for future performance evaluations of a wide range of real- and non-real-time streams. Furthermore, possible evaluation scenarios include investigations of point to multipoint 'trading-sessions' and therein the scalability and performance of the QoS signalling.

6. Conclusions

In this article an OOD based approach for design of API/PPIs to facilitate reconfiguration of protocol stacks for software-radios has been introduced and described. The importance role of active interface implementations using object oriented design mechanisms in this context has been emphasised. The OPtIMA architecture is implemented in Java. The Java platform was selected as it supports code mobility, OO properties, serialisation, inheritance and encapsulation properties. Based on the proposed OPtIMA architecture/framework two representative implementations for QoS signalling using different underlying transport platforms, were described. The core of this work has focused on the definition and provision of a library of classes, which can be used to flexibly build reconfigureable protocol stacks. Protocols implemented under this framework may be used to implement the signalling stack for virtually any underlying wireless communication network; the main advantage of OPtIMA is the ability to reconfigure protocol stacks built within the framework. This becomes a vital facility for the next generation of mobile communication networks (3G/4G) in which variable data rates will be delivered for a number of different types of applications, where the need to employ application specific adaptations to the protocol stack may arise. The API/PPIs can work with legacy implementation of protocol layers as well as component-based forms of protocol layer *i.e.*, the framework supports composible protocols. Within the context of soft-radios, protocol reconfiguration would be performed under the control/supervision of a "Reconfiguration Manager" unit. Specification of functionalities and operation of reconfiguration manager is not within the scope of this article.

Furthermore, two different implementation models for the proposed open programmable protocol interfaces have been proposed and assessed. Both offer the required flexibility to support protocol stack re-configuration. The first possible solution proposes an architecture in which even the interfaces are implemented as objects, whereas the second proposal follows the classical definition of APIs, in which the API is merely a formal definition implemented in the underlying layer. Although both implementation strategies can be supported within OPtIMA, the current OPtIMA architecture has been based on the former of the two implementation models. By extending the existing base protocol classes and customisation, it is possible to implement application-specific behaviour *i.e.*, users would be permitted to install and run custom protocol stacks, protocol layer or addition/deletion of components within any given layer, thus tailoring protocol functionality/components to application requirements. It is assumed that Protocol classes are implemented compliant to the appropriate API/PPI, by the vendors. Also the proposed PIs are capable of mapping the application QoS onto network and link-layer QoS parameters, as required. As a fully java-based multimedia application, the trader application required significant amount of memory and CPU resources to operate. But although use of different technologies and software packages does create some considerable overheads, we expect that much of it will diminish with the introduction of new versions of these packages (whereby their API remains the same or becomes extended only, *i.e.*, the code of our implementation will not need to be changed with these updates).

Finally, the model represents a paradigm shift away from purely stratified protocol stacks towards concurrent approach, which promote active programming interfaces to facilitate non-sequential message and packet passing whilst providing an independent signalling interface for applications.

7. Acknowledgements

The authors wish to acknowledge the support of Mobile Virtual Centre for Excellence (MVCE) UK, in the funding of the work reported in this paper.

8. References

[1] J.Mitola III, "The Software Radio Architecture" , IEEE Comm. Mag. May 1995, pp. 26-38

[2] M.Butler et. Al. , "The Layered Radio" , MILCOM, NY, 1998, pp. 174-79

[3] D.Tennenhouse and D. Wetherall, "Towards an Active Network Architecture" , Comp. Commun. Rev. vol.26, No. 2, Apr. 1996

[4] A.Kulkarni and G.Minden , "Active Networking Services for Wired/Wireless Networks" , Proc. INFOCOM 99, vol.3, NY 1999, pp. 1116-23

[5] A.KulKarni and G.Minden, "Composing Protocol Frameworks for Active Wireless Networks" , IEEE Commun. Mag. Mar. 2000

[6] K. Moessner, S. Vahid, R. Tafazolli, "Reconfiguration Management for Software Radio Networks", in preparation.

[7] K. Moessner, R. Tafazolli, "Terminal Reconfigurability – The Software Download Aspect", IEE Int. Conf. On 3G 2000, Mar. 2000, London UK.

[8] A S Tanenbaum, Computer Networks 3rd ed. , Prentice-Hall, 1996

[9] A Park, J Meggers, On The Move, http://www-i4.informatic.rwth-aachen.de/Research/MMCommunication/Projects/OnTheMove.html .

[10] M Meyer, (ed.), Design of MASE V1.0 Deliverable D17, ACTS AC034, http://www. sics.se /~onthemove/docs/OTM_d17.doc , 1996

[11] Mobile Station Application Execution Environment (MExE), Service description, GSM 02.57 V7.0.0, ETSI, 1998

[12] D Beyer et. al., API Framework for Internet Radios, http://www.rooftop.com

[13] P Lin, et. al., Programming Interfaces for IP Routers and Switches, an Architectural Framework Document, IP Subworking Group IEEE P1520, 1999

[14] B Krupczak, K L Calvert, M H Ammar, Implementing Communication Protocols in Java, IEEE Communication Magazine, October 1998

[15] Review of DSP design requirements: Task T1.13, Mobile VCE - Terminals Group, 1999

[16] Coad and Yourdon, Object Oriented Analysis, 2d edition, Englewood Cliffs, N.J.: Yourdon Press, 1991.

[17] G Booch, Object Oriented Analysis and Design With Applications, 2d Ed., Benjamin/Cummings Publishing, Redwood City, Ca., 1991

[18] C Aurreocoechea, et.al., A Survey of QoS Architectures, ACM/Springer Verlag Multimedia Systems Journal, Special Issue on QoSX Architecture, Vol. 6, No. 3, pp. 138-151, May 1998

[19] http://www.selfpage.com

[20] K. Moessner, S. Vahid, R. Tafazolli, "Performance Evaluation of Signalling Protocols on different Middleware Platforms for Service Provision in

Distributed Mobile Communication Networks", WPMC'99 (VTC99), September 21-23, Amsterdam, Netherlands.

Part 6

Enabling Technologies

Front End Architecture for a Software Defined Radio Base Station

Wolfgang König[1], Gerd Wölfle[1], Christian Fischer[2], Tim Hentschel[3]

[1] Alcatel Corporate Research Center, D-70430 Stuttgart, Germany
[2] University of Karlsruhe, Institut für Höchstfrequenztechnik und Elektronik, D 76128 Karlsruhe.
[3] Dresden University of Technology, Mannesmann Mobilfunk Chair for Mobile Communications Systems, D-01062 Dresden

Abstract: This paper is aiming at the concept for the frontend design of a flexible platform running Software Radio as it is investigated in the SORBAS base station project. It briefly outlines the Software Radio idea and the motivation for a software defined radio base station. The global architecture of such a base station is described including system boundaries and restrictions given by the technology. Emphasis is laid on technology for multiband antennas, specific analogue issues (power amplifier linearisation) and processing issues for a digital front-end providing up- down conversion, channelisation, filtering *etc*.

1. Introduction

It is obvious that the Software Radio idea has triggered world-wide R&D activities, most of them related to application within a mobile station. However, some of these emerging technologies can well be applied for the base station of a mobile communication network. This paper reports about activities undertaken within the scope of the SORBAS (**So**ftware **R**adio **B**ased **A**ccess **S**ystem) –Base Station Project which is one of the projects associated to UMTSplus, a new system concept sponsored by the German Ministry for Education and Research (BMBF) aiming at "Universality and Mobility in Telecommunication networks".

The activities of this project are concentrated around the following main issues:

- analogue front-end (filters, receivers, transmitters, Intermediate Frequency - IF- local oscillators);
- power amplifiers (efficiency, linearity and linearisation algorithms);
- A/D and D/A-conversion at high frequencies;
- digital front-end (up- and down-conversion, sample rate adaptation, channelisation)
- high speed baseband processing hardware (complexity of the flexible platform, computing capacity, algorithms).

At the present stage, the development of a Software Radio system forms a very challenging issue because the basic idea to move the point of digitisation as close as possible to the antenna and to open the flexibility to handle different air interface standards put tremendous requirements on the hardware platform to be used.

2. Motivation and Basic Architecture

The mobile communication market is one of the fastest growing market in the world. Though recognising the benefits of a common world-wide unique standard for future mobile systems, world wide industrial competition shows up a difficult environment for standardisation, leading to a 'family' of third generation standards that has been approved by ITU.

As a consequence, the "Software Radio" idea is emerging as a pragmatic solution to overcome the standardisation issues aiming at a software controlled configuration and re-configuration of the mobile user equipment and radio base stations:

- enable terminals to adapt dynamically the radio environment in which they are currently located;
- enable corresponding base station to accept remotely controlled or automatically the radio environment it is expected to operate in.

The Software Radio in its final stage must not be constrained to a particular standard but able to offer services of any already standardised system or future ones on any radio frequency band. Its compatibility with any defined radio mobile is guaranteed by its re-configurability. For the base station side another important aspect is, that all work in the field (maintenance, upgrade, *etc.*) is a very costly part and every remotely executed measures are welcome. Therefore such a Software Radio base station receiving its configuration during the production cycle and remaining open for remote upgrade or re-configuration may bring operational benefits.

The advantages of a Software Radio solution can mainly be summarised in two areas:

Development / Production

- generic scalable platform for different applications;
- reduction of development effort;
- shortening of development cycles;
- larger production quantity at lower production costs.

Application/Operation

- eases introduction of new services / features and on-line error recovery;
- simplified evolution to more efficient systems.

However for a commercial introduction at a certain point in time, a careful trade off between flexibility and at least initially higher costs for a generic platform has to be done. As it will be shown in the next chapters, there are also a couple of technological restrictions, that will not allow a 'real' software defined radio basestation in the near future. However as technology improves and costs for todays high end signal processing components come down, a steady migration towards more and more flexible solutions will take place.

The basic architecture of a software defined radio basestation is shown in Figure 1.

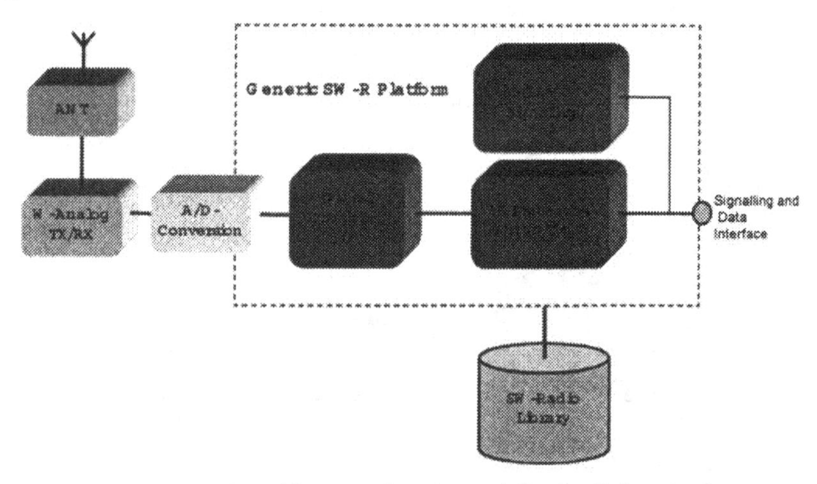

Figure 1. Basic architecture of a software defined radio base station

The above figure represents the up-link (*i.e.*, receiving) and the down-link (transmitting) part of a software defined radio base station. Within the SORBAS project a multistandard solution covering UMTS FDD and GSM 1800 is addressed. The block "Analogue Front-End" (AFE) includes all the necessary analogue Radio Frequency (RF) and Intermediate Frequency (IF) receiver and transmitter components, such as duplexers, amplifiers, mixers, filters, *etc.* The next block, the AD/DA-converter unit is built of wide-band A/D and D/A-converters. The Digital Frontend accomodates the necessary functions for digital filtering, digital down- respectively up-conversion, sample rate adaptation and channelisation. Baseband (BB) -Processing unit includes all functions required for the baseband processing including, modulation, demodulation, framing, coding *etc.* The O&M module stands for all functions related to operation and maintenance as well as network signalling. Digital Frontend, BB-Processing and O&M modules are considered to be based on a generic hardware platform driven by software application modules which were provided by a so called software radio library.

3. Multiband Antennas

For a wireless telecommunication system like a software based radio system the antenna performance is of high significance. The main challenge for an antenna design as it is required for the described system is the high bandwidth from 1710MHz to 2170MHz covering GSM 1800 and UMTS FDD frequency bands.
One of the major research topics of the Institut für Höchstfrequenztechnik und Elektronik (IHE), Universität Karlsruhe, is the design and measurement of antennas for mobile communication applications as well is the development of the necessary simulation tools for such antennas.

Extensive studies at IHE leaded to the result, that the most suitable technology for a low-cost multiband antenna for base stations is the aperture coupled microstrip-patch-antenna. Figure 2 shows a single patch element.

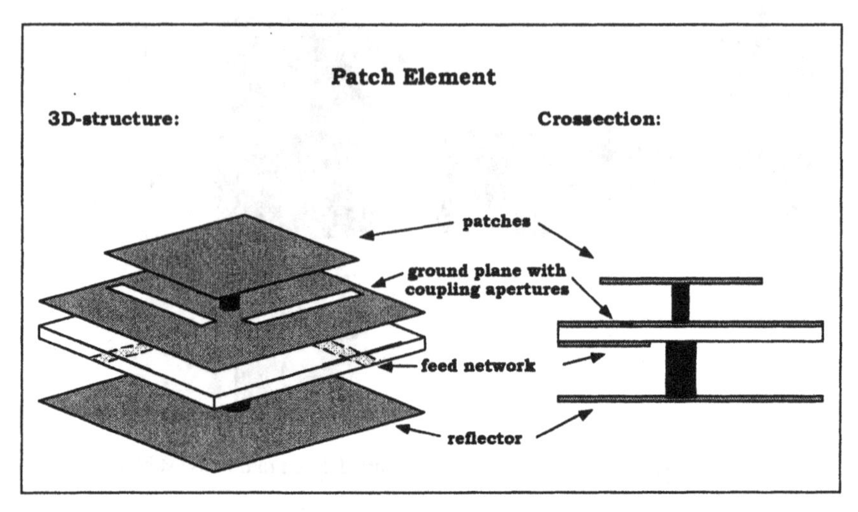

Figure 2. Structure of a single patch element

This structure unifies several positive aspects from the manufacturing and design point of view:

The design is light weight but still very stable. All components are commercially available, therefore of low cost. The used materials are easy to machine.

The antenna provides dual linear polarization, allowing the usage of polarisation diversity. In the design process, the planar structure can be treated well by accurate method-of-moment (MOM) based simulation tools.

By introducing a second radiating patch element above the main patch, it is possible to improve the bandwidth without enlarging losses or reducing the gain of the antenna. Stacked patch element were fabricated and measured at IHE and showed excellent performance concerning return loss, polarization coupling and pattern.

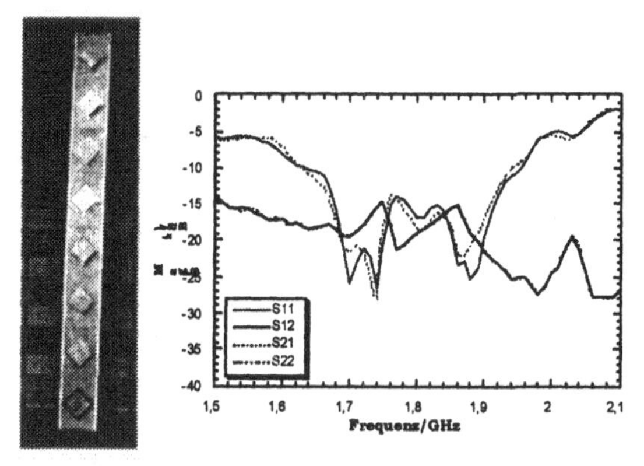

Figure 3. Front view, return loss (S_{11}, S_{22}) and polarization coupling (S_{12}, S_{21}) of a 8×1 stacked patch array

To reach a vertical 3dB-width of 7° (±1°), which is often required for base station antennas, the single elements have to be combined to an 8×1 stacked patch array. Figure 3 shows a realized 8×1 array and its measured return loss and polarization coupling. The measurements show very good performance of the array concerning antenna pattern, sidelobes and input match. The return loss is within the expected level of less than –14dB over a frequency bandwidth of more than 15%. The measured performance of the single elements and of the 8×1 array indicate, that the full required bandwidth can be realized in a further prototype.

4. Analog Signal Processing

4.1 The Analogue Front-End

The Analogue Front-End (AFE) is still necessary in current Software Radios because it is up to now not possible to digitise the signal directly at the antenna. In the AFE receiver the RF signal passes a Low Noise Amplifier (LNA) and a band filter before its transformation to the IF. Because of the wideband components the AFRE receiver can be implemented for multi-standard/multi-carrier operation.

In the transmitter the IF-signal is transformed to the RF and amplified with the power amplifier (PA). Depending on the linearisation technique the AFE contains also the chains for reference or feedback signals. In this paper a hybrid linearisation with feedforward and adaptive digital predistortion is presented (see Section 4.3), so a feedback for the predistortion and a reference for the feedforward is required as shown in Figure 4.

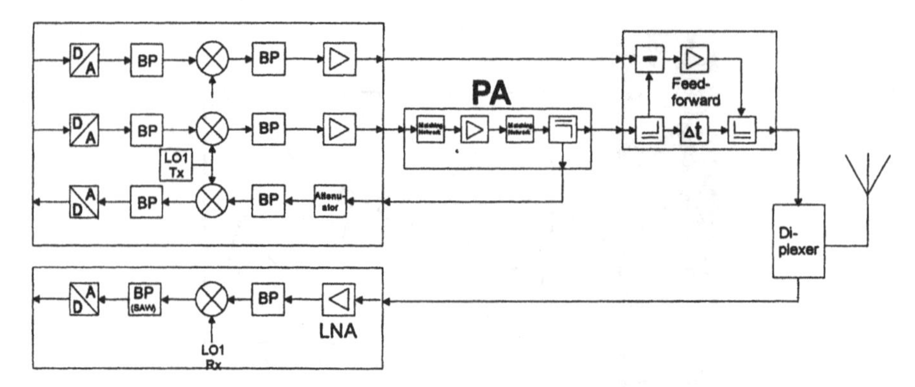

Figure 4. Modules of the Analogue Front-End

The components used in the AFE have only a very limited frequency bandwidth. This leads to problems if different standards must be served with the same AFE, especially when the RF carrier frequency is very different. Solutions for GSM900 and DCS1800 or for GSM900 and UMTS UTRA-FDD are currently not possible with a single AFE. A common AFE for DCS1800 and UMTS-FDD is possible because the total required frequency bandwidth can be handled with most components, except of the linearised power amplifier (see Section 4.3).

4.2 Analog Components for Software Radio Solutions

The mixers and pre-amplifiers are wideband-compatible and can be used for multi-standard or multi-carrier AFEs. The filters must be selected according to the standard with the widest bandwidth. Also the LNA can serve a wide band and support multi-standard/multi-carrier operations.

The only band-specific component is the power amplifier (PA). Due to the high requirements concerning linearity the linearisation methods are up to now limited to a bandwidth of 25–30 MHz. So several PAs must operate in parallel if the supported carriers/standards require a wide bandwidth.

4.3 Linearised Power Amplifier

As mentioned before, the bandwidth of the linearised PA is limited to max. 30 MHz. The PA transistor in the 2 GHz range has a bandwidth of approx. 100 MHz (due to internal matching of the PA), so two independent transistors for DCS1800 and UMTS UTRA-FDD are necessary. It is up to now not possible to use a single PA transistor for a multi-standard AFE. The external matching of the PA reduces again the bandwidth. So the PA module without linearisation is limited to a bandwidth of 70 MHz.

The linearity of the PA depends on the specification of the air interface. For UMTS UTRA-FDD at least -50 dBc are required. For multi carrier DCS1800 more than –70 dBc are necessary. PA transistors today have a linearity of max. –40 dBc (with back-off leading to reduced efficiency), so an external linearisation method must be applied to fulfil the standards. In this paper two methods are presented and a proposal for the combination of both methods is shown. The first approach is analogue feedforward which is applied to the RF signal. In contrast to this approach the second one is digital predistortion which is implemented in the baseband. A combination of both methods leads to the required linearisation of −70 dBc. Feedforward alone leads to an improvement of 10..15 dB which is sufficient for single carrier UMTS UTRA-FDD PAs, if the above mentioned transistors with a high linearity are chosen. But for multi-carrier applications both linearisation methods must be combined.

4.3.1 Analogue Feedforward

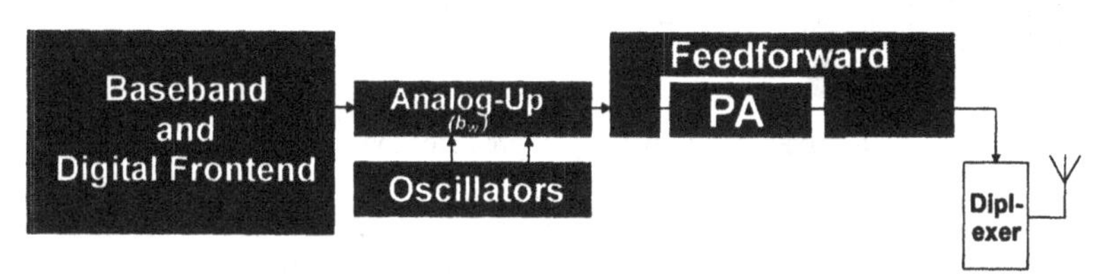

Figure 5. Analogue feedforward

The principle of analogue feedforward is shown in Figure 5. The feedforward approach is a real-time linearisation, *i.e.*, the output signal after the power amplifier is compared to the original (reference) signal before the PA and the difference between the two signals (error signal) is amplified, inverted and added to the output signal (see [1]).

Critical points for the feedforward linearisation are the compensation elements for the delay (and phase shift) in the two amplifier transistors. I-Q-Vector modulators can be used for a variable shift in combination with fixed delay lines for the basic delay. The adjustment of the modulators is realised with a pilot signal close to the band or in the band of the signal. This pilot is feed to the signal before the PA (after the coupler) and if the pilot is still in the output signal of the PA, the settings of the phase shifters must be adjusted.

A limitation of the feedforward is the linearity of the error amplifier. If the error signal needs a high amplification, non-linear signals are added in the error amplifier and the total performance decreases. Very often a second feedforward loop is added to reduce these effects. With a single loop 20 dB in linearity can be achieved.

4.3.2. Digital Predistortion

To eliminate 3rd order intermodulation products with digital predistortion at least three times of the signal bandwidth is required in the digital (after the predistortion) and analogue processing of the signal. The signal at the output of the PA is fed back to the Digital Frontend (DFE) where the predistortion is applied to the signal (see Figure 6). [1] shows an example for the implementation of an adaptive digital predistorter.

The output of the PA is fed back to the predistorter where the difference between reference signal (input signal) and output signal is determined. Depending on this difference a new value for the Look-Up-Table (LUT) is determined and stored in the memory. So the update of the LUT works independent of the predistortion of the signal.

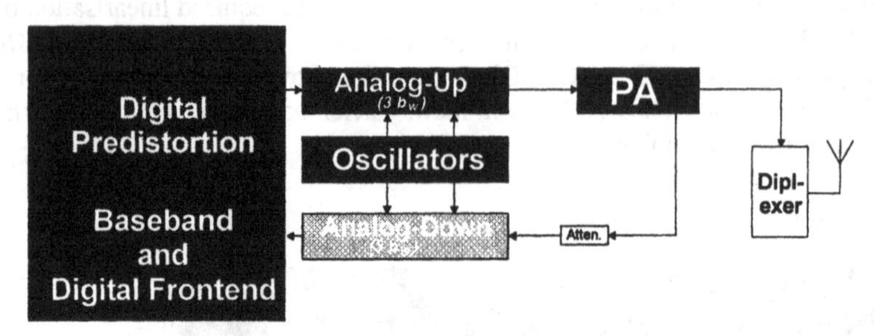

Figure 6. Adaptive digital predistortion

But due to the processing of the signal in the predistorter and due to the limited sample frequency of the A/D-converter (see section 4.4) in the feedback chain, the digital predistortion is limited to a channel bandwidth of 15 MHz (for 3rd order elimination this would lead to a processing of 45 MSamples/s).

4.3.3 Hybrid solution

Both linearisation methods can be combined to achieve high linear output signals. While each method leads to an improvement of 20 dB in linearity, a combination achieves up to 30 dB. Multi-carrier applications require always a hybrid linearisation to fulfil the specifications in the standards.

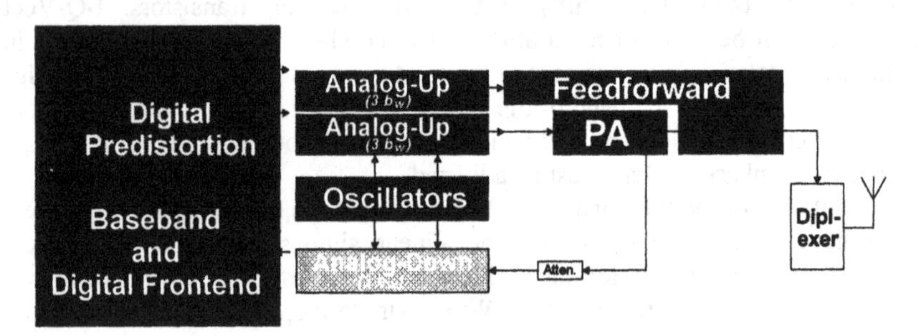

Figure 7. Hybrid linearisation

The feedforward approach is based on the comparison between the input (reference) and output signal of the PA. If predistortion is applied, the input signal is predistorted and therefore not suitable as (original) reference signal for the comparison with the output signal.

To solve this problem a further reference chain is added.

The approach presented in Figure 7 combines the advantages of both linearisation methods. The predistortion is responsible for the basic linearisation and the feedforward adds a post-processing to the remaining non-linearity. So the feedforward has a very high efficiency and linearity because of the small amplitude of the error signal. One loop is sufficient even for high linearity applications. The slow predistortion adoption is sufficient because the fast fine-tuning is made with feedforward and so the predistortion must not detect small changes in the behaviour of the PA.

5. Digital Signal Processing

5.1 The Digital Front-End

The Digital Front-End (DFE) is the very part of the front-end of a software defined base-station where front-end functionalities are realized by means of digital signal processing. It bears the required flexibility and parameterizability of a software defined base-station which can be adopted to different air-interfaces. Based on a common hardware platform and standard-specific software the DFE typically performs digital down-conversion, channelization, matched filtering, and sample rate conversion in the receive branch. Digital up-conversion, pulse-shaping, and sample rate conversion are part of the transmit branch. The DFE interfaces the analog/digital interface (A/D converter in the receive branch, and D/A converter in the transmit branch) on one side, and the base-band processing on the other side.
Since the sample-rate at the analog/digital interface is usually very high, the effort for realizing a software defined DFE goes beyond the capabilities of digital signal processors. As an example, digital down-conversion should be mentioned which is done by means of multiplying the incoming signal with a complex rotating phasor. It can be realized by means of two separate multiplications of each input sample with samples of a sine and a cosine function. With an input sample rate of e.g., $f_S = 65$ MSps this would require 130 million multiplications per second, which cannot be realized with digital signal processors in an efficient way. Therefore, it is one of the main tasks to design a common parameterizable hardware platform on which the functionalities of a DFE can be implemented with lowest possible effort.

5.2 Digital Up-/Down-Conversion

Digital up-conversion in the transmit branch and digital down-conversion in the receive branch is done by multiplying the signal with a complex rotating phasor as shown in Figure 8 where the input bandpass signal $s_{BP}(k)$ is down-converted to I-Q base-band. This approach requires two multipliers and a relatively large look-up table holding all samples of the respective sine and cosine functions. Even if hard-wired the structure of Figure 8 is fully parameterizable and can be adapted to any carrier frequency f_0 by simply selecting the appropriate samples of the complex rotating phasor from the look-up table. Still, it is not an efficient approach. Several ways of avoiding the large look-up table have been suggested. One of them is to use the CORDIC algorithm for calculating respective samples of sine and cosine functions.

The CORDIC algorithm [10] is an iterative method for converting between cartesian and polar coordinates by just using additions and shift-operations. Feeding it with a number, it calculates the cartesian coordinates of this number interpreted as an angle *i.e.*, it calculates the sine and the cosine of the angle. Thus, it can be used to calculate the required complex rotating phasor and can replace the look-up table. Still, the CORDIC algorithm can also be used for rotating the coordinates of a vector. This enables to directly feed it with the signal to be up- or down-converted which is shown in Figure 9 for the case of down-conversion. The great advantage of the CORDIC algorithm is that it avoids a large look-up table. Like the conventional approach of Figure 8 it is fully parameterizable and thus, can be adapted to any carrier frequency. For further details on applying the CORDIC algorithm for digital up- and down-conversion see [4].

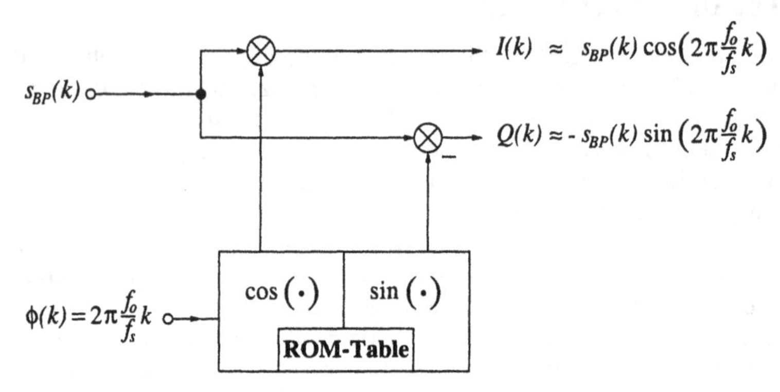

Figure 8. Conventional I-Q Down-Conversion

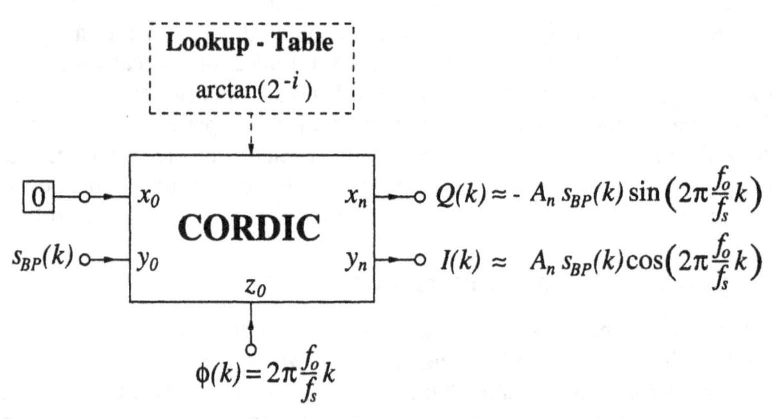

Figure 9. I-Q Down-Conversion with CORDIC

5.3 Filtering

Filtering is required in the DFE for many purposes e.g., for channelization, matched filtering, and pulse shaping. Often, these filtering tasks are intermeshed with decimation and interpolation tasks. Strictly spoken, decimation and interpolation are sample-rate conversion tasks which will be dealt with in the next section. Still, the combination of filtering with decimation or interpolation enables very efficient implementations e.g., as polyphase filters. Otherwise, conventional digital filters must be implemented.

The main problem encountered with digital filters is that the number of coefficients depends on the filter characteristics. Since different air-interfaces usually require different filters it is not possible to have a common digital filter platform with a fixed number of coefficients stored in memory. A way-out to this problem is to calculate the necessary coefficients at the time they are required. This can be done by implementing a description of a continuous-time impulse-response of a filter, and calculating the samples of this impulse response. In the next section this approach will be dealt with in the context of sample-rate conversion.

5.4 Sample-Rate Conversion

Different air-interfaces are based on different chip-/symbol-rates. Thus, it is necessary to process the digital signals at standard-specific clock-rates. The simplest way of obtaining digital signals at certain clock-rates is to clock the A/D converter according to these clock rates. Still, the generation of arbitrary clock signals with high quality (e.g., low jitter) is very costly. Therefore it is sensible to clock the analog/digital interface with a fixed master clock, and to perform digital sample rate conversion (SRC) [3].

SRC is a process of resampling and thus, requires filtering in order to avoid aliasing [3]. As mentioned in the previous section the implementation of a continuous-time impulse response of a filter and the calculation of the required samples of this impulse response is a very flexible solution for digital filtering in the context of software radio. In order to keep the effort low for calculating the samples of the impulse-response, polynomial filters are a favored class of filters. Combining them with SRC and exploiting the polyphase framework leads to the well-known Farrow-structure [8]. For further details on the application of the Farrow-structure to SRC see e.g., [9].

5.5 Implementation Issues

The suggested solutions for digital up-/down-conversion, filtering, and SRC enable an implementation as an ASIC. Since these solutions are fully parameterizable the ASIC – although hard-wired – would be adaptable to arbitrary carrier frequencies, sample rates, and channel spacings, and thus, to different air-interfaces. Another means of implementation are FPGAs providing flexibility that is independent from any algorithm. Thus, FPGAs can serve as a basis for implementing algorithms in software defined applications where parameterization is not sufficient or possible.

6. Conclusion

Work undertaken so far in the SORBAS project leads to the conclusion that the Software Radio philosophy looks promising with respect to an evolutionary approach for practical implementation. A reasonable amount of fundamental Software Radio techniques can be commercially applied today, such as modular algorithms and procedures in the base band processing area, digital IF conversion and filtering, *etc.* It is the objective of the SORBAS project to identify those techniques and to develop them further with respect to applicability in next generation mobile radio base stations and to demonstrate the benefits that can be achieved using such techniques. On the other hand it is clearly not the intention to show the feasibility of a 'pure' Software-Radio solution irrespectively of the required implementation effort being far away from applicability in mid-term future products. Although, technological evolution moves quit fast especially with respect to availability and cost of processing power for chip and symbol rate processing, there are some limitations that will need significant technological improvements before a real broadband software defined radio, standard compliant base station becomes feasible. This mainly concerns A/D-D/A converters that must be able to handle, at reasonable costs, demanding requirements within a base station concerning sample rates, bandwidth, spurious free dynamic range *etc.* as well as the high power transmission path of the analog front where dedicated solutions seem to be unavoidable for the coming years. However this will not block introduction of more and more flexible solutions taking benefit of rapidly advancing signal processing techniques.

7. Acknowledgement

Work on this project is partly funded by BMBF within the UMTSplus system concept.

8. References

[1] Wolfgang König, Gerd Wölfle, Holger Herbig, "Software Radio Based Access System (SORBAS) Base Station," *Proc. 1th Karlsruhe Workshop on Software Radios,* pp. 51-57, Karlsruhe, Germany, March 2000.

[2] T. Hentschel, G, Fettweis, "Sample Rate Conversion for Software Radio", Proc. 1th Karlsruhe Workshop on Software Radios, pp. 13-18, Karlsruhe, Germany, March 2000.

[3] Hentschel, T. Henker, M. and Fettweis, G.: The Digital Front-End of Software Radio Terminals, *IEEE Personal Communications*, August 1999, Vol. 6, No. 4, pp. 40-46.

[4] Löhning, M., Hentschel, T., and Fettweis, G.: Digital Down-Conversion in Software Radio Terminals, *European Signal Processing Conference (EUSIPCO)*, Tampere, Finland, 2000.

[5] H. Herbig, L. Lundheim, N.K. Rossing, T. Hentschel, M.-H. Silly, Y. Rosmansyah, R. Thiruvathirai "The SORT Project-Software Radio Demonstration", *Proc. 1th Karlsruhe Workshop on Software Radios,* pp. 59-67, Karlsruhe, Germany, March 2000.

[6] Joseph Mitola III, "Cognitive Radio: Agent-based Control of Software Radios", *Proc. 1th Karlsruhe Workshop on Software Radios,* pp. 35-43, Karlsruhe, Germany, March 2000.

[7] G. de Boer, R. Mann Pelz, M. Benthin, "Architecture for a Software-Configurable Mobile Terminal based on the GSM, UMTS and HIPERLAN/2 Standard", *Proc. 1th Karlsruhe Workshop on Software Radios,* pp. 45-50, Karlsruhe, Germany, March 2000.

[8] Farrow, C.W.: A Continuously Varying Digital Delay Element. *IEEE Int. Symp.*

on Circuits and Systems (ISCAS) 1988, pp. 2641-2645, Espoo, Finland, June 1988.

[9] Ramstad, T.: Fractional Rate Decimator and Interpolator Design, *European Signal Processing Conference (EUSIPCO)*, Rhodes, Greece, 1998, pp. 1949-1952.

[10] Volder, J. E.: The CORDIC trigonometric computing technique, *IRE Transactions on Electronic Computers*, vol. EC-8, pp. 330-334, September 1959.

DSP-based CDMA Satellite Modem: CNIT/ASI Project

THE WP3 GROUP:, Andrea Conti[1], Annalisa Verdoliva[5], Claudio Sacchi[2], Davide Dardari[1], Fabrizio Argenti[3], Fabrizio Frescura[4], Giacinto Gelli[5], Luca Ronga[3], Mario Tanda[5], Paolo Antognoni[4], Piero Castoldi[6], and Roberto Corvaja[7]

[1] Università di Bologna
[2] Università di Genova
[3] Università di Firenze
[4] Università di Perugia
[5] Università di Napoli
[6] Università di Parma
[7] Università di Padova

Abstract. Multimedia services require a large amount of bandwidth for the delivery of the video and audio signal with adequate quality of service. Hence, the bandwith allocation has to efficiently managed. CDMA is a flexible multiple access technique since it allows a dynamic configuration of the resources allotted to the users of the system. The paper describes a CDMA-based satellite multimedia system, detailing the DSP implementation of the satellite modem.

1 Architecture

Multimedia Services require a large bandwidth to be supported. On costly media, like satellite links, an efficient and flexible radio interface access scheme has to be provided. Within the scope of the ASI/CNIT project "Multimedia Services on Heterogeneous Network connected via Satellite", a CDMA-based satellite network is investigated.

A block diagram of the architecture of the DS/CDMA modem considered in the present dealing is depicted in figure 1.

The modem architecture schematised here can provide a maximum upstream bit-rate equal to 64 Kb/s $\times$ 6 = 384 Kb/s (corresponding to $m = 6$ encoders enabled to transmit).

The multiple access protocol chosen is the fully asynchronous CDMA, with upstream transmission allowed to each user without any bandwidth or time restriction. Such a choice is motivated by the intrinsic asynchronicity of the multimedia application considered for the actual use of the modem (i.e. interactive video-conference). The selected spreading factor N is equal to 63, corresponding to an occupied signal bandwidth equal to about 4.12 MHz.

The hardware architecture has two different sides or sections:

- Custom ASIC devices side;

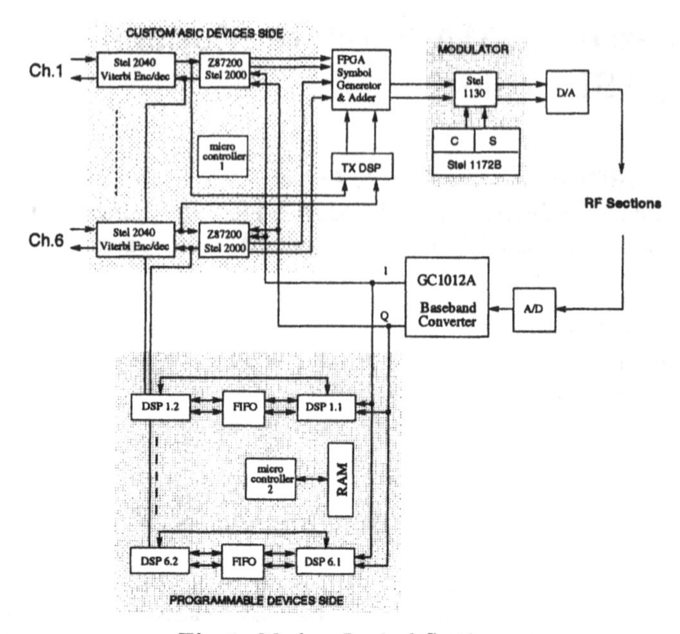

Fig. 1. Modem Logical Sections

- Programmable devices side;

The first side, with a fully operating transmitter and receiver, will allow the development of a high performance, DSP device side. In this way, limited performance single user (conventional) receiver, implemented by the Stel 2000A Spread Spectrum transceiver, will be sensibly improved by the multi user detection algorithms hosted on the DSPs.

As previously stated, the first side is based on the transceiver chip Stel 2000A. This device, based on a differential PSK modulation (binary or quaternary), has a programmable processing gain ranging from 11 and 64 chips. It can operate at a maximum clock frequency FCLOCK=45.056 MHz. The table below summarizes the Stel 2000A performance capabilities.

CLOCK	Proc. gain	BPSK data-rate	QPSK data-rate
45.056 MHz	11	1.024 Mbps	2.048 Mbps
45.056 MHz	64	176 Kbps	352 Kbps

The Stel 2000A is present in each of the 6 channels, together with a convolutional Viterbi encoder/decoder.

The programmable device side, based on DSP, supports both conventional (single user) and multiuser detection. As you can see in the figure 1, 14 DSPs are present in this board:

2 DSPs will be used to give higher flexibility to the transmitter side, (variable length preamble, and other future developments).
12 DSPs, (2 in each of the 6 channels),are employed to implement the multi user algorithms.

The two sides receive the same baseband signal, in order to compare the different performance of the ASICs and the DSPs.

On both sides a μ-controller is available, performing the indicated tasks:

- Custom side: configure the ASIC devices; communicate with a Host PC through a USB or a RS-432 link, and a custom software driver;
- Programmable side: exchange data among the DSP; send and receive data with the custom side

The modem device operates in three different modes: a *custom mode*, a *hybrid mode* and a *DSP mode*. In the first one the transmitter and the receiver are implemented by the Custom side. In the hybrid mode the custom side, and the 2 DSPs in the transmitter, are used to produce a flexible modulated signal while the other DSPs implement a multi user detector. In DSP mode the DSPs work both to transmit and to receive.

2 Code Set and System Capacity

According to the above mentioned constraints about the spreading factor (i.e.,$N = 63$), and to the use of a fully asynchronous CDMA multiple access protocol, the code set choice choice has been performed considering the maximisation of the system capacity. In order to ensure a satisfactory quality of service to the multimedia application to which the modem is addressed (i.e. interactive video-conference), the upper bound for system BER was settled to 10^{-6} after Viterbi decoding, which approximately corresponds to an uncoded BER equal to 10^{-3}.

From a BER evaluation point of view, the modem system represented in figure 1 can be decomposed into independent DS-SS conventional QPSK modulator/demodulator, each of one transmitting information asynchronously over the channel at a bit-rate of 64 Kb/s. An analytical performance evaluation in terms of BER for an asynchronous DS/CDMA system, with M-PSK modulation in reported in [8], using tight upper and lower bounds on the actual BER. Despite the good results yielded, the bounding procedure shown in [8] is very complex. For sake of simplicity it can be replaced by the standard Gaussian approximation on the DS/CDMA BER, whose expression for QPSK-modulated systems can be derived by [8] and [9]:

$$P_b = \mathrm{Q}\left[\frac{1}{\sqrt{2}}\left(\frac{1}{6N^3}\sum_{k=2}^{K}\rho_{k,l} + \frac{N_0}{4E_b}\right)^{-1/2}\right] \tag{1}$$

where: $\frac{E_b}{N_0}$ is the bit signal-to-noise ratio (SNR), $\rho_{k,l}$ is a quadratic term related to the cross-correlation of the PN sequences [8], and hence linked directly to the MAI variance. Further on, the results provided by (1) have been validated through simulations. It can be noticed from (1) a 3dB asymptotic performance degradation factor with respect to a BPSK-modulated DS/CDMA system. This is due to the statistical dependency of complex MAI components affecting the two transmission branches I and Q [8].

From a theoretical point of view, the set of binary 63-length PN codes optimal with respect to the cross-correlation properties is the Kasami set [7], which is able to meet the Welch bound on cross-correlation [7]. Unfortunately, Kasami set is too small for our application as it contains only 8 sequences, whereas the maximal number of users to be served by the satellite modem system is equal to 24 (corresponding to four station transmitting at full rate). For this reason, binary Gold codes, which are sub-optimal in the case of 63-length have been firstly experimented. Gold codes are characterised by a well-known favourable regularity of the cross-correlation pattern, which can assume only three values for each time shift [7]. The cardinality of the Gold set is equal to $N + 2$, meaning in our case that M = 65 codes are available for users' transmission.

Other kind of sequences with favourable cross-correlation properties have been experimented in order to improve the BER performances of the system and therefore its capacity. In particular, the performances provided by four-phases sequences, such the Even-Odd-Equivalent (EOE) Gold sequences [10], and the Frank-Zadoff-Chu (FZC) sequences [5] have been tested. These kind of sequences show very favourable cross-correlation properties, better than binary sequences, in few-user BPSK-modulated DS/CDMA systems.

It can be observed that the use of quaternary PN spreading codes, theoretically characterised by better cross-correlation properties than the binary ones, doesn't provide any improvement in the system capacity (the same result of Gold codes is achieved by EOE-Gold ones, whereas the use of FZC sequences allows one to support the asynchronous transmission of 11 users @ SNR = 20 dB, one user less than Gold and EOE-Gold sequences). This fact is not a surprise: indeed it is known from Spread Spectrum theory that the influence of different PN choice on the performance of a DS/CDMA system decreases as the number of asynchronous users increases. The simulation results about system BER, shown in Table 1, have substantially confirmed the validity of the analytical Gaussian evaluation in the considered case of noticeable number of asynchronous users.

¿From all the consideration made above, it has been stated to consider the use of binary Gold codes for the actual implementation of the DS/CDMA modem, as other PN code sets do not allow one to achieve any significant improvements in the system performances and capacity, although involving an increased algorithmic complexity.

Table 1. PN Code Performances

Number of users	Employed sequences	SNR	BER (simulation)	BER (Gaussian approx)
K = 12	Binary Gold	20 dB	9.3*10-4	8.77*10-4
K = 12	EOE-Gold (4 phases)	20 dB	10-3	7.78*10-4
K = 12	FZC (4 phases)	20 dB	3.3*10-3	2.2*10-3
K = 14	Binary Gold	25 dB	2*10-3	1.9*10-3
K = 14	EOE-Gold (4 phases)	25 dB	2.5*10-3	1.6*10-3
K = 14	FZC (4 phases)	25 dB	4.6*10-3	3.9*10-3

3 Synchronization and tracking

3.1 System description and data frame format

The Multi-user detection techniques that will be proposed in the next section need an input signal without frequency offset and perfectly synchronized. In this section we propose a flexible frame format for the CDMA satellite system under investigation. The proposed frame format allows an efficient timing acquisition, causing a negligible transmission overhead. Moreover, a joint low-complexity scheme for symbol acquisition and tracking with frequency offset recovery is proposed, and some preliminary results are presented.

Canonical synchronization algorithms based on the use of matched filter are not suitable for the hardware platform under development. First, the DSP platform is not able to manage such as complexity required to implement a matched filter working at the sampling rate. Second, the synchronization block has to work before the Multi-User detection block, where the interference level does not allow a reliable synchronization detection. This justifies the transmitted frame structure, shown in Fig. 2, that is composed by a continuous data stream with dedicated synchronization symbols, inserted every N_{synch} symbols, characterized by a different spread code and a power h time higher than for data symbols. As stated by the system specifications, each station has six spreading codes available for the transmission of its data. The codes sent from the same station are synchronous, but, due to the lack of synchronization among the stations, the overall CDMA system must be regarded as asynchronous. The proposed data frame envisions a periodic insertion (every $N_{synch} = (T_{training} + T_{data})/T_s$ symbols, T_s being the symbol interval) of a synchronization symbol, which is the sequence of unmodulated chips allocated to the specified station for the purpose of synchronization. The choice of a Gold sequence with the same spreading factor of the data is necessary to avoid destruction of the cyclostationarity of the received signal and consequently, to avoid a sudden transient in the convergence of the adaptive receiver. Furthermore, the synchronization symbol must be transmitted with a higher power with respect to that devoted to data. The ratio, h, between the power level for synchronization and data symbols must be determined as a trade off between the performance of the correlation acquisition algorithm and the impairment of the induced near-far effect on the data channels.

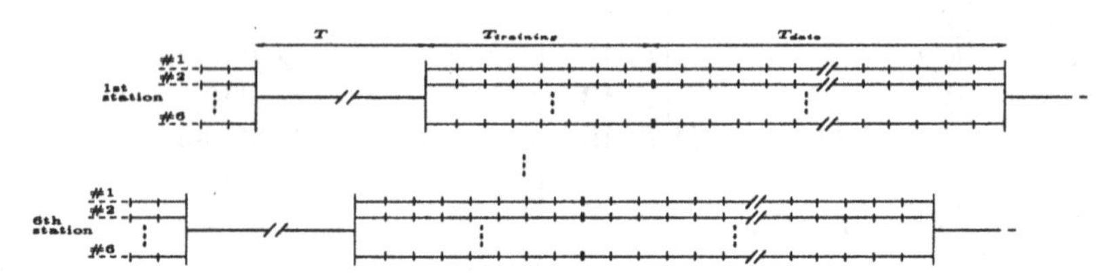

Fig. 2. Data frame with time-multiplexed synchronization symbol, $T = T_s$ is the time interval dedicated to the synchronization symbol

The complex envelope of the received signal, referred to the useful synchronization symbols, can be written as

$$r(t) = \sum_{i=-\infty}^{\infty} \sum_{n=0}^{N-1} A\, c_n\, g(t - nT_c - iN_{synch}T_s) + d(t), \tag{2}$$

where $A = 1 + j$ is the transmitted synchronization symbol, $\{c_n\}$ is the code employed for the synchronization by the useful station, $N = 63$ the spreading factor and $g(t)$ is a rectangular pulse with duration equal to the chip time $T_c = T_s/N$. The component $d(t)$ contains all the contributions due to the other stations, thermal noise and data symbols of the useful station which represent an interference term for the synchronization unit. It could also take into account the nonlinear effects due to High Power Amplifier. In [5,6] is shown that nonlinear disturbance can be considered as an additive noise uncorrelated to other noise and useful components.

3.2 Acquisition

The synchronization scheme has to acquire and track time reference and recover the frequency offset through the detection of the synchronization symbols.

The received signal is sampled each $t_s = \frac{T_s}{N\,F_s}$ seconds where F_s is the oversampling factor.

Each $T_{synch} = N_{synch}T_s$ seconds (the time separation between two synchronization symbols) $NF_s + N_{corr}$ received samples are put into a memory buffer corresponding to a time window slightly greater than one symbol time T_s. During the $T_{synch} - T_s$ remaining time N_{corr} correlations are performed, using different sets of NF_s samples from the memory buffer, each one shifted of one sample.

The generic correlation output, z_k, is

$$
\begin{aligned}
z_k &= AN \sum_{i=-\infty}^{\infty} \Phi(kt_s - iT_{synch}) \\
&+ A \sum_{i=-\infty}^{\infty} \sum_{k,n=0;k\neq n}^{N-1} c_n c_k^* \Phi(kt_s - T_c(n-k) - iT_{synch}) \\
&+ d_{z_k}
\end{aligned}
\tag{3}
$$

where $\Phi(t)$ is the auto-correlation function of $g(t)$ and d_{z_k} is the response of the correlator to the interfering term $d(t)$ for $t = k\ t_s$. The second term of equation (3) also represents the auto-interference due to the spreading code auto-correlation tails (because we are transmitting isolated synch pulses). In the case of heavy loaded system this term could be neglected.

The N_{corr} correlation outputs are processed by a peak detector whose output is $y_k = \frac{|z_k|}{|A|\ N\ \Phi(0)}$. These values y_k are then compared with a suitable threshold th_{acq}. If one of the outputs exceeds this threshold we enter in the tracking mode. If all the outputs do not overcome the threshold, the time reference is shifted forward by N_{corr} samples. The drawback of this technique is a long acquisition time that is not a problem in the application under consideration. This time can be reduced by increasing the number of correlations, N_{corr}, that corresponds to an increasing of the complexity.

Another decision algorithm could use a Maximum-based strategy in which the timing reference is chosen by means a maximum search on the correlation outputs. In this case we can write the False Alarm Probability for the worst-case, where the synchronization symbols of the desired station are transmitted simultaneously to the synchronization symbols of the interfering stations. Then, the false alarm probability P_{FA} is given by the probability that the output of correlator for any value of $k \neq 0$ exceeds the value for $k = 0$, which corresponds to the alignment condition

$$
P_{FA} = \mathrm{Prob}\left\{ \bigcup_{\mathrm{k=1}}^{\mathrm{N_{corr}}} (\mathrm{y_k} > \mathrm{y_0}) \right\}. \tag{4}
$$

In Fig. 3 the false alarm probability is shown as a function of the SNR (signal to thermal noise ratio), with one and six stations transmitting.

The results are evaluated with Gold codes and asynchronous stations, in the worst case condition, where the synchronization symbol of the desired station is transmitted simultaneously to the synchronization symbols of the interfering stations.

As a further investigation, the results in terms of false alarm probability can be obtained with the alignment strategy that the alignment is decided for the most frequent index which happens over M slots. In this case the alignment condition is not decided after the first synchronization symbol is received, but it is delayed after M synchronization symbol are received, that is ML symbols. The false alarm probability can be evaluated with the same assumptions as before, since we assume a static channel, at least over ML symbols.

In the same figure the false alarm probability is shown as a function of the SNR, for decision taken after 1, 4, 8 and 16 slots.

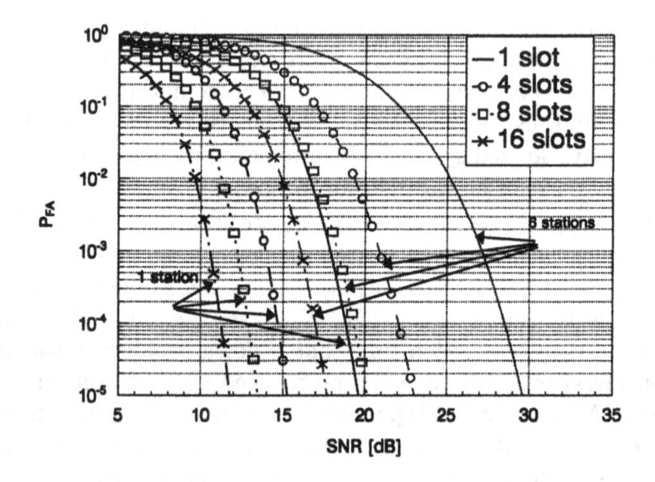

Fig. 3. False alarm probability for different stations simultaneously transmitting

It can be clearly seen the improvement achieved by delaying the synchronization decision, however it is not worthwhile increasing the delay over 4 or 8 slots, since the improvement in not so large afterwards.

3.3 Tracking

The tracking mode is similar to the acquisition one with reference to the correlation processing and differs only in the decision policy. In this case the time reference is shifted in the range $[-N_{corr}/2, N_{corr}/2]$ samples units according to the maximum output level found. If this level becomes below a suitable threshold, th_{tr}, we return to the acquisition mode.

An upper bound of the probability to exit from the tracking mode is

$$P_{tr\rightarrow acq} < 1 - Q(\frac{\sqrt{h}}{\sigma}, \frac{th_{tr}\sqrt{h}}{\sigma}) \tag{5}$$

where $\sigma^2 = E[d_{z_k}^2]/(2|A|^2 N^2 \Phi^2(0))$. $Q(\cdot,\cdot)$ denotes the Marcum Q-function. When the system is interference limited $\sigma^2 \simeq \frac{K-1}{3N}$ where K is the total number of stations.

The choose of the threshold should be a compromise between the probability to exit from the tracking mode and the probability of false tracking, that is, the probability that we remain in the tracking mode when we should come back to acquisition mode. A good approximation of this probability is

$$P_{Ftr} \simeq \exp\left\{-\frac{th_{tr}^2 h}{2\sigma^2}\right\} \tag{6}$$

$P_{tr\to acq}$ and P_{Ftr} slightly depend on N_{corr}, but expressions (5) and (6) have been developed considering only the effect of the higher correlation output. Due to the auto-correlation function behavior, this is the dominant one and the use of the expressions is justified.

The loose of tracking condition is a dangerous event due to the long acquisition time, so it is necessary to keep low the probability to exit from the tracking mode $P_{tr\to acq}$. As will be shown in the results, it is convenient to delay this decision after the observation of two or more synchronization symbols with a peak level below the threshold.

In Fig.4, $P_{tr\to acq}$ and P_{Ftr} against the threshold $th_{tr_{rel}}$ is shown for $K = 16$ and $h = 4, 16$.

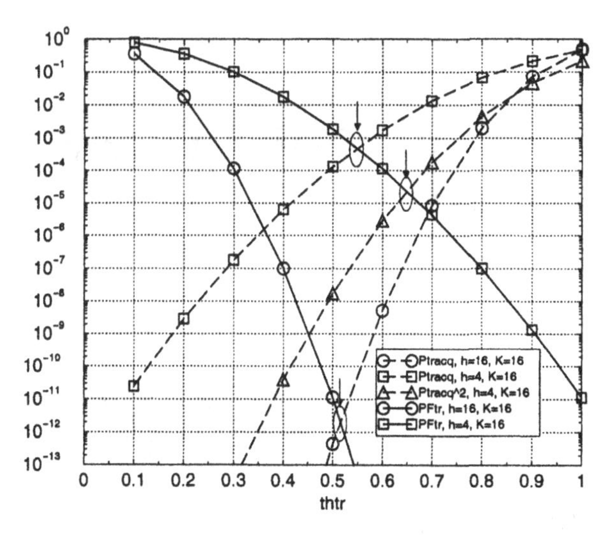

Fig. 4. Optimum value for th_{tr} with respect to $P_{tr\to acq}$ and P_{Ftr}

In this case thermal noise and auto-correlation tail interference are negligible with respect to the multiple access interference. The optimum tradeoff is about $th_{tr} = 0.5$ In the case of a tracking algorithm that delays the decision over two synchronization symbols, the probability to exit from the tracking mode is squared and the optimum tradeoff becomes $th_{tr} = 0.65$ for $h = 4$. As can be noted this simple strategy allows an relevant gain in terms of minimum $P_{tr \to acq}$.

3.4 Frequency offset estimation

As regards frequency recovering we refer to a modified version of the algorithm proposed in [4], here working only during the synchronization symbol and once the tracking mode has been achieved. Only two additional correlations per synchronization symbol are required.

In Fig.5 a transient sample of $max\{y_k\}$ behavior and the frequency offset estimation $\hat{\delta f}$ normalized to the symbol rate are shown as a function of the number of synchronization symbols processed. These curves have been obtained by simulation for $N_{corr} = 63$, $N_{synch} = 1000$, $K = 16$, $h = 16$ and normalized frequency offsets $\delta f T_s = 0.3, 0.5$. In this case the synchronization algorithm reaches the tracking mode after 16 synchronization symbols. The frequency tracking scheme starts when the synchronization section is in tracking mode, and a reliable frequency acquisition is achieved after about 30 synchronization symbols. It has been checked that the frequency tracking algorithms is able to work with frequency deviations in a range of $[-0.5/T_s, 0.5/T_s]$. In the case of $1/T_s = 64$ Kbit/s the tracking window is 64 KHz.

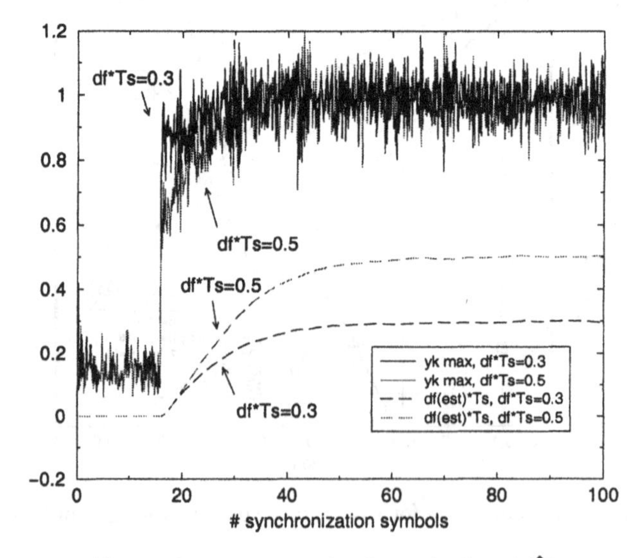

Fig. 5. Transient sample of $max\{y_k\}$ and $\hat{\delta f}$

3.5 Hardware implementation

The complexity of this scheme is very low and essentially depends on the average number of correlations needed for symbol time. In fact we have to perform $N_{corr}+2$ in a T_{synch} time, with could be in the order of hundreds or thousands of the symbol time. Some experimental tests have been carried out on a single DSP board based on Texas Instruments TMS320C62x working at $160MHz$. Fixing $N = 63$, $F_s = 4$, $1/t_s = 16 \cdot 10^6$ (which corresponds to a symbol rate about $64Ksymb/s$), up to 2500 CPU cycles are available for processing if I/O operations are performed via DMA clocked with the sampling frequency. A complex correlation operation can be performed in about 600 CPU cycles using parallel assembler coding. In the case of $N_{corr} =$ 63 and $N_{synch} = 1000$ the algorithms proposed lead to a CPU utilization of about only 1.6%. By increasing N_{corr} the acquisition time can be reduced at the expense of an increasing of CPU utilization. We have to keep in mind that a matched filter synchronization scheme requires one equivalent correlation per sampling time t_s, which is not implementable with actual DSP.

4 CDMA Reception

4.1 Algorithms for data detection and their performance

The system capacity is heavily limited by interference, due to the relatively low processing gain and the asyncronous MAI interference. The use of multiuser detection is necessary in order to maintain the performances at an acceptable level.

Several different multi-user detection schemes have been analysed for the DSP implementation. Both theoretical and simulated results are described here. The performance of these algorithm has been assessed by computer simulation with the aim to validate their performance in the presence of non-ideal working condition such as inaccurate timing. Indeed, this is the case since the acquisition algorithm can not reach a resolution larger than the sampling interval.

4.2 Parallel Interference Cancellation

In this section the performance analysis concerning the application of a single parallel-interference-cancellation (PIC) stage [12] is considered. It is known that the computational complexity of this multi-user receiver increases with the number of stages, and in general it is equal to $O(KN)^m$, where m is the number of detection stages ($m = 1$ corresponds to the conventional matched filter stage), K is the number of active codes and N is the number of bits to be detected. The performance of the PIC algorithm is strongly influenced by the MAI reconstruction performed by each stage [12]. The task of the i-th stage is influenced by: the correct symbol estimation at the $(i-1)$-th stage, the resolution in the estimation of the timing of each user and the quality of the signal amplitude estimation.

Under the reasonable assumptions of absence of near-far effect, perfect knowledge of the amplitudes and AWGN channel, some BER results concerning a single-stage PIC multi-user receiver have been derived. The computer simulations, not reported here, have been carried out for $K = 18$ and $K = 24$ asynchronous transmitting users. The BER characteristics provided by the single-stage PIC receiver satisfies the upper bound of 10^{-3} for SNR >10dB, when $K = 18$ asynchronous users populate the system. For $K = 24$ users, the BER characteristic provided by the single-stage PIC receiver is not strictly compliant with the above mentioned BER requirements; it is quite close to the upper bound of 10^{-3} (i.e. BER = $2.4 * 10^{-3}$ @ SNR = 25 dB). The involved performance degradation in terms of system BER after Viterbi decoding can be eventually accepted in such case. For this reason, it can be concluded that the single-stage PIC receiver can be regarded as a valuable alternative for a sub-optimal low-complexity multi-user detection scheme with respect to other blind and adaptive receivers presented in literature.

4.3 Trained Adaptive Algorithms

The trained adaptive detector does not require knowledge of any spreading sequence but it needs a training sequence (provided by the proposed data frame). Denoting by $a_k(m)$ the k-th user data ($k = 1, 2, \ldots, K$) and by $\mathbf{y}(m)$ the observation vector, the receiver updates the vector $\hat{\mathbf{t}}_k(m)$ of its FIR coefficient according to the following LMS adaptation rule

$$\hat{\mathbf{t}}_k(m+1) = \hat{\mathbf{t}}_k(m) + \mu e_k(m)^* \mathbf{y}(m),$$

where $e_k(m) = a_k(m) - \hat{\mathbf{t}}_k^H(m) \cdot \mathbf{y}(m)$ and

$$a_k(m) = \begin{cases} a_k^{(t)}(m) \text{ in the } \textit{training} \text{ mode} \\ \hat{a}_k(m) \text{ in the } \textit{decision directed} \text{ mode} \end{cases}$$

Soft decision are obtained as $x_k(m) = \hat{\mathbf{t}}_k^H(m) \cdot \mathbf{y}(m)$ and hard limiter provides hard decision $\hat{a}_k(m) = \text{quant}[\ x_k(m)\ ]$.

Fig. 6 reports the simulation results regarding the performance of receiver operating with five asynchronous interfering stations. We can note that the trained detector is robust to timing offset.

4.4 Blind adaptive MMSE detection

The blind adaptive detector requires knowledge of the desired spreading sequence $\mathbf{c}_k$ ($k = 1, 2, \ldots, K$), but it does not need any training sequence. Denoting by $a_k(m)$ the k-th user data and by $\mathbf{y}(m)$ the observation vector,

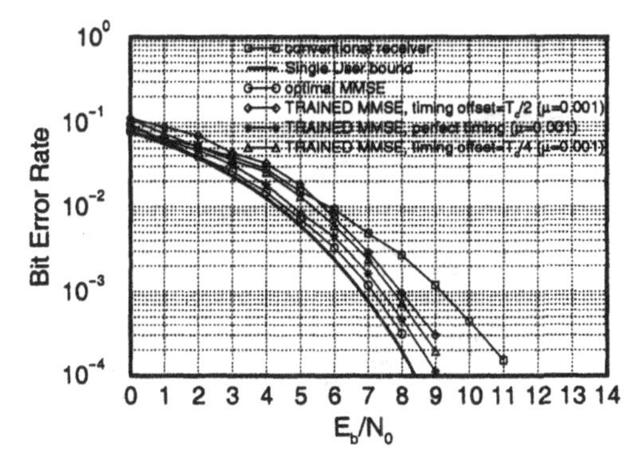

Fig. 6. BER performance of the trained MMSE detector

the receiver updates the vector $\hat{\mathbf{t}}_k(m) = \mathbf{c}_k + \mathbf{x}_k(m)$ $(\mathbf{x}_k^H \cdot \mathbf{c}_k = 0)$ of its FIR coefficient, using the the following LMS adaptation rule

$$\hat{\mathbf{x}}_k(m+1) = \hat{\mathbf{x}}_k(m) - \mu[\hat{\mathbf{t}}_k^T(m) \cdot \mathbf{y}^*(m)]\left(\mathbf{y}(m) - [\mathbf{y}^T(m) \cdot \mathbf{c}_k^*]\mathbf{c}_k\right)$$

which updates only $\mathbf{x}_k$ (since $\mathbf{c}_k$ is fixed and known).

If we test the receiver performance under non ideal condition (e.g. timing offset) a mismatch between the nominal desired code $\tilde{\mathbf{c}}_k$ and the actual code $\mathbf{c}_k$ may appear. To correct this problem, a constraint on the surplus energy $\chi = ||\mathbf{x}_k||^2$ is required, to avoid the cancellation of the useful signal. Especially for high signal to noise ratio a constraint on the surplus energy must be explicited (by trial and error). The modified LMS adaptation rule (χ depends on ν) becomes:

$$\hat{\mathbf{x}}_k(m+1) = \hat{\mathbf{x}}_k(m) - \mu[\hat{\mathbf{t}}_k^T(m) \cdot \mathbf{y}^*(m)]\left(\mathbf{y}(m) - [\mathbf{y}^T(m) \cdot \mathbf{c}_k^*]\mathbf{c}_k\right) - \mu\nu\hat{\mathbf{x}}_k(m)$$

Fig. 7 reports the simulation results regarding the performance of receiver operating with five asynchronous interfering stations. We notice that the blind detector suffers a large performance degradation due to timing offset and that a constraint on the surplus energy is beneficial for the receiver performance specially for high signal-to-noise ratios.

4.5 Robust Multiuser Detection

Recently, a robust multiuser detector (MUD) for (differential) noncoherent reception of DPSK signals in flat-fading non-Gaussian channels, has been proposed in [13]. At the receiver, the resulting complex discrete-time signal corresponding to the ith signaling interval is given by

$$\mathbf{r}_n(i) = \sum_{k=1}^{K} \mathbf{g}_k(i) b_k(i) a_n^k + \mathbf{w}_n(i), \quad n = 1, \ldots, N \tag{7}$$

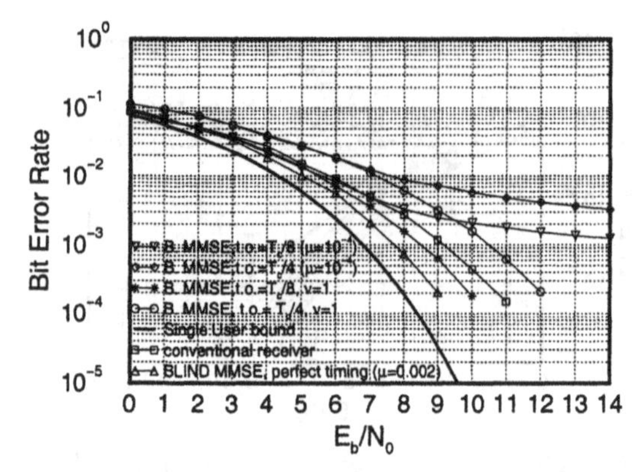

Fig. 7. BER performance of the blind MMSE detector

where N is the processing gain, $a_1^k, a_2^k, \ldots, a_N^k$ is the signature sequence of the kth user and $g_k(i)$ is the kth channel fading complex coefficient. Equation (7) can be written in matrix notation as

$$\underline{r}(i) = \underline{H}\ \underline{\theta}(i) + \underline{w}(i)\,, \tag{8}$$

where $\underline{r}(i)$, $\underline{w}(i)$ and $\underline{\theta}(i)$ are real vectors.

The basic idea of robust MUD is to recover the symbols in (8) by first estimating $\underline{\theta}(i)$, and then extracting the symbols from these continuous estimates. The required estimates of $\underline{\theta}(i)$ are obtained by using a structure belonging to the class of M-estimators proposed by Huber, which minimize a function $\rho(\cdot)$ (called the *penalty function*) of the residuals:

$$\hat{\underline{\theta}}(i) = \arg \min_{\underline{\theta}(i) \in R^{2K}} \sum_{j=1}^{2N} \rho\left(r_j(i) - \sum_{l=1}^{2K} h_{jl}\theta_l(i) \right). \tag{9}$$

Once estimated the vector $\underline{\theta}(i)$, the symbols are given by

$$\hat{b}_l(i) = \mathrm{sgn}\left\{ \Re\left[\hat{\boldsymbol{\theta}}_l(i) \hat{\boldsymbol{\theta}}_l^*(i-1) \right] \right\},$$

where $\hat{\boldsymbol{\theta}}_l(i) \triangleq \hat{\theta}_l(i) + j\hat{\theta}_{l+K}(i)$. Note that when $\rho(x) = x^2/2\beta$ ($\beta > 0$) and the user signature waveforms are linearly independent, the solution of (9) is the well known least-squares estimator and the corresponding detector is a differential form of the standard decorrelator. The results of computer simulations (not reported here) have shown that the robust MUD clearly outperforms the decorrelating detector for moderate and high values of the SNR. Furthermore, as the SNR becomes very large both detectors present the same error floor due to the fading rate.

5 References

1. P. Castoldi and H. Kobayashi, "Low complexity Group Detectors for Multirate Transmission in TD-CDMA 3g Systems", submitted to *Globecom 2000.*
2. T. J. Lim, S. Roy, "Adaptive filters in multiuser (MU) CDMA detection", Wireless Networks 4 (1998), pp. 307-318.
3. S. Haykin, *Adaptive filter theory*, 2nd ed., Englewood Cliffs, NJ: Prentice Hall, 1994.
4. A.Q.Hu, P.C.K.Kwok, T.S.Ng, "MPSK DS/CDMA carrier recovery and tracking based on correlation technique," IEE Electronics Letters, Feb. 1999, vol. 35, no.3.
5. A.Conti, D.Dardari, V.Tralli, "Analytical Characterization of Non-linear Effects in Satellite CDMA Systems," Proc. of Fifth European Conference on Satellite Communications, Nov. 1999, Toulouse, France.
6. A.Conti, D.Dardari, V.Tralli, "An Analytical Framework for CDMA Systems with Nonlinear Amplifier and AWGN," submitted to IEEE Trans. on Communications.
7. D. Sarwate, M. B. Pursley, "Correlation properties of pseudo-random and related sequences", Proceedings of IEEE, Vol. 68, No. 5, May 1980, pp. 593-619.
8. F.M. Ozluturk, S. Tantaratana, A.W. Lam: "Performance of DS/SSMA Communications with MPSK Signalling and Complex Signature Sequences", IEEE Trans. on Comm. Vol. 43, No. 2/3/4, February 1995, pp.1127-1133.
9. M.I. Irdish, I.S. Salous, "Bit Error Probability for Coherent M-ary PSK systems", IEEE Trans. on Comm. Vol. 39, March 1991, pp. 349-352.
10. H. Fukumasa, R. Kohno, H. Imai, "Design of Pseudonoise Sequences with Good Odd and Even Correlation Properties for DS/CDMA", IEEE Journal of Select. Areas of Comm., Vol. 12, No. 5, June 1994, pp. 828-836.
11. D.C. Chu, "Polyphase Codes With Good Periodic Correlation Properties", IEEE Trans on I.T, July 1972, pp. 531-532.
12. S. Moshavi, "Multi-User Detection for DS-CDMA Communications", IEEE Communication Magazine, October 1996,Vol. 10, No6, pp.124-136.
13. H. V. Poor and M. Tanda, "Multiuser detection in fading non-Gaussian Channels," in *Proc. of 33rd Annual Conference on Information Sciences and Systems,* The Johns Hopkin University, Baltimore, MA, March 1999.
14. G. Gelli, L. Paura, and A.M. Tulino, "Cyclostationarity-based filtering for narrowband interference suppression in direct-sequence spread-spectrum systems," *IEEE J. Select. Areas Commun.*, vol. 16, no. 9, pp. 1747–1755, Dec. 1998.

A Round Robin Protocol for the Integration of Video and Data Bursty Traffics in Wireless ATM Networks

Alessandro Andreadis, Giuliano Benelli, Giovanni Giambene, Francesco Partini

Dipartimento di Ingegneria dell'Informazione - Università degli Studi di Siena
Via Roma, 56 - 53100 Siena, ITALY

Abstract: Future wireless systems will allow a mobile access to *Asynchronous Transfer Mode* (ATM) networks. This paper proposes a novel *Medium Access Control* (MAC) protocol based on a round robin policy. A token bucket scheme has been adopted, so that a source enabled to transmit can send a maximum number of packets according to the number of tokens in the related bucket. We have considered *real time-Variable Bit Rate* (rt-VBR) traffics and bursty *Available Bit Rate* (ABR) traffics. The scheduler estimates the congestion of the rt-VBR buffers and accordingly adapts the cycle order. The obtained MAC scheme, called *Adaptive Token Bucket – priority based Round Robin* (ATB-RR), has permitted to obtain satisfactory quality of service levels for both rt-VBR and ABR traffics.

1. Introduction

Today there is an increasing interest for providing multimedia services to mobile users. The scenario envisaged in this paper is a *Wireless Asynchronous Transfer Mode* (WATM) - *Local Area Network* that allows a wireless access to the public ATM infrastructure for users with *Mobile Terminals* (MT), for instance, *Personal Digital Assistants* (PDAs) [1]. Small *Access Points* (APs) provide the coverage over few hundreds of meters (pico- and micro-cells); the frequency band at 5.2 GHz has been considered. A channel bit-rate B_c = 10 Mbit/s has been envisaged in both uplink and downlink according to a frequency division duplexing scheme (a symmetric traffic scenario has been assumed). Each packet (*i.e.*, wireless ATM cell) has a length L = 54 bytes with a payload of L_p = 48 bytes [2].

Several types of *Medium Access Control* (MAC) schemes have been proposed for WATM [2, 3], but the identification of an efficient MAC protocol, able to guarantee suitable *Quality of Service* (QoS) for different traffics, is still an open research issue. Aim of this paper is to propose an effective MAC scheme that is able to manage isochronous and bursty traffics. In particular, we envisage the *real-time Variable Bit-Rate* (rt-VBR) class to support videoconferencing traffic and the *Available Bit-Rate* (ABR) class for WWW surfing traffic. Video cells have a deadline within which they must be transmitted. The WWW surfing traffic has a bursty nature with no special requirement on the cell transfer delay.

In this paper a centralized assignment scheme is proposed where MTs (*i.e.*, traffic sources) are cyclically authorized to transmit by the AP. We use a round robin scheme in order to reduce the variance of the cell transfer delay with respect to a *First-Input First-Output* discipline. Moreover, rt-VBR sources are prioritized with respect to ABR sources. A *token bucket* scheme [4] has been adopted to control the number of cells that a source can transmit at each cycle.

A token bucket scheme is typically used to regulate the data flow (*traffic policer* and *traffic shaper* [5]). In this paper, the scheduler at the AP is based on a token bucket regulator: for each source, tokens are put into a suitable bucket at rate r tokens/s; the bucket has a capacity of B tokens. If the bucket is full, newly arriving tokens are discarded. The cells from a traffic source are queued at the related MT in a suitable buffer. When a given source is enabled to transmit according to the round robin policy, the polling message contains the number n of tokens in the related bucket at the AP. Hence, this source can transmit up to n cells to the AP (see Figure 1). If the number of cells in the buffer exceeds n, the remaining packets will be served in the next cycle. The largest burst a source can transmit to the AP is equal to the bucket size, B. When a traffic source sends the last packet in its buffer or the last packet for which it has the permission to transmit (*i.e.*, tokens are exhausted), this source sets in the header a field that denotes the number of remaining cells in its buffer. This information is used by the AP to adapt the priority order for the service of video sources.

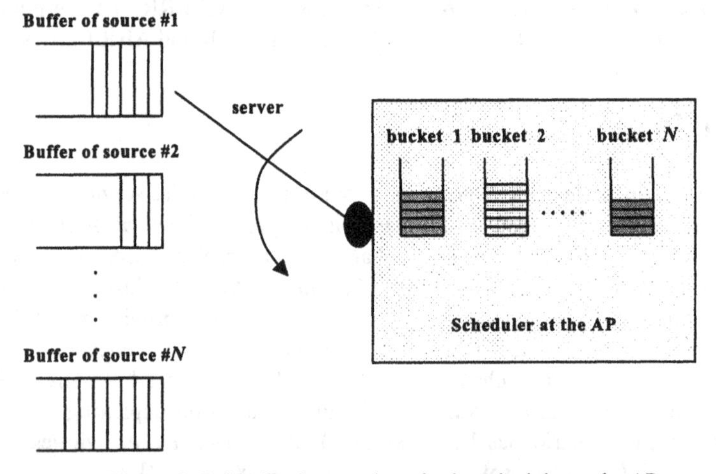

Figure 1. Token bucket regulator in the scheduler at the AP

The rate r is given by the long-term average traffic in the case of an ABR source (*i.e.*, the *Sustainable Cell Rate*, SCR, declared by the ABR source at the session set-up). For video rt-VBR sources, the token rate r is given by a measure of the local cell generation rate taken at the AP on a suitable time window. Indeed, a video source can be modeled as a multi-state source, where a different bit-rate is associated with each state (see Section 2.2) [6]. Since a video source produces a correlated traffic, there are some intervals where it emits at a greater bit-rate than the mean value and other intervals where the produced bit-rate is much lower than the mean value. Hence, the measure taken at the AP permits to estimate the current state of the video source and to adapt the token rate, consequently. The bucket capacity B has been introduced in order to control the service unbalance in favor of a given source; this is important especially for video sources where cells must be served within a deadline. For data sources, we have not considered a specific B value, since they do not require a real-time service.

In our MAC scheme, the AP estimates the congestion of the buffers of video sources; hence, the cycle order is adapted so as to serve first the video sources with a higher buffer congestion [5]. For all these reasons, the proposed MAC scheme has been called *Adaptive Token Bucket – priority based Round Robin* (ATB-RR) protocol.

Our round robin protocol is similar to a scheduler that manages slot allocations according to a variable frame length *Time Division Multiple Access* (TDMA) scheme; hence, comparisons will be made with the results shown in [2].

2. Traffic Models

Within MTs we consider M_v *Videoconferencing Terminals* (VTs) and M_w *WWW Surfing Terminals* (WTs). The models for these traffic sources are detailed below.

2.1 Videoconferencing Terminals

We have considered the model proposed in [6] to characterize videoconferencing sources taking into account the bit-rate variability of the video frame compression algorithm. Each video source is modeled as a *Discrete-time Markovian Arrival Process* (D-MAP). The cell transmission time (*i.e.*, a *slot*) is $T_{slot} = 8L/B_c$.

The bit-rate produced by a VT can be considered as the aggregated output of M independent minisources, each alternating between OFF and ON states. Each minisource in the ON state produces traffic at the constant rate of A bit/s; whereas in the OFF state no traffic is generated. The time intervals (in slots) spent in ON and OFF states are geometrically distributed with parameters $\alpha = 1/p$ and $\beta = 1/q$, where p (and q) is the mean time spent in ON (and OFF) in slots. A minisource in the ON (or OFF) state makes a transition towards the OFF (or ON) state at the end of a slot with probability α (or probability β). The minisource activity factor is $\psi_v = p/(p + q)$. We have considered that in a slot at most one minisource can make a transition from ON to OFF or vice versa. Hence, there are not sudden traffic variations for a VT (*i.e.*, videoconferencing source case). A VT can be modeled through the discrete-time Markovian modulating process described in Figure 2. Parameters p , q and A of a minisource can be obtained as follows [6]:

$$A = \frac{\mu}{M} + \frac{\sigma^2}{\mu}\left[\frac{bit}{s}\right],\ q = \frac{1}{aT_{slot}}\left(1 + \frac{M\sigma^2}{\mu^2}\right)[s],\ p = \frac{1}{aT_{slot}}\left(1 + \frac{\mu^2}{M\sigma^2}\right)[s] \quad (1)$$

where μ is the mean bit-rate produced by a VT, σ^2 is the variance of the bit-rate produced by a VT and parameter a characterizes the slope of the autocovariance function of the bit-rate produced by a VT (if a decreases, a more correlated cell generation process is obtained).

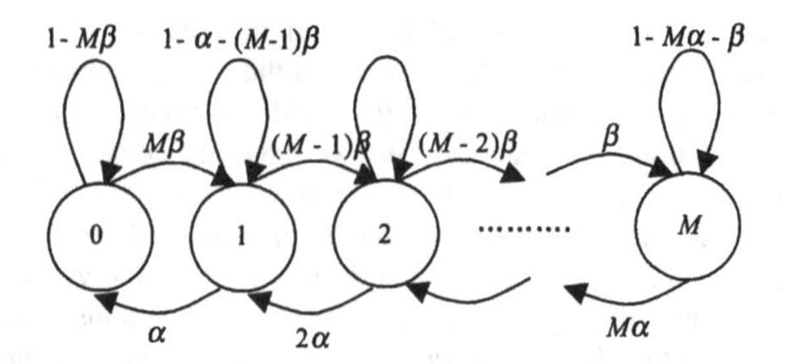

Figure 2. Modulating process for the cell generation of a VT (D-MAP model)

The state probability of the modulating process in Figure 2 is binomial:

$$Prob.\{state = i\} = \binom{M}{i} \psi_v^i (1-\psi_v)^{M-i} \quad i = 0, \ldots, M \tag{2}$$

In this model, a bit-rate iA is produced when the VT is in the state i. Hence, the mean bit-rate produced by the VT (*i.e.*, SCR) is $\mu = M\psi_v A$ bit/s and the *Peak Cell Rate* (PCR) is MA. We can consider $5 \leq M \leq 10$ to model a single video source [6]. We have assumed: $\mu = 512$ kbit/s, $\sigma = 256$ kbit/s, $M = 10$ minisources/VT and $a = 3.8\ s^{-1}$ (also lower a values are considered in [2, 6] for videoconference). On the basis of these values, the minisource activity factor ψ_v is equal to 0.28.

The video service belongs to the rt-VBR traffic class: if a video cell is not transmitted within D_{VMAX}, it is discarded at the source. The video *Cell Loss Ratio* (CLR) can not exceed a maximum value, CLR_{MAX}. The following requirements have been assumed: $D_{VMAX} = 90$ ms and $CLR_{MAX} = 10^{-4}$ [6, 7].

2.2 WWW Surfing Terminals

We refer to the WWW browsing traffic model proposed in [8]. In particular, we consider a simplified model (with heavier load conditions), where a WWW browsing session follows immediately the previous one. An MT generating WWW traffic produces *packet calls* separated by *reading times*. The number of datagrams generated per packet call is geometrically distributed with expected value $m_{Nd} = 25$; the datagram interarrival time is exponentially distributed with mean value $m_{Dd} = 0.125$ s. The reading time between packet calls is exponentially distributed with mean value $m_{Dpc} = 4$ s. The WT activity factor $\psi_w = m_{Nd} \times m_{Dd} / (m_{Nd} \times m_{Dd} + m_{Dpc}) \approx 0.439$. Each datagram has a length in bytes $l_{w\_byte} = \lfloor x \rfloor$, where $\lfloor . \rfloor$ is the *floor function* and x is a random variable with the following truncated Pareto *probability density function* (pdf) with parameters $\gamma = 1.1$, $k = 81.5$ bytes and $m = 6666$ bytes:

$$pdf(x) = \frac{\gamma k^{\gamma}}{x^{\gamma+1}} [u(x-k) - u(x-m)] + \omega\, \delta(x-m) \tag{3}$$

where $u(.)$ is the unitary step function and $\omega = (k/m)^{\gamma}$.

Datagrams are packetized with a payload L_p = 48 bytes [2]. From (3), the distribution of the datagram length in cells, l_w, is:

$$Prob.\{l_w = j\ cells\} = \begin{cases} 1-\left(\dfrac{k}{jL_p+1}\right)^{\gamma}, & j=2 \\ \left(\dfrac{k}{(j-1)L_p+1}\right)^{\gamma} - \left(\dfrac{k}{jL_p+1}\right)^{\gamma}, & 2<j<L_{w\_max} \\ \left(\dfrac{k}{(j-1)L_p+1}\right)^{\gamma}, & j=L_{w\_max} \end{cases} \tag{4}$$

where $L_{w\_max} = \lceil m/L_p \rceil$ and $\lceil . \rceil$ is the *ceiling function.*

According to distribution (4), the mean datagram length in packets is $L_w \approx 8.2$ cells and the mean squared length of a datagram is $L_{wq} \approx 343$ cells^2. The QoS parameter considered for WTs is the mean *Cell Transfer Delay* (mean CTD).

The burstiness degree of a traffic source is given by the peak-to-mean traffic ratio. A VT has a burstiness degree $\mathcal{B}_v = 1/\psi_v$; whereas a WT has a burstiness degree $\mathcal{B}_w = 1/\psi_w$. For the values assumed in this paper we have: $\mathcal{B}_v \approx 3.5$ and $\mathcal{B}_w \approx 2.27$; hence, a VT produces a more bursty traffic than a WT.

2.3 System Stability

Under the assumption of stability, the total data throughput, η_w, must be equal to the input traffic: $\eta_w = \psi_w T_{slot} M_w L_w / m_{Dd}$ erlangs. Whereas, the video throughput, η_v, must account for the cell rate $M_v \mu/(8L_p)$ produced by M_v VTs and for the loss of video cells with probability *CLR*. Since VTs and WTs use the same resources, the sum of η_v and η_w (thereafter referred to as η_{tot}) must be lower than one erlang:

$$\eta_{tot} = M_v \frac{\mu}{8L_p}(1-CLR)T_{slot} + \frac{\psi_w}{m_{Dd}} L_w M_w T_{slot} < 1 \quad erlangs \tag{5}$$

This is the stability condition to be used for a proper sizing of the *Connection Admission Control* (CAC) protocol. Moreover, assuming a good management of video sources, we can neglect the video *CLR* in (5) and we can derive an upper bound for the traffic produced by WTs as a function of M_v, μ, L_p, T_{slot}.

3. ATB-RR Protocol Description

We consider the following taxonomy for the MAC protocols:

1. *Fixed access* protocols that grant permission to send only to one terminal at once, avoiding collisions of messages on the shared medium. Access rights are statically defined for the terminals.

2. *Demand-adaptive* protocols that grant the access to the network on the basis of requests made by the terminals. This class encompasses reservation and token-based schemes.
3. *Contention-based* protocols that give transmission rights to several terminals at the same time. This policy may cause two or more terminals to send simultaneously and their messages to collide on the shared medium.

Fixed access schemes are not efficient with bursty traffics, because they can not adapt to varying traffic conditions. Moreover, contention-based schemes are not adequate to manage correlated and heavy traffics. An advantage of demand-adaptive protocols over most contention-based ones is the existence of an upper bound on the transmission delay. Hence, in the presence of video correlated traffics and data bursty traffics, we propose a demand-adaptive protocol that takes advantage of the variability of the video source generation to multiplex also data traffic when capacity is available. In particular, we envisage a round robin scheme where the AP cyclically enables the transmissions of MTs by sending them a polling message that permits to transmit up to a maximum number of cells (*i.e.*, the number of tokens in the related bucket). If the MT has no cell to transmit, it immediately sends a packet with a release command so that the AP consequently transmits the polling message to the next MT in its polling list.

The requirements for an efficient polling scheme are: (*i*) a negligible round trip propagation delay with respect to the cell transmission time; (*ii*) low overhead due to polling messages; (*iii*) a stable traffic stream produced by sources. In our case, previous point (*i*) is fulfilled due to the use of pico- and microcells and high bit-rate transmissions (in our numerical examples, $T_{slot} \approx 4.32\ 10^{-5}$ s, whereas the maximum round trip propagation delay within a cell of radius 300 m is about equal to $2\ 10^{-6}$ s). Moreover, points (*ii*) and (*iii*) are met, because VTs produce a varying but generally heavy traffic (due to the priority order, the probability that the VT has an empty buffer when it is served is negligible). At the end of the *k*-th cycle the scheduler at the AP evaluates the parameters in Table 1.

Table 1. Description of the round robin variables

Symbol	Definition
T_i^k	Cell generation rate estimated for the *i*-th source in the *k*-th cycle
R_i^k	Residual number of cells in the buffer of the *i*-th source at the end of the *k*-th cycle (*i.e.*, the number of cells left in the buffer due to the exhaustion of tokens)
τ_i^k	Time elapsed from the service end of the *i*-th source in the *k*-th cycle to the beginning of the new *k*+1-th cycle
t_i^k	The service time for the *i*-th source in the *k*-th cycle
A_i^k ($\tilde{A}_i^k$)	Number of cells transmitted (generated) by the *i*-th source in the *k*-th cycle
δ_i^k	The interrogation time for the *i*-th source in the *k*-th cycle (*i.e.*, all overhead related to the *i*-th source in the *k*-th cycle)

According to Table 1, we have:

$$t_i^k = T_{slot}\ A_i^k \quad [s] \tag{6}$$

where $T_{slot} = 8L/B_c$ is the cell transmission time.

Source indexes are organized so that index i corresponds to VTs for $1 \le i \le M_v$ and corresponds to WTs for $M_v + 1 \le i \le M_v + M_w$. According to Table 1, the duration of the n-th cycle, σ^n, and the time elapsed up to the k-th cycle, Σ^k, are:

$$\sigma^n = \sum_{i=1}^{M_v+M_w} t_i^n + \delta_i^n \quad [s] \quad \text{and} \quad \Sigma^k = \sum_{n=1}^{k} \sigma^n \quad [s] \tag{7}$$

The cell rate T_i^k is estimated by the scheduler for each source, as

$$T_i^k = \begin{cases} \dfrac{\sum_{n \in \Omega_W^k} \tilde{A}_i^n}{\Gamma_\Delta^k}, & \text{for VTs} \\ \dfrac{\sum_{n=1}^{k} \tilde{A}_i^n}{\Sigma^k}, & \text{for WTs} \end{cases} \quad \left[\frac{cells}{s}\right] \tag{8}$$

where Γ_Δ^k is the time on which the generation rate is evaluated and depends on the cycle index k and Δ, the mean time for which the video source emits at a constant rate, as follows: $\Gamma_\Delta^k = \sum_{n=h}^{k} \sigma^n \ : \ \sum_{n=h}^{k} \sigma^n \le \Delta < \sum_{n=h-1}^{k} \sigma^n$, $h \le k$. Moreover, Ω_W^k is the set of cycle indexes that characterize the time Γ_Δ^k (*i.e.*, from k back to h).

Parameter Δ is the *coherence time* of a video source and, according to Section 2.2, it is the mean sojourn time in a state of the chain in Figure 2:

$$\Delta = \sum_{j=0}^{M} \frac{1}{j\alpha + (M-j)\beta} \, Prob.\{state = j\} \quad [s] \tag{9}$$

In our numerical examples, according to the parameter values assumed for VTs, we have obtained $\Delta = 0.065$ s for $a = 3.9\ s^{-1}$. The AP updates at end of the k-th cycle all the parameters in Table 1 for each traffic source on the basis of (6)-(9). Hence, the AP uses T_i^k as the token generation rate for the sources during the k+1-th cycle. Moreover, the AP computes the priority order P_i^{k+1} for VTs in the k+1-th cycle as:

$$P_i^{k+1} = R_i^k + T_i^k \tau_i^k, \quad 1 \le i \le M_v \tag{10}$$

In the k+1-th cycle, VTs are served in order of decreasing P_i^{k+1} value. Note that parameter P_i^{k+1} represents an estimation[1] (made by the AP) of the number of cells in the buffer of the i-th VT at the beginning of the k+1-th cycle. According to this priority order, we try to serve first the VT buffer with a greater congestion, which is expected to be the most critical case for the fulfillment of cell deadlines.

After having served all VTs, the scheduler enables the transmission of WTs according to the following criterion. Due to the maximum bucket size B_v of each VT, at most B_v M_v video cells can be transmitted in a cycle. In order for the video traffic to experience a QoS insensitive to the presence of data traffic, we consider that WTs can be served in the k+1-th cycle at most for the time B_v M_v T_{slot} - $\sum_{i=1}^{M_v}\left(t_i^{k+1}+\delta_i^{k+1}\right)$. We have not considered a priority order among WTs (data cells have not stringent delay constraints to be fulfilled). Hence, the AP enables the transmission of a WT by sending it the number of tokens, which is the minimum between the token bucket value and the remaining cells to reach the maximum cycle duration of B_vM_v cells. If not all the WTs can be served in a cycle, the service order is restared in the next cycle from the interruption point.

A rough estimate of the optimal choice of the maximum token bucket size B_v for VTs can be given by imposing that the maximum cycle duration does not exceed the video cell deadline (hence, it is possible to serve in time a cell arrived at a VT soon after this VT has released the service):

$$B_v M_v T_{slot} = D_{VMAX} \tag{11}$$

With the assumed parameter values, we have B_v = 300 cells for M_v = 7 VTs.

4. ABR Performance Analysis

Owing to the extreme complexity of the video source model, we limit this analytical study to the case of ABR data traffic only: resources are shared among all the WTs by means of a round robin policy and there is no constraint on the maximum cycle duration. Let us neglect all overheads. If the arrival process at each WT was Poisson distributed, the system behavior could be studied by means of a global *M/G/1* queue (the Pollazchek-Khintchine formula does not depend on the service discipline [9]). In our case, we still consider a global queue, but the total arrival process due to WTs is not Poisson distributed. We approximate the datagram arrival process of a WT by means of a 2-state *Markov-Modulated Poisson Process* (MMPP); this is possible if we make an exponential approximation for the time a WT sojourns in the packet call state. Accordingly, we have adopted the following model to study the system: $\sum_{M_w} 2 - MMPP^{[P]}/D/1$,

[1] This estimation is reasonable since the traffic produced by a video source is correlated (see Figure 2).

where $\sum_{M_w} 2-MMPP^{[P]}$ stands for the aggregation of M_w WTs each of them approximated by a 2-state *MMPP* arrival process of datagrams with truncated Pareto distribution, D is the deterministic cell service time, "1" means that only one cell can be transmitted at once.

In order to study this system we have modified the approach proposed in [10], by taking into account that cells have a compound arrival process due to both the generation process of datagrams and their variable length in cells. Hence, the mean *CTD* can be obtained as follows [10]:

$$mean\ CTD = \left\{1 + \frac{\eta_w\left(1-\frac{1}{M_w}\right)+\frac{\lambda_1''(1)M_w}{\eta_w}}{2\left[1-\eta_w\right]} + \frac{\xi_{M_w}'(1)}{\eta_w}\right\}T_{slot} \quad [s] \tag{12}$$

where η_w is the traffic intensity produced by WTs and parameters $\lambda_l''(1)$ and $\xi_{M_w}'(1)$ have complex expressions that can be found in [10].

5. Results

The ATB-RR scheme has been implemented in OPNET® [2]. Table 2 shows the values used for numerical evaluations. Very long simulation runs have been performed to achieve reliable results. System stability has been verified by (5).

Table 2. System parameter values

Parameter	Value
Channel bit-rate (both uplink and downlink)	B_c = 10 Mbit/s
Packet format	Payload/cell length = 48/54 bytes
Number of minisources (video)	M = 10 minisources/VT
Mean video source bit-rate	μ = 512 kbit/s
Standard deviation of the bit-rate (video)	σ = 256 kbit/s
Autocovariance parameter (video)	$a = 3.9\ s^{-1}$
Mean reading time of the WWW browsing source	m_{Dpc} = 4 s
Mean number of datagrams per packet call	m_{Nd} = 25 datagrams/packet call
Mean datagram interarrival time in the packet call	m_{Dd} = 0.125 s
Maximum distance in a cell from the AP	300 m
Polling message length	6 bytes

Figure 3 presents the comparison between simulation and analytical results with ATB-RR in the case of WTs, where the channel bit-rate B_c is reduced to 220 kbit/s. According to (5), the maximum number of WTs is 17 in this case. These results show an acceptable agreement that improves with the number of WTs. Indeed, the

[2] OPNET® is a registered trademark by *OPNET Technologies, Inc.*

greater the number of WTs, the better the approximation of the total arrival process as sum of 2-state *MMPP* arrival processes.

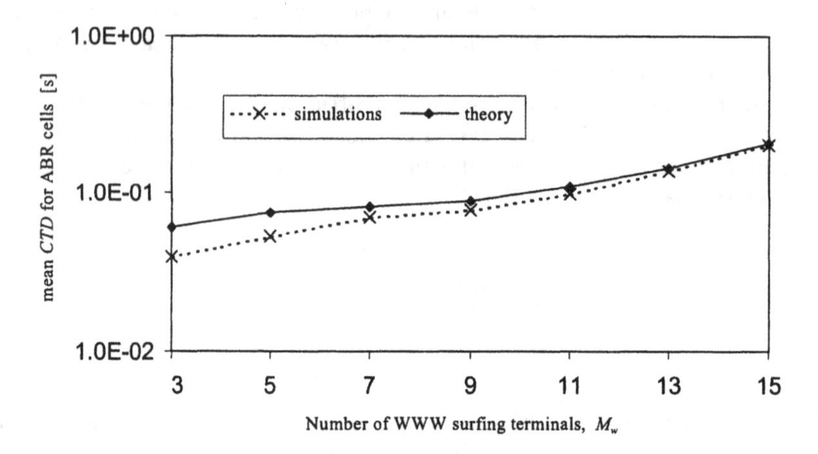

Figure 3. Comparison between theoretical and simulation results (WWW surfing traffic)

Figure 4 presents the video *CLR* comparison between ATB-RR, PRAS (*Prioritized Regulated Allocation Scheme*) [2] and a simplified version of ATB-RR, denoted as TB-RR (*Token Bucket-Round Robin*), where no adaptive priority is used for VTs. These results show that the TB-RR scheme can not guarantee a good performance. Whereas, the ATB-RR protocol achieves a performance improvement with respect to PRAS. It is worth noting that the ATB-RR protocol allows a simple implementation and that the video *CLR* performance strongly depends on the choice of parameter *a* (*i.e.*, lower video *CLR* values are expected if *a* increases).

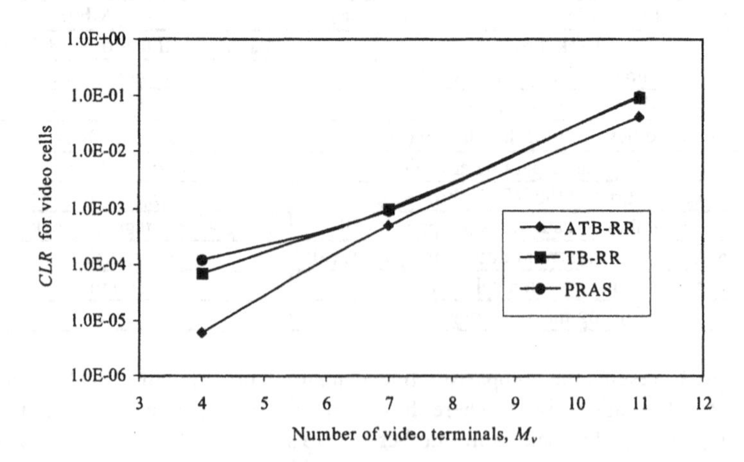

Figure 4. Performance comparisons (video traffic)

Finally, Figs. 5.3 and 5.4 respectively show the video CLR and the mean data CTD as functions of the number of WTs for M_v = 4 VTs. We note that the video traffic QoS is slightly sensitive to the presence of WWW browsing traffic. Moreover, the ABR traffic is managed with short delays if the number of WTs is sufficiently low.

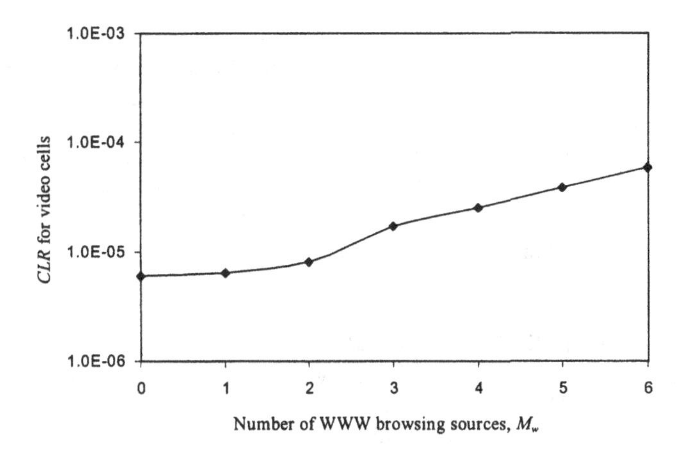

Figure 5. Video CLR as a function of the number of WTs, M_w, for $M_v = 4$ VTs

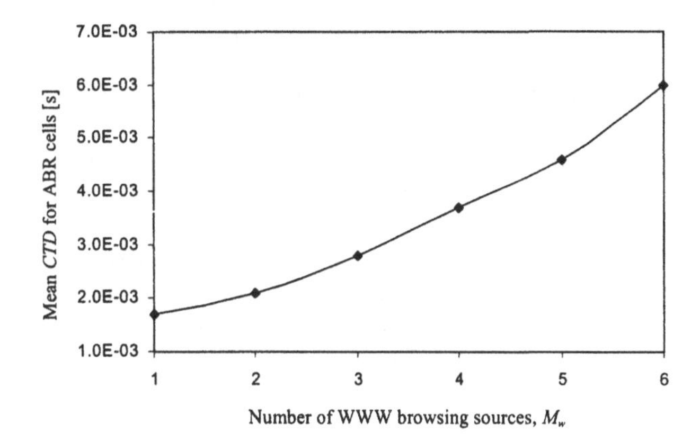

Figure 6. Mean *CTD* as a function of the number of WTs, M_w, for $M_v = 4$ VTs

6. Conclusions

This paper has proposed a novel round robin MAC scheme, called ATB-RR, to support video sources and data bursty traffics in a WATM network. A scheduler has been envisaged that uses a token bucket scheme to determine the maximum service time for a given source at each cycle and that adapts the service order of video sources on the basis of their estimated congestion. The preliminary results shown in this paper have highlighted that the ATB-RR scheme outperforms the

PRAS scheme and permits to manage efficiently video sources also in the presence of data bursty traffics.

7. References

[1] Osama Kubbar and Hussein T. Mouftah, "Multiple Access Control Protocols for Wireless ATM: Problems Definition and Design Objectives", *IEEE Comm. Mag.*, pp. 93-99, Nov. 1997.

[2] N. Passas, L. Merakos, D. Skyrianoglou, F. Bauchot, S. Decrauzat, "MAC Protocol and Traffic Scheduling for Wireless ATM Networks", *Mobile Networks and Applications* (1998) pp. 275-292.

[3] J. Sanchez, R. Martinez, M. W. Marcellin, "A Survey of MAC Protocols Proposed for Wireless ATM", *IEEE Network*, vol. 11, no.6, Nov.-Dec. 1997, pp. 52-62.

[4] J. NG, "On Traffic Burstiness and Priority Assignment for the Real-Time Connections in a Regulated ATM Network", *IEICE Trans. Comm.*, Vol. E82-B, No. 6, pp. 841-850, June 1998.

[5] O. Altintas, Y. Atsumi, T. Yoshida, "Urgency-Based Round Robin: A New scheduling Discipline for Multiservice Packet Switching Networks", *IEICE Trans. Comm.*, Vol. E81-B, No. 11, pp. 2013-2021, November 1998.

[6] C. Blondia and O. Casals, "Performance Analysis of Statistical Multiplexing of VBR sources", *Proc. of INFOCOM'92*, pp. 828-838.

[7] "WAND design requirements", CEC Deliverable, n. 1D3, available at http://www.tik.ee.ethz.ch/ ~wand.

[8] ETSI, *Selection Procedures for the Choice of Radio Transmission Technologies of the UMTS (UMTS 30.03 Version 3.1.0)*; ETSI, Sophia-Antipolis, Cedex, France, Nov. 1997.

[9] L. Kleinrock, *Queuing Systems*. J. Wiley & Sons, N.Y., 1976.

[10] B. Steyaert, H. Bruneel, G. H. Petit and D. De Vleeschauwer, "A Versatile Queueing Model Applicable in IP Traffic Studies", *COST 257 Project, TD* (00)-02, Barcelona, January 2000.

An Efficient Frequency Sampling Design of FIR Raised-Cosine Filters for Software Radio Applications

Marina Mondin, Letizia Lo Presti, Massimiliano Laddomada

Dipartimento di Elettronica, Politecnico di Torino,
C.so Duca degli Abruzzi, 24, 10129 Torino (Italy).
email:{mondin,lopresti,laddomada}@polito.it

Abstract: An important requirement in the design of data transmission filters is the minimization of intersymbol interference (ISI), which is zero if the overall impulse response (transmitter filter, channel and receiver filter) satisfies the Nyquist criterion, i.e. it has uniformly spaced zero–crossings.

An important class of transfer functions, satisfying the Nyquist criterion, is the so–called raised–cosine filter family. In this paper, a design method is given to find the coefficients of couples of linear–phase transmitter/receiver FIR filters, that cascaded have raised–cosine amplitude.

The design is based on Frequency Sampling techniques, and the filter parameters are chosen in order to obtain maximum stopband attenuation. The filter coefficients can be easily evaluated and the optimal filter parameters can be obtained with tables. The design method is very simple and suited for non–filter–oriented users.

The main advantages of this design approach concerns its simplicity, flexibility, and efficiency by which the receiver and transmitter filters can be generated. Furthermore, the design method may be efficiently developed in the software radio systems, based on architectures using general-purpose processors, such as DSP or FPGA platforms.

1 Introduction

One of the main advantages of software radio terminals consists in the fact that, replacing analog with digital signal processing, it's possible to reconfigure the systems via software. Thereby, with a dedicated hardware, such as FPGA or DSP, any filter, which realizes a generic transmit waveform, may be completely reconfigured according the requirements of the current transmission standard ([3],[4],[5]).

Different requirements are imposed to the transmitter and receiver filters in the design of a data transmission system: the transmitter filter is used to band–limit the signal spectrum to the Nyquist bandwidth, while the receiver filter must reject the out–of–band noise or the side–channels, and it needs therefore to have high stopband attenuation.

The stopband attenuation is in fact a very important constraint, and it must be minimized in order to reduce the interchannel interference (ICI). Besides, the cascade of the transmitter and receiver filters must satisfy the Nyquist criterion, in order to avoid intersymbol interference (ISI). Various techniques have been proposed for designing digital filters satisfying the Nyquist criterion. A very common solution is the use of FIR filters, which have been successfully designed using linear programming (LP) techniques ([6] and [7]) in [8], [9] and [10], obtaining filters with equiripple stopband behaviour. In particular, filters designed in [8] and [10] can be split into two low-pass (Tx and Rx) filters. Although linear programming is a very flexible and powerful technique for designing digital filters it does suffer from some numerical ill-conditioning problems for high-order filter design. In ([10]) a simple modification to the standard linear programming approach to FIR filters design has been proposed, which avoids the necessity for a dense grid of frequency points. However, the design method can not guarantee the convergence of the proposed algorithm, and such technique assures that the designed filters are optimal only in the sense that they achieve the maximum possible stopband attenuation for a given filter order and stopband edge frequency. A valid alternative to the use of linear programming methods has been proposed in [11], in which an iterative technique for designing equiripple FIR Nyquist filters using a multistage structure is shown. The multistage implementation shown in [11] is very efficient, but the resulting filter cannot be easily split into a transmitter/receiver filters pair, because it is not generally verified that the two separate filters can meet the ICI specifications. Design methods for FIR structures are very common because FIR filters can be easily constrained to have linear phase. On the other side, FIR structures generally require a large number of multipliers to meet the design specifications. Some alternative solutions able to avoid this problem can be found in literature.

In this paper linear phase FIR structures have been considered. The considered evaluation criterion is the sidelobes amplitude in the stopband (which characterizes the amount of ICI). A design procedure will be described, able to give the filter parameters either with simple formulas and design charts, or with a fast computer program. The proposed design method controls the stopband attenuation of the global transfer function, and it constraints the stopband attenuation of both the transmitter and the receiver filters as well. This is a very important requirement, especially when the transmission channel is characterized by a low signal-to-noise ratio. The design method gives a very efficient solution and, with respect to other techniques, has a number of advantages and features. First, it is guaranteed to yield a solution and there are no convergence problems. Furthermore, with respect to the method proposed in [10], it involves a much smaller computational effort because both transmitter and receiver filters have a symmetric impulse response; this latter characteristic allows a reduction of 50% in the number of multipliers. The FIR filters are inherently linear phase filters and so there is no need of group delay equalizers. Finally, with respect to the ICI minimization, we give a design chart whose points may be stored in a ROM memory. In this way, by a programmable platform it is possible to retrieve the optimum combination of parameters in order to design a very efficient filter with a very small number of taps.

2 The Nyquist Criterion in the Digital Domain

Denoting by $h[n]$ the overall system impulse response, the Nyquist criterion in the time domain states that

$$h[n] = \begin{cases} A & \text{if } n = n_0 \\ 0 & \text{if } n = n_0 \pm kN_s,\ k = 1, 2, \ldots \end{cases} \tag{1}$$

where A is a nonzero constant, n_0 is the discrete sampling instant and N_s is the number of samples per period. N_s represents the oversampling factor, as, being T the sampling interval, F the sampling frequency and R_s the baud rate (in symbols per second), we have $F = 1/T$, $R_s = 1/N_sT$, and $N_s = F/R_s$. The Nyquist criterion can be expressed in the frequency domain as well: being $H\left(e^{j2\pi fT}\right)$ the z–transform of $h[n]$ evaluated for $z = e^{j2\pi fT}$, and $H_T\left(e^{j2\pi fT}\right)$ a function equal to $H\left(e^{j2\pi fT}\right)$ for $-\frac{1}{2T} < f < \frac{1}{2T}$ and zero elsewhere, the following condition must hold

$$\sum_{k=-\infty}^{+\infty} H_T\left(e^{j2\pi(f-k/N_sT)T}\right) = AN_s \tag{2}$$

(the frequency range $\left(-\frac{1}{2T}, \frac{1}{2T}\right)$ will be called in the following the "Nyquist bandwidth"). Note that the transfer function has an odd symmetry around the point $\left(\frac{1}{2N_sT}, \frac{AN_s}{2}\right)$.

From equations (1) and (2), we can see that the Nyquist criterion in the digital domain is practically the same as in the analog domain, therefore the digital filter design can be performed starting from analog results, i.e. digitizing some well–known analog Nyquist filters.

3 Raised–Cosine Filters

It is well–known from the analog filter theory that the so called raised cosine transfer function [1], denoted by $R(f_a, \rho, T_s)$, where f_a is the analog frequency, ρ the roll–off, and T_s the symbol period of transmission (i.e. $T_s = N_sT$), satisfies the Nyquist criterion. The expression of a raised–cosine function is

$$R(f_a, \rho, T_S) = \begin{cases} 1 & \text{for } |f_a| \le f_1 \\ \cos^2\left[\frac{\pi}{4\rho}(2|f_a|T_s - 1 + \rho)\right] & \text{for } f_1 < |f_a| \le f_2 \\ 0 & \text{elsewhere} \end{cases} \tag{3}$$

where $f_1 = \frac{1-\rho}{2T_s}$ and $f_2 = \frac{1+\rho}{2T_s}$. The function is commonly shared between transmitter and receiver filters, as $R^\alpha(f_a, \rho, T_s)$ and $R^{1-\alpha}(f_a, \rho, T_s)$, where α is chosen in order to optimize the performances.

As the expressions of the Nyquist criterion in the analog and digital domain are equivalent, we will try to approximate a transmitter and a receiver filter with transfer functions $R^\alpha(f, \rho, T_s)$ and $R^{1-\alpha}(f, \rho, T_s)$ in the Nyquist bandwidth (where f_a is the digital frequency), that will satisfy Eq. (2) when cascaded.

The problem that we consider in this paper is therefore the design of a digital FIR filter with linear phase and transfer function magnitude equal to $R^\alpha(f, \rho, T_s)$ with $0 \le \alpha \le 1$. As we are looking for a causal FIR filter, we introduce a linear phase $\Phi(f)$ equal to $-\pi fT(N-1)$, being N the filter length. Therefore we will start our design from the ideal digital transfer function

$$H_I(f) = R^\alpha(f, \rho, T_s)e^{j\Phi(f)} \tag{4}$$

4 Filter Design

Various FIR design methods have been proposed in literature [2]. We chose the Frequency Sampling (FS) method, which allows the evaluation of the filter coefficients $h[n]$ with a simple direct formula [2]

$$h[n] = \frac{1}{N} \sum_{k=0}^{N-1} H_k e^{j\frac{2\pi}{N}nk} \tag{5}$$

where H_k are the samples of the ideal transfer function $H_I(f)$ at the equally spaced points $f_k = \frac{k}{NT}$, $k = 0, \ldots, N-1$. The FS method produces a digital transfer

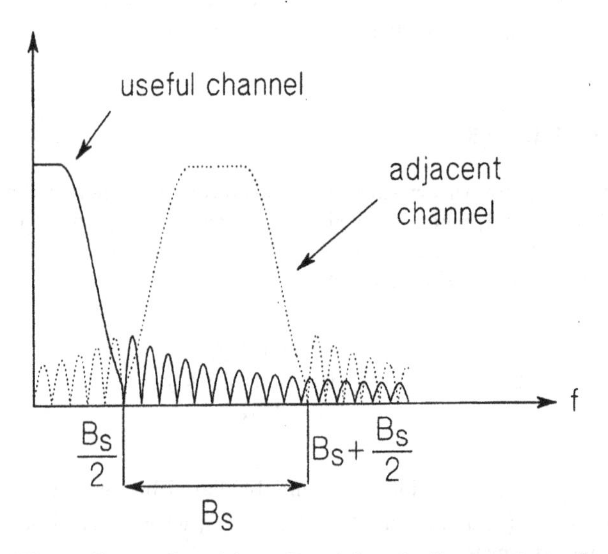

Fig. 1. Spectral position of useful and adjacent channels.

function $H_D(f)$, which is forced to assume the ideal values H_k at the frequencies f_k, but $H_D(f)$ is not directly controlled in all the other frequencies. Despite of the fact that the FS method doesn't give us control on the filter behaviour in the time domain, it has been shown in [16] that FIR raised–cosine filters designed with FS offer good performances in terms of ISI, compared with Time Domain methods. In our design method, the filter length N is set equal to $2N_sE_p$, where E_p is the number of periods on the right (left) side of the impulse response. The value E_p is a real value compatible with the constraint that $2N_sE_p$ must be an integer number. We will examine both integer and non–integer values of E_p. The samples H_k to be

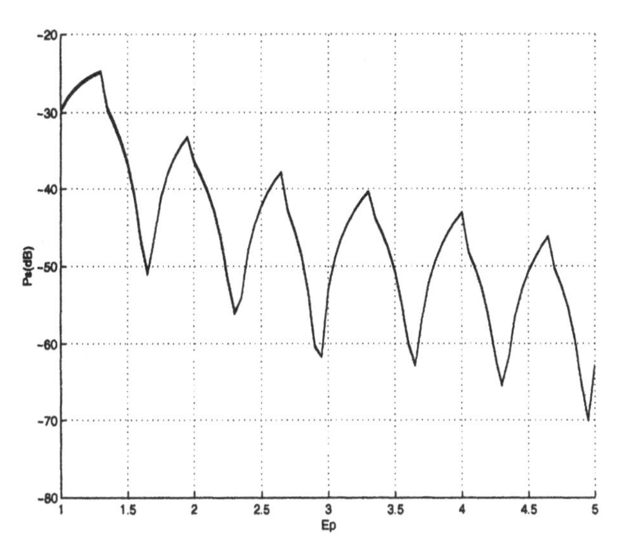

Fig. 2. Value of P_s as a function of E_p for $\rho = 0.5$, $\alpha = 0.5$ and $N_s = 10, 20, 30$

plugged in Eq. (5) can be evaluated from Eq. (3) and Eq. (4):

$$H_k = \begin{cases} e^{-jk\pi\frac{N-1}{N}} \\ \text{for } 0 \leq k < N_A \\ \cos^{2\alpha}\left[\frac{\pi}{4\rho}\left(\frac{k}{E_p} - 1 + \rho\right)\right] e^{-jk\pi\frac{N-1}{N}} \\ \text{for } N_A \leq k < N_B \\ 0 \\ \text{for } N_B \leq k \leq N - N_B \\ \cos^{2\alpha}\left[\frac{\pi}{4\rho}\left(\frac{N-k}{E_p} - 1 + \rho\right)\right] e^{\frac{-j\pi(k-N)(N-1)}{N}} \\ \text{for } N - N_B < k \leq N - N_A \\ e^{-j(k-N)\pi\frac{N-1}{N}} \\ \text{for } N - N_A < k \leq N - 1 \end{cases} \tag{6}$$

where we have set $N_A = \lceil E_p(1-\rho) \rceil$ and $N_B = \lceil E_p(1+\rho) \rceil$.

5 ICI Performances

In order to evaluate the effects of ICI, we first define the filter bandwidth B_s as twice the "gross bandwidth" of the useful channel (see Fig. 1), i.e. the frequency range from zero to the first null of the filter transfer function, that will obviously be located in $\frac{B_s}{2}$. In usual applications, the interchannel interference is due to an adjacent channel with the same frequency characteristics of the useful channel, starting at the frequency $\frac{B_s}{2}$ and of bandwidth B_s. (see Fig. 1). In order evaluate the amount of ICI allowed by the filter, we introduce the parameter P_s, defined as the medium lobe in the adjacent channel with bandwidth B_s

$$P_s = \frac{1}{B_s} \int_{\frac{B_s}{2}}^{\frac{3}{2}B_s} |H_D(f)| df. \tag{7}$$

It has been numerically verified that the value of P_s does not depend on the number of samples per symbol N_s, the so called oversampling factor, but only on the roll–off *parameter*ρ and the number of periods E_p, as shown in Fig. 2 in the case of a square–root raised cosine filter ($\alpha = 0.5$). This behaviour can be explained as follows: considering the digital transfer function $H_D(f)$, the first null is generally placed in $k = N_B$ (see Fig. 3), that is at the frequency $f = \frac{N_B}{NT}$. Within the stopband

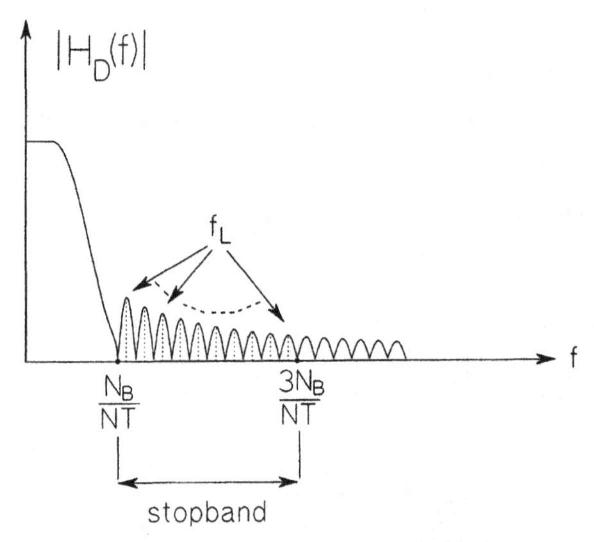

Fig. 3. Magnitude of the digital filter $H_D(f)$

B_s, the maxima of the magnitude sidelobes are placed very near to the middle point between two null samples, that is in the points $f_L = \frac{L+0.5}{NT}$, $N_B \leq L < N - N_B - 1$. The parameter P_s defined in Eq. (7) can be approximated as

$$P_s \approx \frac{1}{2N_B} \sum_{L=N_B}^{3N_B-1} \left| H_D\left(\frac{L+0.5}{NT}\right) \right| = \frac{1}{2N_B} \sum_{L=N_B}^{3N_B-1} |H_D(f_L)| \tag{8}$$

The quantity $|H_D(f_L)|$ can be expressed as in ([2], pag. 107) obtaining the following equation:

$$|H_D(f_L)| = \tfrac{1}{N} \left| \sin\left(\tfrac{\pi L}{N}\right)^{-1} \right. \tag{9}$$

$$+\sum_{k=1}^{N_A-1}(-1)^{-k}\left[\sin\left(\tfrac{\pi(L-k)}{N}\right)^{-1}+\sin\left(\tfrac{\pi(L+k)}{N}\right)^{-1}\right]$$
$$+\sum_{k=N_A}^{N_B-1}(-1)^{-k}R_k\left[\sin\left(\tfrac{\pi(L-k)}{N}\right)^{-1}+\sin\left(\tfrac{\pi(L+k)}{N}\right)^{-1}\right]\Bigg|$$

When $N \gg 1$, as it is generally verified in practical filters, the quantities $\sin(\cdot)$ in Eq. (10) can be approximated with their arguments, obtaining the expression:

$$|H_D(f_L)| \approx \frac{1}{\pi}\left|\frac{1}{L}+\sum_{k=1}^{N_A-1}(-1)^{-k}\left[\frac{1}{L-k}+\frac{1}{L+k}\right]\right.$$
$$\left.+\sum_{k=N_A}^{N_B-1}(-1)^{-k}R_k\left[\frac{1}{L-k}+\frac{1}{L+k}\right]\right| \tag{10}$$

where

$$R_k = \cos^{2\alpha}\left[\frac{\pi}{4\rho}\left(\frac{k}{E_p}-1+\rho\right)\right]e^{-jk\pi\frac{N-1}{N}} \tag{11}$$

Substituting Eq. (11) in Eq. (8), we obtain an approximated expression of P_s that does not depend on the number of samples per symbol N_s and that justifies the behaviour of Fig. 2:

$$P_s \approx \frac{1}{2N_B}\sum_{L=N_B}^{3N_B-1}\frac{1}{\pi}\left|\frac{1}{L}+\sum_{k=1}^{N_A-1}(-1)^{-k}\left[\frac{1}{L-k}+\frac{1}{L+k}\right]\right.$$
$$\left.+\sum_{k=N_A}^{N_B-1}(-1)^{-k}R_k\left[\frac{1}{L-k}+\frac{1}{L+k}\right]\right| \tag{12}$$

where $N_A = \lceil E_p(1-\rho)\rceil$ and $N_B = \lceil E_p(1+\rho)\rceil$.

As it can be seen in Fig. 2, P_s shows some minimum values for certain optimal values of E_p. This situation will repeat for every value of ρ, i.e given a value of ρ, P_s will show a certain number of minima for some "optimal" values of E_p. This behaviour can be verified in Fig. 4, where the values of P_s as a function of E_p are shown for different values of ρ. In this latter figure, it can be observed that the values of P_s decrease when ρ increases. For a given value of α, a proper choice of the design parameters E_p and ρ can therefore result in a filter with very high stopband attenuation.

In software radio applications it is important the number of filter taps is maintained as low as possible, while still maximizing (or locally maximizing) the stopband attenuation. It is also important to have simple filter design methods, possibly

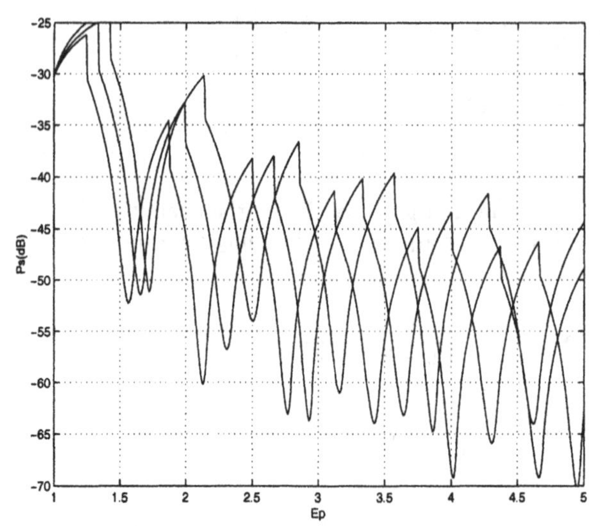

Fig. 4. Value of P_s as a function of E_p for $\rho = 0.4, 0.5, 0.6$ and $\alpha = 0.5$.

completely automatic, that allow us to select good filters without the need for interactive optimization. For this purpose, a graph showing the values of the values (E_p, ρ) in which the maxima stopband attenuations (i.e. the minimum values of P_s) are located is shown in Fig. 5 for three different values of α (i.e. $\alpha = 0.4, 0.5, 0.6$) In particular, the upper curve refers to $\alpha = 0.6$, the lower to $\alpha = 0.4$ and the median curve to $\alpha = 0.5$. Since for a given value of ρ different optimal choices for E_p are possible, the minimum "optimal" value of E_p has been reported in Fig. 5, in order to minimize the number of taps N. The values of the minima of P_s considered in Fig. 5 are all less than -40 dB, and their value decreases as E_p increases. Fig. 5 can therefore be used as design chart: given the values of α and N_s, the values (E_p, ρ) corresponding to a minimum of P_s (that is a minimum of the ICI) can be directly read from the chart (paying attention to the constraint that $2N_sE_p$ must be an integer number). The designed filter will have the minimum possible value of E_p, and therefore the minimum possible value of N for the given N_s.

Note that the couple of values (E_p, ρ) shown in Fig. 5 are "optimal" in the sense that they locally minimize the parameter P_s for the selected value of α.

In order to render the design method completely automatic, and therefore more suited to a software radio application, a linear approximation of the curves of Fig. 5 can be obtained. In particular, the values of α, ρ and E_p in Fig. 5 are related to each other by the following approximation:

$$E_p = \begin{cases} 1.945\rho\alpha - 1.611\rho + 1.97 & \text{for } 0 < \rho \leq 0.5 \\ -0.9\rho + 0.5\alpha + 1.83 & \text{for } 0.5 < \rho \leq 1 \end{cases} \tag{13}$$

Equation (13) allows to automatically determine the optimal value of ρ when α and E_p are given (this choice will be optimal for every value of N_s), or to select

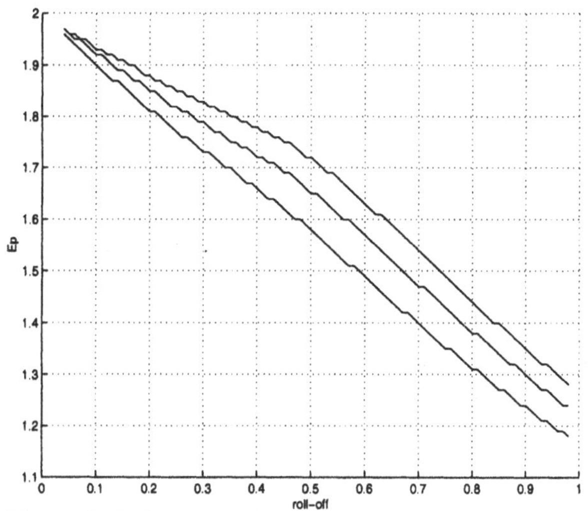

Fig. 5. (ρ, E_p) design table for "optimal" square-root raised-cosine filters. The upper curve refers to $\alpha = 0.6$, the middle curve to $\alpha = 0.5$ and the lower to $\alpha = 0.4$.

a possible value for E_p if ρ and α are given. In this second case, the final value for E_p, denoted $\hat{E}_p$, must be chosen as

$$\hat{E}_p = \frac{round(2E_pN_s)}{2N_s} \tag{14}$$

where round(.) is the function rounding its argument to the closest integer. Eq. (14) selects $\hat{E}_p$ as the value closest to E_p among those satisfying the constraint of having $N = 2\hat{E}_pN_s$ an integer number.

6 Design Example

Example: As an example, we considered the specifications

$$\alpha = 0.5, N_s = 5, \rho = 0.11$$

A local maximum of the stopband attenuation can be obtained by choosing $E_p = 1.9$ from the Fig. 5, leading to a filter containing 19 (symmetrical) coefficients, whose transfer function is shown in Fig. 6.

As it can be observed from Fig. 6, a minimum stopband attenuation of 40 dB has been achieved in this design.

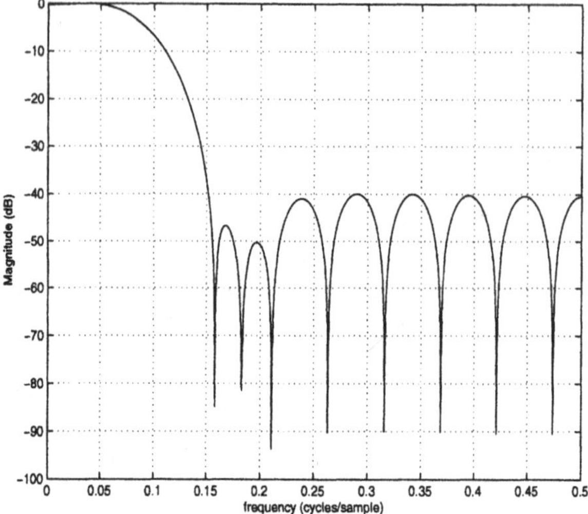

Fig. 6. Transfer function magnitude of the square–root raised–cosine filters described in the Design Example.

7 Conclusions

A simple and efficient design method for low ICI raised cosine FIR filter pairs based on Frequency Sampling techniques has been described. The proposed design method separately controls the stopband attenuation of both the transmitter and the receiver filters. Simple design tables for square–root raised–cosine digital filters have been presented (see also [17]). Moreover, optimal filters can be obtained for every value of the raised–cosine exponent α, which can be an important design parameter in presence of non–linear transmission channels. A design program is available to obtain the minimum length FIR filter able to meet fixed specifications having N_s, ρ, E_p and α as parameters. An example have illustrated that the proposed method can be used as an alternative to more complicated and time consuming FIR design techniques.

References

1. S. Benedetto, E. Biglieri, V. Castellani, "*Digital Transmission Theory*", Englewood Cliffs, N.J., Prentice Hall, pp.324–333, 1987
2. L. Rabiner, B. Gold, "*Theory and Application of Digital Signal Processing*", Englewood Cliffs, N.J., Prentice Hall, pp.105–107, 1987
3. M. L. Welborn, " *Direct Waveform Synthesis for Software Radios*", IEEE 1999
4. T. Hentschel, M. Henker, G. Fettweis " *The digital Front-End of software radio terminals*", IEEE Personal Communications, pp.40–46, Aug. 1999
5. H. R. Karimi, N. W. Anderson, P. McAndrew " *Digital signal processing aspects of software definable radios*", IEE, Savoy Place, London, 1998
6. L. Rabiner, "*Linear Programming Design of FIR Digital Filters*", IEEE Trans. Audio and Electroacoustics, Vol. AU–20, p.280, Oct. 1972

7. K. S. Steiglitz, "*Optimal design of FIR filters with monotone passband response*", IEEE Trans. on ASSP, Vol. ASSP–27, pp.643–649, Dec. 1979
8. A. C. Salazar, V. B. Lawrence, "*Design and implementation of transmitter and receiver filters with periodic coefficients nulls for digital systems*", Proc. IEEE Int. Conf. in Acoustics, Speech, and Signal Processing, Paris (France), pp.306–310, May 1982
9. J. K. Liang, R. J. P. De Figuereido, F. C. Lu, "*Design of optimal Nyquist, partial response, Nth band, and nonuniform tap spacing FIR Digital Filters using linear programming techniques*", IEEE Trans. on Circuits and Systems, vol. CAS–32, pp.386–392, Apr. 1985
10. H. Samueli, "*On the Design of Optimal Equiripple FIR Digital Filters for Data Transmission Application*", IEEE Trans. on Circuits and Systems, vol. CAS-35, pp.1542–1546, Dec. 1988
11. T.Saramaki, Y. Nuevo, "*A Class of FIR Nyquist (Nth band) Filters with Zero Intersymbol Interference*", IEEE Trans. on Circuits and Systems, vol. CAS-34, pp.1182–1190, Oct. 1987
12. H. Samueli, "*On the design of FIR digital data transmission filters with arbitrary magnitude specifications*", IEEE Trans. on Circuits and Systems, Vol. 38, p.1563, Dec. 1991
13. F. M. De Saint-Martin, P. Siohan, "*Design of optimal linear-phase transmitter and receiver filters for digital systems*", Circuits and Systems, 1995. ISCAS '95., 1995 IEEE International Symposium on
14. H. Baher, M. O'Malley, "*Design of FIR sampled-data filters for data transmission*", Circuits and Systems, 1990., IEEE International Symposium on
15. M. Renfors, T. Saramaki, "*Pulse-shaping filters for digital transmission systems*", GLOBECOM '92. Communication for Global Users., IEEE
16. L. Lo Presti, "*FIR design of Digital Filters*", 8th European Conference on Electrotechnics, June 1988, Stockholm (Sweden)
17. L. Lo Presti, M. Mòndin, "*Design of optimal FIR raised-cosine filters*", Electronics Letters, pp.467–468, March 1989

High Computational Capacity Flexible Architecture for the Realization of Wide Band Digital Communication Systems for Radio Relays

Fabrizio Giani

Laboratorio Ponti Radio, Marconi Communications S.p.A. – Mobile – Defence Division, Via A. Negrone 1A 16153 Gevova, Italy

Abstract: According to an authoritative research institute[1] Software Radio represents the meeting between Digital Radio and software technologies. The system we are developing, that is the object of this presentation, is characterized by an high computational capacity (about 20 GFLOPS), necessary to the development of wideband (about 30 MHz) communication systems in Radio Relay applications. It is fully implemented on a flexible and highly integrated hardware platform (FPGA-Field Programmable Gate Array). This work describes a specific case and tries to highlight potentiality the system presents, beyond the specific application, in order to allow generic Software Radio architectures.

1. Introduction

Object of this presentation is a multi-rate multi-mode modem suited to wideband Radio Relay applications. Its feature is full digital implementation of carrier and timing synchronization, AGC (Automatic Gain Control), adaptive equalization. Algorithms can be dynamically updated or substituted by simply reprogramming of the FPGAs. The circuit is integrated with other functional blocks: the Framer-Deframer and source Encoders-Decoders (in the specific application the universal multi-rate Framer-Deframer and the completely programmable FEC (Feedforward Error Corrector) & Interleaver-Deinterleaver).

2. Transmit Path

The transmit path of the system is implemented using a standard architecture (Figure 1).

After the Framer there are the programmable Reed-Solomon Encoder and the programmable convolutional Interleaver. A waveforms synthesizer follows, whose output is fed to an I&Q modulator employing 4× interpolating root raised cosine filters. Inverse Sinc filtering is added prior to DAC (Digital to Analog Converter). Operating DAC frequency is about 80 MHz and resolution is 14 bits.

---

[1] Roke Manor Research

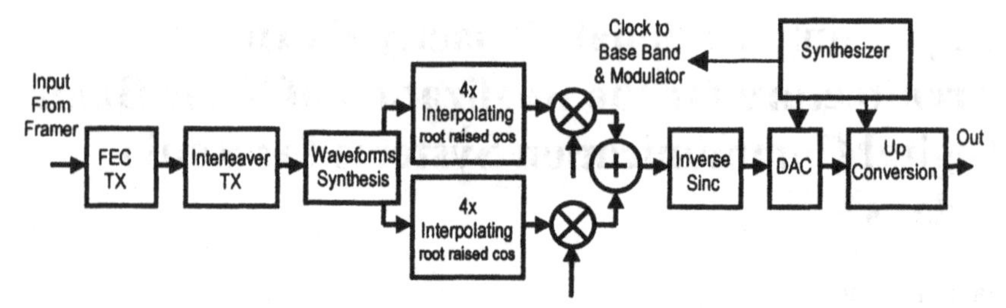

Figure 1. Transmit Path.

3. Receive Path

A schematic of the receiving system follows (Figure 2):

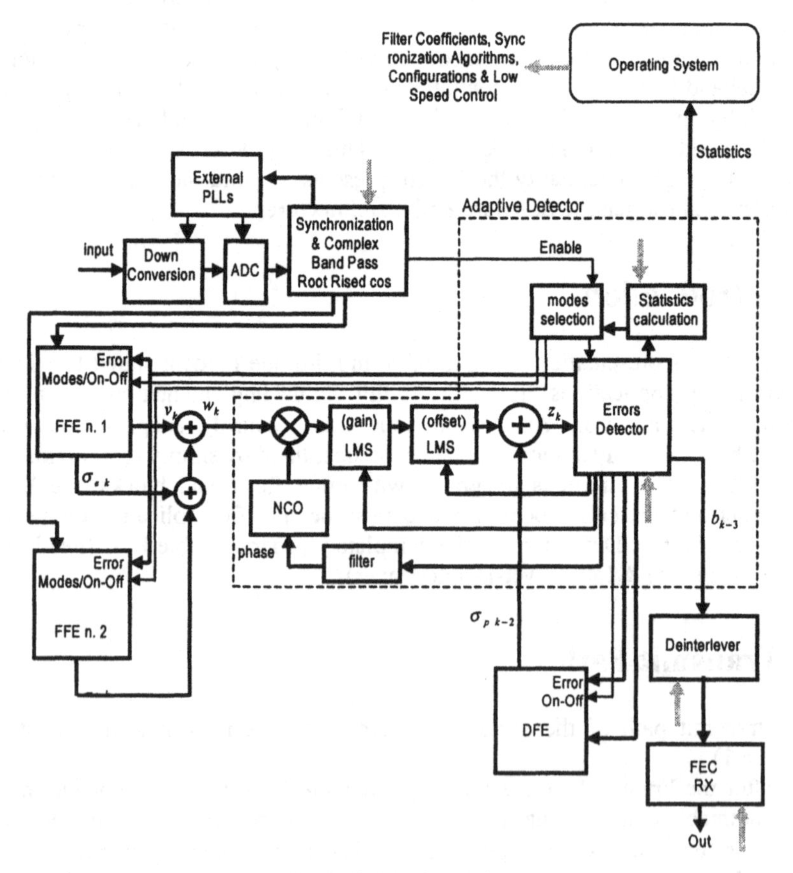

Figure 2. Receive Path

3.1 Filtering and Timing Synchronization

The received signal is digitized with a 12 bits resolution at a rate around 80 MHz and pass band match filtered (root raised cosine). The result is the analytical representation of the demodulated signal.

Choosing of the right sampling time is made according to a modified form of the Gardner's [1] algorithm. The first modification consists in bringing the Gardner's approximation from the complex envelopes domain to the analytical signals domain. The second modification is made adding further filtering stages to better average the algorithm timing error.

3.2 Adaptive Detection

Besides being corrupted by noise and distorted by ISI (Inter Symbol Interference), the constellation exiting the matched filter exhibits a random average radius. The constellation also rotates with a random speed (i.e., due to wrong carrier reconstruction) around an offset point whose position depends, itself, on system asymmetries and on the nature of possible interferers. The adaptive decision process removes these uncertainty factors extracting a final constellation on which actually make the decision.

In absence of undesired amplitude modulations, the average radius of the constellation remains practically constant over a long period of time (respect to symbol period). In this case the constellation amplitude normalization parameter (the gain) can be drawn using an LMS (Least Mean Square) [2] algorithm applied to the difference between the modulus of the constellations before detection (z_k) and after detection (b_k).

The phase difference between z_k and b_k has short period variations (respect to symbol period), so LMS algorithm has to be modified replacing its update term: let's call γ_k this new term. This is drawn through an Integral-Proportional transformation of a term, ε_k, representing the derivative of the phase error, during frequency acquisition, or the actual phase error, during the phase lock acquisition. A description of what has just been affirmed about phase recovery follows Equations 1-6.

$$\mu_k = z_k b_k^* \tag{1}$$

$$\varepsilon_k = \begin{cases} \mathit{Phase\ Detection}: \vartheta_k = \angle z_k - \angle b_k \approx \mathrm{Im}\{\mu_k\} \\ \mathit{Frequency\ Detection}: (d\vartheta/dt)_k \approx \mathrm{Im}\{\mu_k\ \mu_{k-1}^*\} \end{cases} \tag{2}$$

$$\gamma_{I\ k} = \gamma_{I\ k-1} + \delta_I\ \varepsilon_k, \qquad \delta_I \in \Re^+ \tag{3}$$

$$\gamma_{P\ k} = \delta_P\ \varepsilon_k, \qquad \delta_P \in \Re^+ \tag{4}$$

$$\gamma_k = \gamma_{P\ k} + \gamma_{I\ k} \tag{5}$$

$$\phi_k = \left[\phi_{k-1} - \delta_\phi \gamma_{k-1}\right]_{\mathrm{mod}(0+2\pi)}, \qquad \delta_\phi \in \Re^+ \tag{6}$$

Assuming that only a monochromatic interferer located at the center of the received spectrum exists, also the constellation offset can be considered being practically constant over a long period of time (respect to symbol time). In the preceding hypothesis the term of depolarization of the constellation is given by Equation 7:

$$e_k = z_k - b_k, \quad \eta\text{:}\min(e^2) \Rightarrow \eta_k = \eta_{k-1} - \delta_\eta 2e_{k-1}, \quad \delta_\eta \in \Re^+ \tag{7}$$

3.3 Adaptive Equalization

Let's point out the temporal index of the constellation entering the Adaptive Detector as Present Instant. Removal of ISI (Inter Symbol Interference) contributions takes place adding to the signal at the Present Instant, a linear combination of the Signal History. The Signal History is defined as the collection of symbols received across Present Instant along time. This linear combination has to be done with a proper set of coefficients through adaptive equalization. This is performed using an FS-FFE (Fractionally Spaced-FeedForward Equalizer), made up by two FFEs, and a DFE (Decision Feedback Equalizer). Signal History and the respective equalization coefficients are defined as follows (Table 1):

Table 1. Signal history and coefficients

Signal	Coefficient
v_k (Present Symbol)	1 (Unitary Coefficient)
$v_{even\,k}$ (Precursors Vector)	$c_{even\,k}$ (FFE n.1 Coefficients Vector)
$v_{odd\,k}$ (Fractional Time Precursors Vector)	$c_{odd\,k}$ (FFE n.2 Coefficients Vector)
v_{p_k} (Postcursors Vector)	c_{p_k} (DFE Coefficients Vector)

Taking into account the terms for ISI removal, the constellation z_k takes the following form (Equation 8):

$$z_k = w_k\, e^{j\phi_k}\, g_k + \eta_k + \sigma_{p\,k} \tag{8}$$

where, g_k, ϕ_k and η_k represent, respectively the normalization term (gain), the phase and the offset reconstructed by the Adaptive Detector and w_k, $\sigma_{p\,k}$ are given by the following expressions (Equations 9-11):

$$w_k = v_k + \sigma_{e\,k} + \sigma_{o\,k} \tag{9}$$

$$\sigma_{e,o\ k} = c_{e,o\ k} \bullet v_{e,o\ k} = \sum_{i=0}^{N_f-1} c_{e,o\ i,k}\ v_{e,o\ i,k} \quad (FFE\ n.1\ or\ n.2) \tag{10}$$

$$\sigma_{p\ k} = c_{p\ k} \bullet v_{p\ k} = \sum_{i=0}^{N_p-1} c_{p\ i,k}\ v_{p\ i,k} \quad (DFE) \tag{11}$$

N_f and N_p represent respectively the precursors equalization depth and postcursors equalization depth (measured in symbol periods).

Supposing FFEs are disabled until PLL (internal to the detector) has locked and only correct decisions are taken, update of c_e, c_o coefficients is purely DD (Decision Directed). LMS algorithm, applied to the complex error e_k, provides (Equations 12-15):

$$e_k = z_k - b_k \tag{12}$$

$$c_{e,o} \min \frac{|e|^2}{2} \rightarrow c_{e,o\ k} = c_{e,o\ k-1} - \delta_{e,o} \left(\nabla_{c_{e,o}} e^*\right)_{k-1} e_{k-1},\ \delta_{e,o} \in \Re^+ \tag{13}$$

$$\left(\nabla_{c_{e,o}} e^*\right)_{k-1} = v^*_{e,o\ k-1}\ e^{-j\phi_{k-1}}\ g_{k-1} \tag{14}$$

from which:

$$c_{e,o\ i,k} = c_{e,o\ i,k-1} - \delta_{e,o} v^*_{e,o\ i,k-1}\ e^{-j\phi_{k-1}}\ g_{k-1}\ e_{k-1},\ \ \delta_{e,o} \in \Re^+. \tag{15}$$

To speed up the convergence in the update of c_e, c_o coefficients, FFEs are activated before the aforementioned PLL has locked, with a CMA (Constant Modulus Algorithm) [3] algorithm.

The DFE is activated exclusively when all other algorithms have reached a good degree of convergence and therefore only in the DD mode (Equations 16-18):

$$c_p \min \frac{|e|^2}{2} \rightarrow \ c_{p\ k} = c_{p\ k-1} - \delta_p \left(\nabla_{c_p} e^*\right)_{k-1} e_{k-1},\ \ \delta_p \in \Re^+ \tag{16}$$

$$\left(\nabla_{c_p} e^*\right)_{k-1} = v^*_{p\ k-1} \tag{17}$$

from which:

$$c_{p\ i,k} = c_{p\ i,k-1} - \delta_p v^*_{p\ i,k-1}\ e_{k-1},\ \ \delta_p \in \Re^+ \tag{18}$$

3.4 Control

In order to avoid instability and false lock of the system it is necessary to guarantee a correct sequence of activation and the correct way of operation of all processes. For such a purpose Adaptive Detector calculates error signals statistics from which some flags are generated. Such flags act on the modes of operation according to a priority scheme and are employed in loading of the right algorithm for the specific phase of the acquisition process. For example, the constellation derotating PLL is activated in Frequency Detection mode or in Phase Detection mode depending on the value of the averaged frequency error. Similarly, FFEs are activated in CMA mode or in DD mode, depending on the value of the averaged phase error. Finally, DFE, like depolarization, is activated, only in DD mode, when the averaged squared decision error is fallen under a particular maximum threshold.

4. Performances and Hardware Optimization

In order to get the maximum in terms of speed from the programmable devices, it is necessary to limit the depth of combinatorial paths introducing pipelining. This technique, because of the introduction of delay taps, in an application like this, where there are many and different feedback loops, can bring to instability. The insertion of a certain number of delays in an update loop, for instance of LMS type, corresponds to assume that the gradient of the quadratic form associated to the algorithm remains almost invariable during a period corresponding to the number of introduced delay taps. This, if others conditions remain invariable, is more true the smaller the algorithm step sizes are. Similar considerations can be made in cases diverging from LMS.

5. Conclusions

The high computational capacity allows the synthesis and the elaboration of complex waveforms with high performances in terms of spectrum efficiency and counteraction of channel mismatches. The type of implementation (FPGA) offers the advantage of reliability and reproducibility of equipment, and the advantage of the simple integration with the control and reconfiguration software.

6. References

[1] F.M. Gardner: "A BPSK/QPSK timing-error detector for sampled receivers", *IEEE Transaction on Communications*, Vol. 34, No. 5, 1986.

[2] B. Widrow, M. E. Hoff Jr. "Adaptive switching Circuits", IRE Wescon Conv. Rec., part 4, August 1960.

[3] M.G. Lagrimore, J.R. Treichler "Convergence Behavior of the Constant Modulus Algorithm" *Conf. Record, 1983 IEE Int. Conf. On Acoust., Speech Processing*, Boston.

The Software Radio Technique Applied to the RF Front-end for Cellular Mobile Systems

Anna Marina Badá, Marcello Donati

New Tech. Group, Radio Technologies, Mobile Networks, Siemens ICN, 20019 Castelletto di Settimo Milanese, Italy

Abstract: The novel software radio concept is a promising technique to develop new RF front-end architectures. It has been used to implement a transceiver for cellular mobile radio Base Station equipment. The most important technical issues are discussed and the prototype developed in Siemens R&D labs is presented. Results of the measurement campaign are shown.

1. Introduction

It is general opinion that the future cellular mobile networks will use different standards.

In the past, the first generation of analogue mobile systems didn't foresee any compatibility between different countries, while the second and next generations see the extension of standards in more countries (*i.e.*, GSM, UMTS).

Anyway, the third generation UMTS standards are constituted by different versions: UTRA TDD, UTRA FDD, CWTS, *etc.*

Even though these UMTS versions follow the same philosophy (all of them are based on the Code Division Multiple Access method, CDMA), the differences between them are substantial.

In fact, the radio access interface (Frequency Division Duplex FDD or Time Division Duplex TDD), the operating RF bandwidth, the mo-demodulation and co-decoding techniques, and the procedures for the network access are quite different. These considerations impose a new approach to designing the radio frequency front-end of the Base Transceiver Stations (BTS), in orderto obtain standard-indepentdent equipment.

The software Radio (SWR) technique performs a new idea as to implement a flexible radio interface for the BTS and also for the Mobile Station (MS) of a cellular mobile system: by re-programming the hardware of the equipment for different standards thorugh an opportune software procedure.

It is important to clarify at once that the SWR hits partially the target of a complete equipment reconfiguration, since some hardware modifications are needed in every case. Anyway, the reliability and the cost reduction for equipment with different standards and access methods are improved by this architecture.

Even if the "Software Radio" term is sometimes used to define generically the software techniques for the managing of the air interface, this paper will be focalised only on the hardware architecture of a SWR transceiver.

In addition, some considerations of the BTS hardware implementation will be presented.

The state of the art will be evaluated and, in the end, will be presented a GSM transceiver prototype with the RF power amplifier, implemented in the Siemens R&D laboratories in Castelletto di Settimo Milanese.

The results obtained, some measurements and the future evolutions will conclude the paper.

2. An Overview of the SWR Technique

Although the physical channel is different according to the standard, it is possible to implement a general RF front-end structure that is independent of the kind of physical channel adopted.

The traditional architecture of the radio interface of a BTS with N carriers is shown in Figure 1: it is characterised by a set of N transceivers. Every receiving part down-converts, amplifies, discriminates and demodulates the up-link signals tuned at the n-th channel frequency *f*.

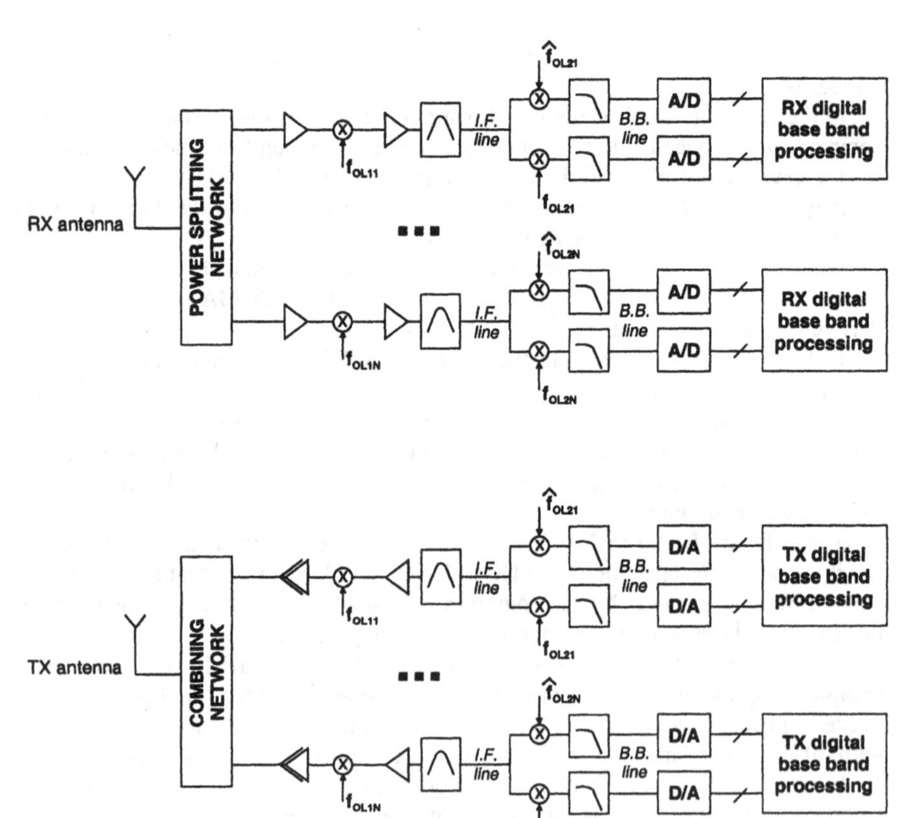

Figure 1. Schemes of traditional RF front-end architecture

Every transmitting part modulates, up-converts, and amplifies the down-link signal at the n-th channel frequency f (for TDD systems) or $f + \Delta f$ (for FDD systems).

Every transceiver has a RF synthesizer used as local oscillator that is programmed in order to up- or down-convert the current channel frequency.

All the transceivers' outputs must be combined to be sent to the antenna through a RF combining network with N inputs and a single output.

Being a RF network, this combiner has to be adapted (typically at 50Ω); when the number of carriers is high, the loss of a wide-band combining network becomes unacceptable. In this case it is necessary to implement a selective combining network that has a typical loss within 3÷4dB independently by the number of carriers (with N<10÷15). These networks are expensive and critical, however.

The SWR architecture, on the other hand, has a single wide-band multi-carrier transceiver directly connected to the antenna (see Figure 2) and a cheaper IF digital combining node that substitutes the RF network [1][2][3]. In addition the frequency channelisation implemented through digital up- and down-conversion further simplifies the analogue circuitry.

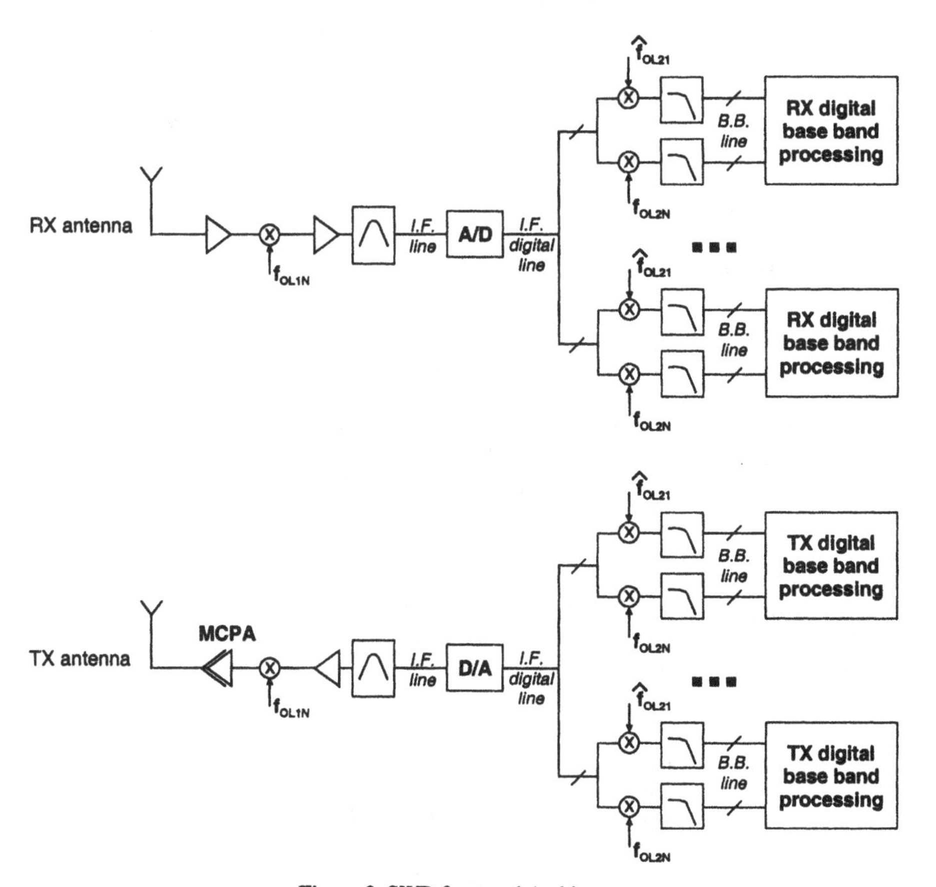

Figure 2. SWR front-end Architecture

A single fixed RF synthesiser instead of N tuned local oscillators constitutes another advantage of the SWR technique.

A further fundamental difference in respect of the traditional solution is the positioning of the Analogue-to Digitial Converter (ADC) and Digital-to-Analogue Converter (DAC) along the chains.

In an ideal case, they are positioned near the antenna connector but the present state of the technological art doesn't permit the digital conversion at RF band so a compromise is chosen and these conversions are placed at intermediate frequency. It follows that the choice of the converters constitutes a critical point in the design of the wide-band SWR: it is necessary to select a particular family of DAC and ADC devices with very tight characteristics.

In fact, the wide-band SWR architecture is typpically characterised by the absence of theAutomatic Gain Control (AGC) on the analogue receiving part, while the standards foresee a high dynamic of the received signals (about 90÷100dB), so the ADC has to suport this value completely.

The DAC has a similar problem, even if the value of the dynamic range is more relaxed (spurs at 75÷80dBc typical).

For the same reasons, the multi-carrier power amplifier (MCPA) constitutes another bottleneck of the SWR architecture. Appropriate structures of MCPA have to be evaluated.

In order to reduce the output power of the MCPA, it is possible to use the SWR architecture in a BTS with Smart Antenna functionality. Moreover, a further improvement can be obtained with the remotisation of the SWR transceiver nearby the antenna.

3. Analogue to Digital Conversion

The most important parameter of an ADC used for SWR implementation is the dynamic linearity, defined by harmonic performance and intermodulation products level.

In order to improve the dynamic linearity of the DC, a dithering technique can be used, randomizing the Differential Non-linearity (DNL) errors [4].

ADC thermal noise could also be an issue, but it is usually well below the noise of the analogue front-end stage, so it doesn't contribute to the overall receiver sensitivity.

Finally, the signal-to-noise ratio (SNR) depends on the number of bits with which the digital samples are represented at the output of the ADC. This value is about 6dB/bit but in our application, the SNR is improved by the processing gain defined by the ratio between the bandwidth of a single modulated carrier and the bandwidth of the Nyquist zone.

In a practical way, the processing gain will be improved by the digital filtering following the ADC conversion.

As stated earlier, the ADC works at intermediate frequency. This can be implemented using the under sampling technique, that is, sampling at a frequency less than half of the analogue input signal frequency (and more than twice the bandwidth of the signal). Note also that the under sampling serves a function quite similar to a frequency shift because the input signals are aliased into the first Nyquist zone as they were in this zone originally [5][6].

The undersampling technique has a technological problem owing to the fact that the ADC device must also guarantee its linear performance over the first Nyquist sone.

Today, the market offers a new generation of 14 bits ADCs that are typically able to work with a sampling rate of 65MHz and have a Spurious Free Dynamic Range (SFDR) of 90dBFS (level indB with respect to the Full Scale).

Latest devices guarantee this performance in the second and third Nyquist zones as well.

4. Digital to Analogue Conversion

The digital to analogue conversion for the Software Radio application has similar problems to those of analogue to digital conversion.

Different architectural structures can be implemented; in any case, it is desirable to have a high intermediate frequency at the DAC output in order to filter the image band and the local oscillator easily after IF to RF up-conversion.

The classical approach forsees the selection of the first Nyquist zone at the output of the DAC by means of a reconstruction low-pass filter. This makes it almost impossible to obtain a high IF frequency value. In fact, this requirement asks for a too-high input sampling rate that is, one that approaches or exceeds the maximum speed of state-of-the-art digital devices interfaced to the DAC.

Another possible method consists of recovering the analogue signal into higher Nyquist zones through a reconstruction band-pass filter [7]. This implementation solves the previous troubles but reduces the SNR owing to the *sin x/x* effect.

Generally, both methods need a high-order analogue reconstruction filer when used for SWR application. For example, with an intermediate frequency f_{if} = 50MHz, a bandwidth BW = 20MHz and a sampling clock at 65MHz, the order of the reconstruction filter for GSM standard has to be $n \geq 12$.

To overcome this, a novel technique for the DAC architecture providing input interpolation filtering for digital cancellation of the adjacent Nyquist zones can be used. The Nyquist zone desired is sampled with a frequency that is a multiple of the clock frequency. The order of the interpolation establishes the number of rejected Nyquist zones and the impact of the *sin x/x* effect.

In this way, the analogue reconstruction filter can be reduced in complexity or even eliminated.

It represents an improvement to the flexibility and reprogrammability of the equipment as the SWR concept requires.

The market already offers some DACs with both digital interpolation and low-pass filtering integrated on chip. Typical figures forsee orders of interpolation M = 4 and stop-band attenuation of –75dBFS with shaping factor = 1.37.

5. Multi-carrier Power Amplifier

The design of the MCPA of the transmitting chain can be very difficult, as stated before.

GSM900 and DCS1800 require the intermodulation spurious level to be at least –75dBc for class 5 BTS (output power P_{out} = 20W per carrier). Supposing the SWR transmitter driven by 10 carriers, the average output power will be 53dBm with output third-order Intercept Point $IP3_{\_out}$ = 93dBm, that represents a prohibitive level.

The required output power of the BTS transmitter can be reduced dramatically combining together the SWR technique with the Smart Antenna solution [8][9] (see Figure 3).

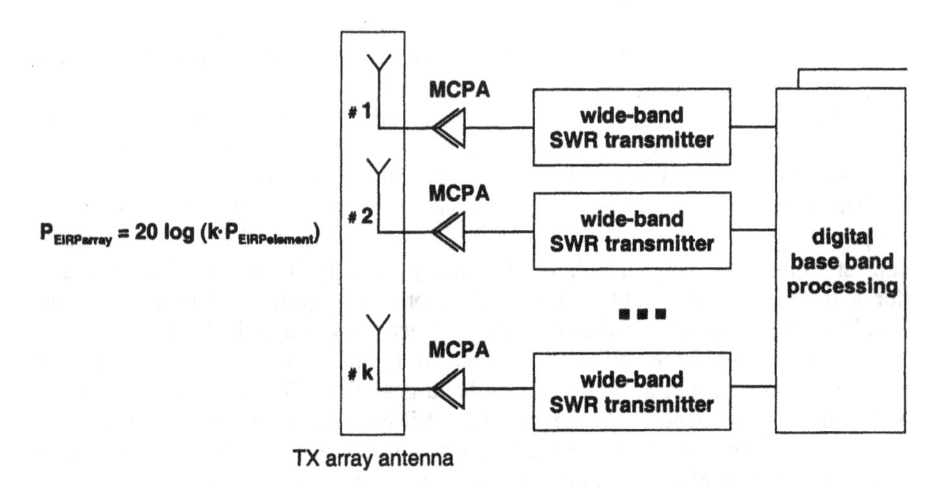

Figure 3. SWR technique applied to Smart Antenna solution

In this way, the output power is distributed between the *k* elements of the array antenna and then phase-combined on air. In this case, if we consider the previous example, with $k = 8$ elements of array antenna, the average output power required becomes only 35dBm and the $IP3_{\_out}$ reduces at 75dBm.

This value is still high but now it is possible to satisfy the requirements by designing a linearised amplifier, for example, in order to increase the efficiency.

In the R&D laboratories of Casteletto di Settimo Milanese a prototype of a double loop feed-forward linearised amplifier was also implemented [10]; its block diagram is depicted in Figure 4.

The amplifier was designed to the GSM900 standard so that it could be used for Base Transceiver Stations with the Smart Antenna functionality.

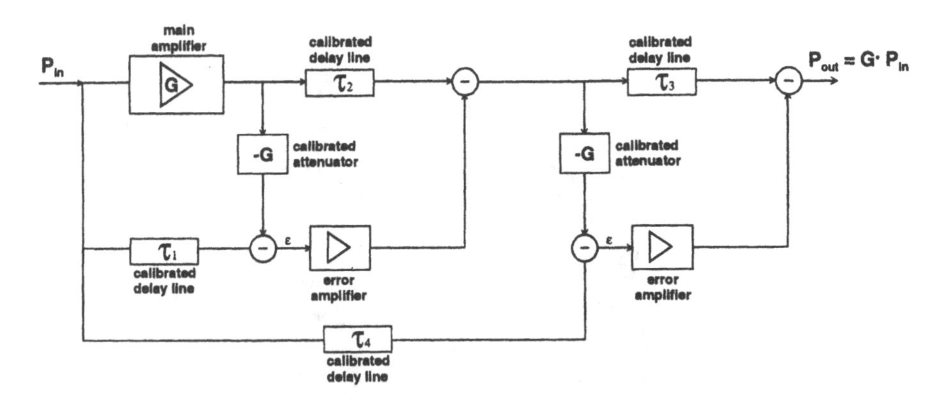

Figure 4. Double loop feed –forward linearised amplifier

The operating frequency band is 925÷960MHz; the typical average output power is 34dBm (2.5W) with an efficiency of 6%; the intermodulation attenuation is 80dBc. The measured output of the amplifier driven with two tones is shown in Figure 5.

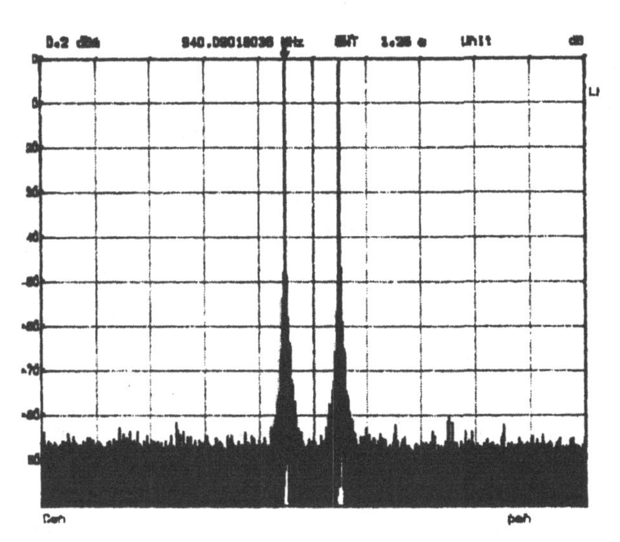

Figure 5. Measured MCPA output driven with two tones (each tone at 0dBm)

The output of the amplifier is able to provide up to 10W but in this case, the intemodulation attenuation degrades.

A photograph of the prototype is shown in Figure 6.

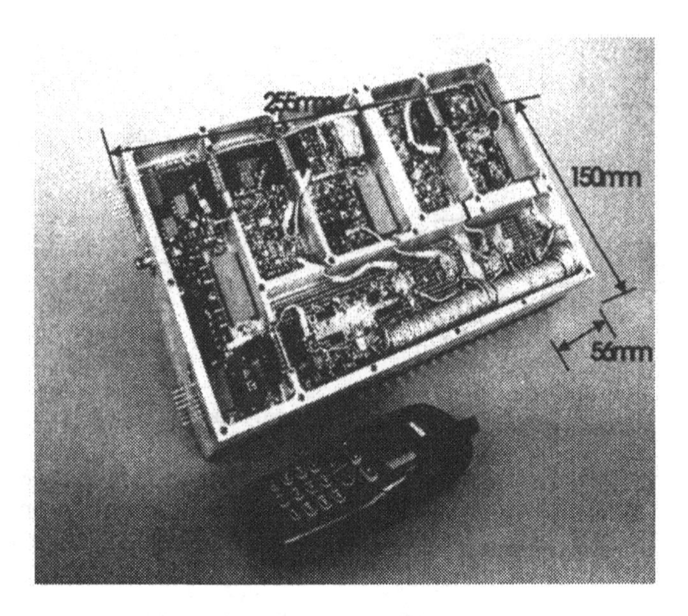

Figure 6. MCPA prototype

6. Remotisation of the Analogue Part

The remotisation of the analogue part of the SWR transceiver and the ADC and DAC nearby the antenna is a way to reduce the ouput power of the MCPA further. In this way, it is possible to recover the loss of the cable between the RF front-end and the antenna element [11].

This remote circuitry can then be connected to the digital part of the BTS through an optical serial link (Figure 7).

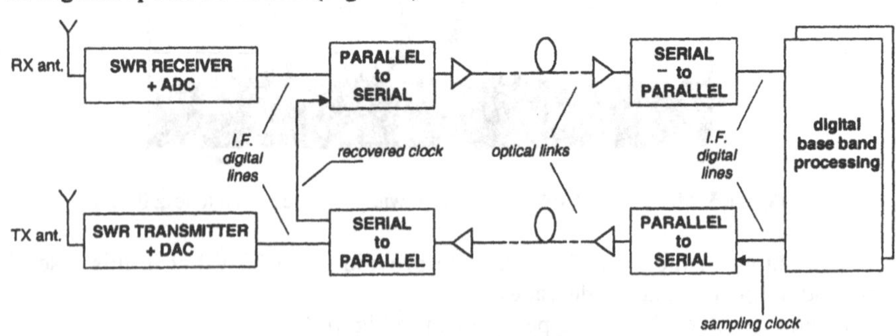

Figure 7. Remotisation scheme

Typically, it is desirable to have only a high-stability reference signal next to the local interface transmitter into the BTS digital side. The clock and data recovery circuits of the remote interface receiver also extract the synchronisation signal for the remote interface transmitter.

Even if the optical link must assure a good transfer quality, the recoverd clock is typically too noisy to be used directly as a sampling source, so it is necessary to filter this signal to reduce the jitter noise.

7. Prototype of the Transmitter

An experimental prototype developed for the GSM system has been implemented as well [12]. The output power is 0dBm for each carrier and the gain of the analogue part between the DAC output and the MCPA input is 13dB; the bandwidth is 10MHz.

The DAC used is a 14 bit device operating in the second Nyquist zone, with sampling rate at 34MSPS and analogue output (intermediate centre frequency) at 26MHz.

A picture of the prototype of the DAC board and the RF front-end is shown in Figure 8 in which it is possible to note the small dimensions of the analogue part.

To show the quality of the results obtained, a measured eye diagram of the GMSK-modulated signal at the transmitter output and the vector envelope are depicted in Figures 9 and 10.

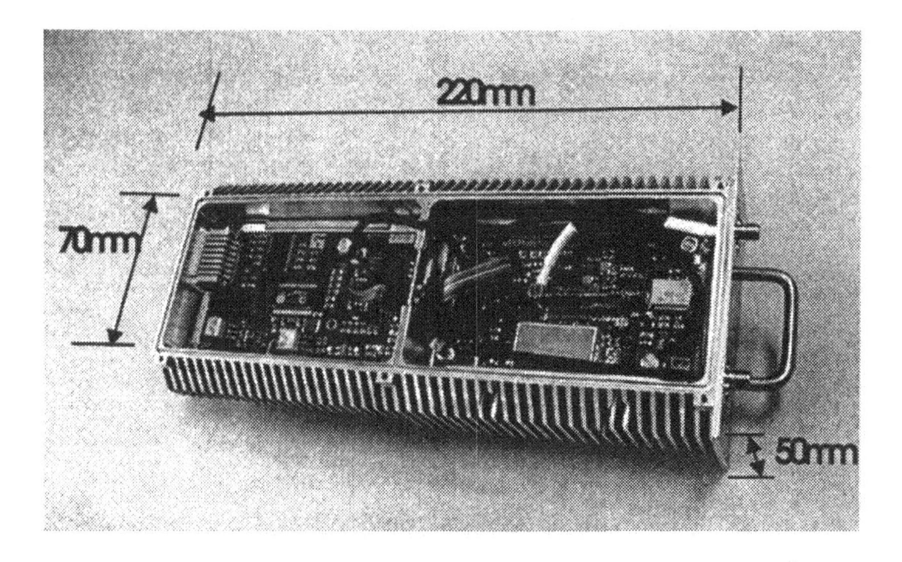

Figure 8. Transmitter prototype

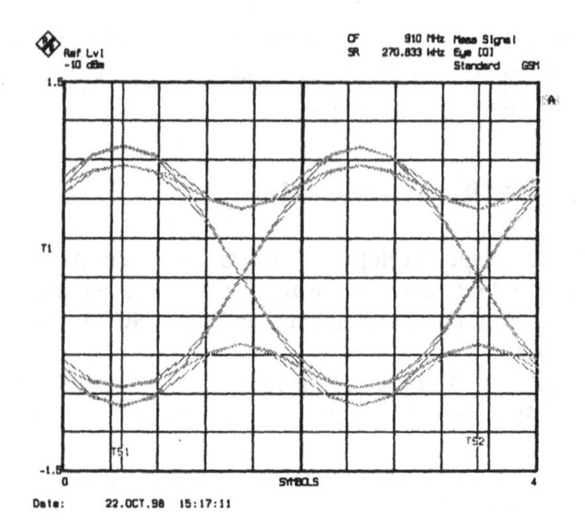

Figure 9. Measured GMSK eye diagram

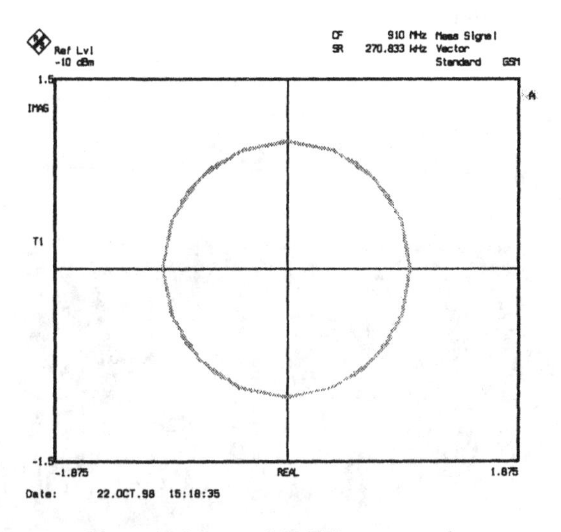

Figure 10. Measured GMSK vector envelope

8. Prototype of the Receiver

The wide-band nature and the absence of AGC of the RF, IF and ADC sections lead the receiver to have high dynamic range requirements.

The prototype implemented makes use of a high linear ADC device with 12 bits used in under sampling mode [13]. The IF centre frequency of the receiver is 70MHz, while the clock frequency is 40MHz. To improve the dynamic range of the ADC, a dithering technique was used.

The gain of the analogue part between the antenna and the output of the ADC is 35dB; the bandwidth is 10MHz and the noise figure is 5dB. The spectrum at the output of the ADC when a monochromatic signal of –40dBm at 908MHz and a –101dBm weak GMSK-modulated signal at 910MHz are present at the antenna input is shown in Figure 11.

To simplify the analogue part, a single down-conversion architecture was chosen. The RF unwanted side band is partially eliminated by means of a RF filter and a Surface Acoustic Wave (SAW) filter placed at intermediate frequency. The channelisation is performed digitally by means of digital down-converters.

Also, the receiver has been implemented in small dimensions: a case similar to that for the transmitter prototype was used to allocate it.

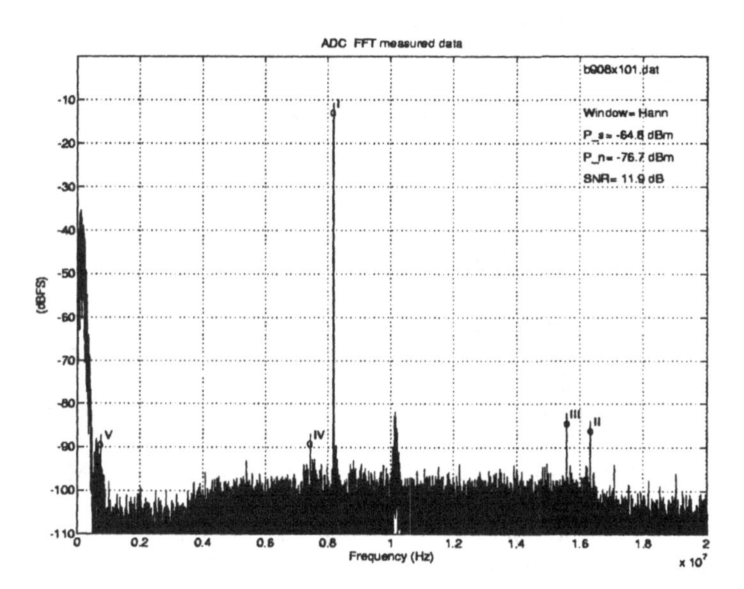

Figure 11. FFT measured spectrum at ADC output

9. Conclusions

The Software Radio concept applied to mobie radio Base Stations was evaluated and a transceiver implemented with wide-band SWR technique was presented. The prototype was developed for GSM standard in order to compare it with existing apparatus. This work was carried out in conjunction with a project development that concerns the implementation of a Smart Antenna Test Bed. It had contributed to prove once more the symbiotic marriage between SWR technique and Smart Antenna syatems.

In addition, thanks to the great flexibility of the SWR architecture that permits the apparatus to be easily reprogrammed, new studies are in progress to adapt the implemented prototype toward the third-generation standards.

10. References

[1] M. Donati, "La Technica Software Radio Applicati ai Sistemi Radiomobili", Alta Frequenza, Luglio-Agosto '96.

[2] Analog Devices Inc., "Software Radio", Microwave Journal, February 1996, pp 128-136.

[3] H.R. Karimi and B. Friedrichs, "Wideband Digital Receivers for Multi-Standard Software Radios", IEE '98.

[4] B. Brannon, "Overcoming Converter Nonlinearities with Dither", Analog Device, Application Note AN-410.

[5] R. Groshong and S. Ruscak, "Undersampling Techniques Simplify Digital Radio", Electronic Design, May 23, 1991, pp 67-78.

[6] G. Hill, "The Benefits of Undersampling", Electronic Design, July 11, 1994, pp 69-79.

[7] Int. Patent App. N. PCT/EP98/04967 "Broadband TX for a Signal Consisting of a Plurality of Digitally Modulated Carriers".

[8] J. Kennedy and M.C. Sullivan, "Direction Finding and Smart Antennas Using Software Radio Architectures", IEEE Communications Magazine, May 1995, pp 62-68.

[9] G. Bucci, A. Colamonico, M. Donati, M. Politi and A. Picciriello, "Experimental GSM Test bed for Adaptive Array Antenna System", MMT '99, Venezia, October 1999.

[10] D. Diamanti, "Double Loop Feed-forward Amplifier" Siemens Internal Technical Report, February 2000.

[11] C. Savazzi, "High Speed Optical Data Link for Smart Antenna Mobile Radio System", MMT '99, Venezia, October 1999.

[12] A.M. Badá, and M. Maddiotto, "Design and Realisation of Digital Radio Transceiver Using Software Radio Architecture", VTC '00-s, Tokyo, May 2000.

[13] A.M. Badá, G. Bucci, M. Donati, M. Maddiotto, M. Politi and C. Savazzi, "Multicarrier Software Radio Transceiver for Mobile adio Systems", ICT '99, Cheju Korea, June 1999.

9 781852 333461